Nonlinear Dynamics of Piecewise Constant Systems and Implementation of Piecewise Constant Arguments

Nonlinear Dynamics of Piecewise Constant Systems and Implementation of Piecewise Constant Arguments

Liming Dai

University of Regina, Canada

World Scientific

NEW JERSEY · LONDON · SINGAPORE · BEIJING · SHANGHAI · HONG KONG · TAIPEI · CHENNAI

Published by

World Scientific Publishing Co. Pte. Ltd.

5 Toh Tuck Link, Singapore 596224

USA office: 27 Warren Street, Suite 401-402, Hackensack, NJ 07601

UK office: 57 Shelton Street, Covent Garden, London WC2H 9HE

British Library Cataloguing-in-Publication Data
A catalogue record for this book is available from the British Library.

NONLINEAR DYNAMICS OF PIECEWISE CONSTANT SYSTEMS AND IMPLEMENTATION OF PIECEWISE CONSTANT ARGUMENTS

ISBN-13 978-981-281-850-8
ISBN-10 981-281-850-2

Printed in Singapore by World Scientific Printers

To my family,
Dai JinPu, Wang ZhaoZhi, Xinming, Lillian and Linda

PREFACE

The systematical studies with mathematical models involving piecewise constant arguments were initiated for solving some biomedical problems. The first-order linear differential equations with variable coefficients and piecewise constant arguments and the differential equations with piecewise constant arguments of alternately advanced and retarded type were first studied. The oscillatory and nonoscillatory behavior of the corresponding systems were found and the characteristics of piecewise constant systems were also interested by the researchers in the areas of differential equations and applied mathematics. It was recognized by the earlier researchers that the differential equations with piecewise constant arguments may combine the features of both differential and difference equations. Since the early 1980's, differential equations with piecewise constant arguments have attracted great deal of attention from researchers in mathematical and some of the other fields in science.

Piecewise constant systems exist in a widely expanded areas such as biomedicine, chemistry, mechanical engineering, physics, civil engineering, aerodynamical engineering, etc. In actual physical and engineering systems, the phenomena related to stepwise or piecewise constant variables or motions under piecewise constant forces are common. These systems can usually be described in mathematical forms of first or second-order differential equations, or systems of differential equations with piecewise constant arguments. The piecewise constant systems may also be the ones with linear combination of known continuous and piecewise constant functions, or the systems with given piecewise constant functions and unknown coefficients. Examples in practice include vertically transmitted diseases, machinery driven by servo-motors and elastic systems impelled by Geneva wheels.

I have been working on piecewise constant dynamic systems since 1991. The initiation of my research in this area started with the finding of some interesting phenomena of piecewise constant systems. A completely linear system with a single piecewise constant variable may lead to extraordinary and nonlinear responses of the system; and the vibration of a spring-mass system may be attenuated or vanished with a piecewise constant exertion of sinusoidal form. This really attracted my attention to the piecewise dynamic systems. The piecewise constant dynamic systems can be very complex and highly nonlinear, and the behavior of the piecewise constant systems are distinctive and remarkably rich in comparing with the conventional dynamic systems of continuous variables. Though numerous research works on piecewise constant systems can be found in the current literature and the literature shows a general progress of interest in the properties of solutions to the governing differential equations with piecewise constant arguments, there is still a lack of thorough and systematical approach for effectively describing and analyzing the linear and nonlinear piecewise constant systems. In fact, there is no monographic books on piecewise constant systems or dynamic systems with piecewise constant variables available in the market. This motivates the writing of the present book. The book intends to provide a step by step and systematical approach for introducing the fundamental principles of the piecewise constant systems and their behavior on the basis of the existing research findings and my research results. The useful especially the newly developed approaches and techniques for analyzing the piecewise constant systems will be emphasized. A major portion of this book includes the principles and techniques that have been developed in my research and used as lecture notes for teaching the graduate students. The research results generated in the most recent investigations in this field and those yet to be published will also be included in the book. With the continuous research efforts on the piecewise constant systems, interesting and significant results have been found in the research. A novel piecewise constant argument was first introduced in early 1990's. With implementation of the piecewise constant argument, a new methodology was developed to bridge the gaps between the piecewise constant systems and the continuous systems. A new approach for analytically solving the differential equations of dynamic systems was also established. Extraordinary and complex characteristics of the piecewise constant systems were found in the research. Utilization of the piecewise constant argument leads to the development of a new numerical method which provides efficient numerical calculations with good convergence and higher accuracy over that of Runge-Kutta method. To diagnose the chaotic and other nonlinear behavior from regular periodic responses of the general dynamic systems including piecewise constant systems, a criterion named periodicity ratio was developed. These research results make the foundation of the present book. It is the hope of the author that this book may serve as an introduction to the subject of nonlinear dynamics of piecewise constant systems and provide principal concepts and theoretically and practically sound tools to the beginners and experienced researchers in this area for studying the piecewise constant systems.

The book is organized into seven chapters and three appendices. The contents of the chapters are carefully selected with the anticipation that the readers may comprehend the fundamental concepts and modern developments of the piecewise constant systems in an efficient manner. Chapter 1 starts with a brief discussion of the history and fundamental use of the differential equations in science fields. The importance of piecewise constant systems and physical phenomena influenced by piecewise constant variables are introduced. Prepreparation of the knowledge needed for comprehending and theoretically and numerically analyzing the piecewise constant systems are presented in Chapter 2, together with the terminologies used in piecewise constant system analyses. The basic concepts of linear and nonlinear differential equations governing continuous and piecewise constant dynamic systems are also presented.

A simple pendulum in physics or a linear spring-mass system in engineering would seem to be one of the simplest physical systems. However, the behavior of the systems may be rich and complex when it is subjected to a piecewise constant exertions. These systems are therefore perfect for introducing the piecewise constant systems, which can be facilitated by mathematical models formulated as second-order differential equations with piecewise constant variables. An external piecewise constant exertion acting on such a system in dynamics has intervals of constancy and may vary its magnitude or its magnitude and direction simultaneously at certain points of time. Continuity of the independent variable of the system and its first derivative at a point joining any two consecutive intervals then implies recurrence relation for the solution at such points. In Chapter 3, several of these systems are presented for initiate the study on the piecewise constant systems. Complete solutions are derived in detail for the systems subjected to various types of piecewise constant exertions. The behavior of the systems of linear and nonlinear formats is also evaluated in this chapter. A piecewise constant argument, $[Nt]/N$, is introduced in Chapter 4, where N is a parameter controls the intervals of constancy and provides flexibility in modeling the piecewise constant systems. With the implementation of the argument, linear and nonlinear continuous dynamic systems can be solved approximately. This leads to the establishment of a novel numerical approach for solving nonlinear differential equations. Utilization of the numerical approach in numerically solving linear and nonlinear dynamic systems is discussed in this chapter. As a common practice, construction of the mathematical models usually involves linearization, assumptions and simplifications of ignoring some unimportant or difficult factors. However, the new numerical approach maintains the original information of the system considered to an utmost level therefore generates numerical solutions of high accuracy. A comparison of the new numerical approach with Runge-Kutta method is conducted. The applicability of the numerical approach is also studied. The multi-degree-of-freedom (MDOF) systems with piecewise constant variables and the characteristics of the MDOF systems are investigated in Chapter 5. The methodology of solving continuous MDOF dynamic systems with implementing the piecewise constant argument introduced is provided with considerations of the linear coupling and damping of the systems.

If a variable of a system can be described by a piecewise constant argument of a very small time unit, the effect of the stepwise disturbance can be dramatically reduced. The behavior of a system with the piecewise constant argument of small time unit can then be a good approximation to that of the corresponding system of continuous variable. With the theoretical proof of this fact, Chapter 6 describes the development of a technique by which the two categories of dynamical systems, those with piecewise constant argument and the others which are continuous, may be completely linked together. With the piecewise constant argument introduced, the difference between the two categories of dynamical systems vanishes in the limiting case as N tends to infinity. An infinite sequence of the solutions with a piecewise constant argument is proved to be convergent and the limit leads to an exact solution of the dynamical system considered. This approach with the implementation of the piecewise constant argument for solving the dynamic systems is therefore called piecewise constantization.

Under certain conditions, irregular and unpredictable time evolution may occur in dynamic systems. The irregular and unpredictable behavior of the systems has been known as chaos. The discovery of chaos has changed the understanding of the foundation of physics, and has had an impact on many fields of science and engineering. In analyzing the properties of motion for nonlinear systems, it is essential to distinguish chaos from other types of behavior of the systems. Chapter 7 describes the development of a criterion named Periodicity Ratio for diagnosing chaos from regular behavior of a dynamic system. Periodicity ratio diagrams are also established for analyzing the periodic, quasiperiodic and chaotic behavior of nonlinear dynamic systems with varying system parameters and initial conditions. For demonstrating the characteristics of the Periodicity Ratio, a comparison of the Periodicity Ratio with Lyapunov exponent is provided.

Appendix A provides the mathematical developments and proofs useful for the modeling and developments of the mathematical formulas used in the context. The fundamental concepts of matrix and mathematical manipulations of the matrices needed for performing the developments described in the context of the book are presented in Appendix B. Appendix C lists the computer programs necessary for carrying out the numerical calculations with the new numerical approaches described in the context. The readers may conveniently use the programs to solve their own dynamic problems with or without piecewise constant variables. The programs also help the readers to comprehend the concepts and techniques described in the book.

The main structure of the book consists of four components, concept and solution developments, characteristics, numerical approaches and applications of piecewise constant approaches in dynamic systems. The chapters of the book tend to be arranged in such a way that each topic in the chapters is self-contained and the main concepts in nonlinear dynamics of piecewise constant systems are explained fully with necessary derivatives in details. The readers may thus gain the main concepts of each chapter with as less as possible the need to refer to the concepts of the other chapters. Readers may therefore start to read one or more chapters of the book for their own interests.

I am grateful to acknowledge many students and faculty for their helps with the book, their support and encouragement during the development of this book. Specific thanks to Leo Xu, Paul Wang, Changping Chen, Mansa Singh, Jingbo Wang, Sharat Chandra, James Lou, Shijun Zhou, Fereshteh Mafakheri, Punnamee Sachakamol, Quan Hu, Mahesh Patwardhan, and Huayseen Lee. Thanks also to Mr. Robert Jones for his supports in computer system management and computations involved in the book. I would also like to thank the University of Regina for its encouragement in performing the research involved in this book.

It has been gratifying to work with the staff of World Scientific Publishing Company through the development of this book. The assistances provided by the staff members have been valuable and efficient.

Finally, I wish to thank my wife Xinming Yan and my daughters Lillian and Linda for their love, consistent support and for tolerating some neglect of family responsibilities. A special gratitude is owed to my mother, Wang ZhaoZhi, for her continuous support on my career and her motherly encouragement, affection and sometimes urging on completing this book during the period that she stayed with me.

Liming Dai

CONTENTS

CHAPTER 1

Fundamentals of Conventional and Piecewise Constant Systems

1.1. Preliminary Remarks

Piecewise constant variations can be seen in many of the phenomena in the real world. These phenomena may often be modeled by piecewise constant systems with corresponding differential equations containing piecewise constant arguments. Such systems are usually considered as discontinuous systems and their behavior are usually more complex and richer in comparing with that of the conventional continuous systems governed by continuous differential equations, as will be demonstrated in the subsequent chapters. Comprehension of the piecewise constant systems and their behaviors and familiarization of the tools necessary for studying the systems are therefore practically significant. The piecewise constant systems are unique in terms of modeling, solution development and behaviors with respect to that of conventional continuous systems. However, the core methodology of modeling the conventional continuous systems and the procedures of solving the differential equations governing the continuous systems have very lose relation with the modeling and approaches for studying the piecewise constant systems. It is therefore important to first of all make ourselves familiar with the fundamentals of the conventional systems governed by continuous differential equations and the piecewise constant systems governed by the differential equations consisting of piecewise constant arguments, before the performance of a systematical study on the piecewise constant systems and their behaviors. The chapter begins with a brief description on the history of the studies on the piecewise constant systems. The

1

approaches for studying the conventional continuous differential equations are then outlined with the basic methodology in modeling and analyzing the conventional continuous systems. This methodology is similar to that to be used for piecewise constant systems. The greatest integer functions and essential definitions and concepts of piecewise constant systems are introduced. There follows a presentation of fundamental approaches on modeling and analyzing the piecewise constant systems with examples. This chapter intend to provide some preliminary knowledge on piecewise constant systems and fundamental tools for modeling and analyzing the piecewise constant systems, more ideas on the piecewise constant systems in details will be developed in the subsequent chapters.

1.2. Remarks on the Development and Analyses of Piecewise Constant Systems in History

Theoretically and practically sound study on piecewise constant systems started in the early 1980's. Since then, differential equations with piecewise constant functions or variables have attracted considerable attention from researchers in mathematics, biology, engineering and the other fields. These differential equations are closely related to impulse and difference equations of discrete arguments. A mathematical model involving a piecewise constant argument was first constructed by Busenberg and Cooke (1982) in the context of a biomedical problem. In their work, a first-order differential equation with a piecewise constant argument was developed based on the investigation of vertically transmitted diseases. Following Busenberg and Cooke's research, first-order linear differential equations with piecewise constant arguments were treated at length in several publications by Shah, Cooke, Aftabizadeh, Wiener, Jayasree and Deo (1983, 1984, 1985, 1992). A typical differential equation studied by the above authors is expressible in the following form:

$$y'(t) = a_0 y(t) + a_1 y([t]) + a_2 y([t] \pm a_3) \qquad (1.1)$$

where a_0, a_1, a_2 and a_3 are constants, $y(t)$ represents an unknown function, and $[t]$ denotes the greatest integer function. The initial value problems so defined have the structure of a continuous dynamical system within each of the intervals of unit length.

In general, behavior of a system with piecewise constant arguments is complex in comparing with that of the corresponding continuous regular systems. The literature shows a general progress of interest in the properties of solutions to the governing differential equations with piecewise constant arguments. The system with the variables of retarded type $t-n$ and advanced type $t+n$ was investigated by Cooke and Wiener (1984, 1987) and Wiener and Aftabizadeh (1988). Existence and uniqueness of the solution of this system and the asymptotic stability of some of its solutions were also studied. The oscillatory properties of its solution were reported with detailed analysis later by Cooke and Wiener (1987). Based on the studies given by Cooke and Wiener, Zhang and Parhi (1989) examined the first-order linear differential equations with variable coefficients and piecewise constant arguments and analyzed oscillatory and nonoscillatory behavior of the corresponding solutions. The oscillatory and asymptotic behavior for some first-order differential equations of more general form involving piecewise constant arguments of various types were demonstrated methodically by Aftabizadeh and Wiener (1985, 1988) and Wiener (1993). The first-order differential equations with the piecewise constant arguments of some peculiar forms also attracted the interest of the researchers in the field of differential equations. Huang reported his results in an analysis (Huang 1990) on oscillatory and asymptotic stability of a differential equation with piecewise constant argument in the form of $t-[t+1/2]$. Research on the oscillatory properties of the system of differential equations of several specific forms involving piecewise constant arguments can be found in the articles by Wiener and Cooke (1989) and Jayasree and Deo (1992). The investigations of mathematical approaches are continuously attracting the attention from the researchers for the behaviors of piecewise constant systems, as can be found from the current literature. Examples of such research works in recent years are the studies on existence of periodic solutions of retarded piecewise constant systems (Yuan 2002), existence, uniqueness and asymptotic behavior of piecewise constant system (Papaschinopoulos 2007), and conditions for oscillations of first-order piecewise constant systems (Wang and Yan 2006).

Very few investigations can be found in the current literature on the modeling and analysis of the behavior of physical systems in dynamics with governing equations of second-order piecewise constant differential equations. Leung (1988) studied the steady state response of a linear mechanical system in which the forcing function is represented by a linear combination of known functions that can be continuous or piecewise constant. The solution of the system was assumed in the form of a linear combination of the given functions with unknown coefficients. For a piecewise constant forcing function, the response at the discrete points of time was obtained. Dai and Singh (1991, 1994) studied the motions of several vibration systems disturbed by piecewise constant forces. The governing differential equations with piecewise constant arguments were formulated and analyzed. The solutions corresponding to the equations were found to be continuous everywhere in the time range considered. The response of various mechanical systems subjected to piecewise constant forces was obtained and the oscillatory, non-oscillatory and periodic properties of motion for the mechanical systems were studied.

With the research results found in the investigations in this field, following need to be emphasized for the development and analyses of piecewise constant systems in nonlinear dynamics to be presented in the subsequent chapters.

1. The piecewise constant systems considered in the field of the dynamics of piecewise constant systems generally have the structure of continuous dynamical systems within the time intervals in which the piecewise constant arguments are constant.

2. The complete solutions of the piecewise constant systems are usually based on the continuity of the systems at the points joining the neighboring intervals. Therefore, the solutions of the differential equations governing the piecewise constant systems combine the features of both differential and difference equations.

3. The overwhelming majority of the investigations in the field of the piecewise constant systems are pure mathematical approaches. The concentrated areas of the investigations are the stability, uniqueness, oscillation and existence of the solutions of the systems especially the periodic solutions of the systems.

4. Most of the archived research works in this field are on the first-order linear piecewise constant systems governed by the differential equations in the similar form as that of equation (1.1). There is still lack of a systematic investigation on the modeling and properties of the physical problems in the dynamics of piecewise constant systems that are governed by higher order and/or multi-degree of freedom differential equations.

5. The mathematical approaches in this field are mainly on the differential equations with the piecewise constant arguments of simple forms, typically in the form of $[t]$ to which a continuous system is considered merely over a unit time interval.

Piecewise constant dynamic systems make an important portion of nonlinear dynamics. Many piecewise constant systems can be found in science and engineering practices and their behaviors show uniqueness in comparing with the conventional continuous systems. In the past years, the author and his colleagues have been working on the physical problems in the dynamics of piecewise constant systems which are mainly governed by the second-order differential equations of single and multi-degree of freedom with piecewise constant arguments. In this book, the methodology and techniques for modeling and analyzing the piecewise constant dynamic systems will be presented based on our research. New piecewise constant arguments will be introduced, relations between piecewise constant systems and the corresponding continuous systems will be established, and the behaviors of the piecewise constant systems will be analyzed. Based on our research results, a new semi-analytical and numerical approach implementing piecewise constant arguments will also be presented, as it shows significant advantages over the existing methods.

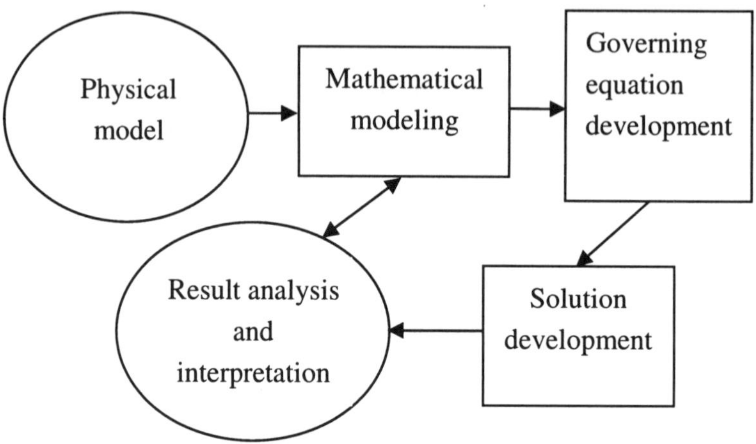

Figure 1.1. Analysis procedures of continuous and piecewise constant systems.

1.3. Modeling and Analysis Procedures for Conventional Continuous and Piecewise Constant Systems

Numerous phenomena or systems in the world involve variations with time or other variables. The variables are usually continuous like those used in conventional continuous dynamic systems; however, for many cases in practices in sciences, the variables can be discontinuous such as piecewise constant. Nevertheless, the key concepts and procedures for modeling and analyzing the conventional and piecewise-constant systems are similar. The primary common procedures for modeling and analyzing the two types of systems can be graphically shown in Figure 1.1 followed by the detailed descriptions with an example.

Physical Model

For analyzing the phenomena or systems existing in the real world, the physical system or the interesting physical components of a system should first be identified, as indicated in Figure 1.1. This identification is necessary for performing the subsequent mathematical modeling. The physical model should contain all the important elements of the system, such that the mathematical model can be consequently established. To illustrate this clearly, consider a pendulum system used in a pendulum clock. Let us say that the motion of the pendulum system of the clock is what we desire to analyze. Imagine the pendulum system is "separated" from the other components connecting with it and identify the system's components, pivot, sleeve, connecting rod and a bob of mass as illustrated in Figure 1.2 as a physical model defined.

The mass of the pendulum is attached at one end of the rod, and the other end of the rod is fixed at the sleeve which may rotate about an axis considered as a pivot, as shown in Figure 1.2. The pendulum system is actually connected with the other components of the clock and can be driven by a balance wheel. These components associated with the pendulum may also be considered if so desire. However, the physical model would then be more involved with an analysis more complex.

Establishment of a proper physical model is therefore very important for the study of a physical phenomenon or the behavior of a system in practice. Proper identification and determination of the system or the necessary components closely related to the main characteristics that one is interested in are the keys. Some simplifications may also necessary.

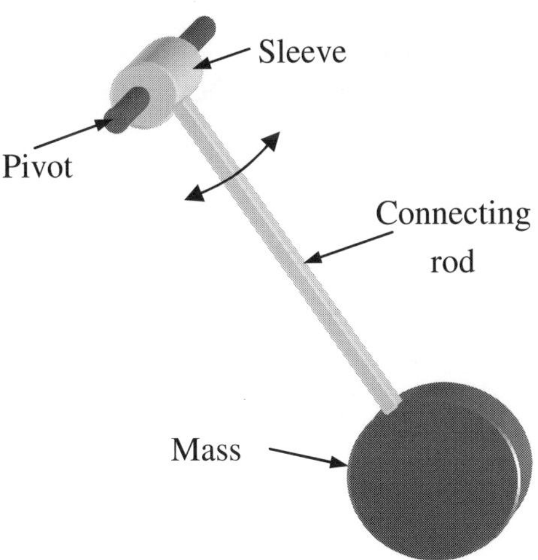

Figure 1.2.   Components of a pendulum system used in a pendulum clock.

Mathematical Modeling

Once a physical model is established, a mathematical model can be generated for the analysis of the system considered. The purpose of the mathematical modeling is to establish a model that includes all the necessary features of the system for the mathematical equations governing the system to be derived. There are generally many factors involved in a phenomenon or a physical system of the real world. In mathematically modeling the phenomena or systems, strictly speaking, it is very difficult and in many cases impossible to consider all the factors. The salient factors may have to be identified first and the applications of limits, hypotheses, simplification or linearization are usually inevitable .... Application of assumptions and simplifications of the physical model with scientific judgments in the field specified are therefore necessary before the derivation of the governing equations.

To get the insight of the mathematical modeling, we again use the pendulum described above. The mathematical model would be too difficult, if not impossible, to develop should we consider all the details of the real system or the physical model developed. A simple mathematical model known as simple pendulum illustrated in Figure 1.3 is therefore established to represent the physical model.

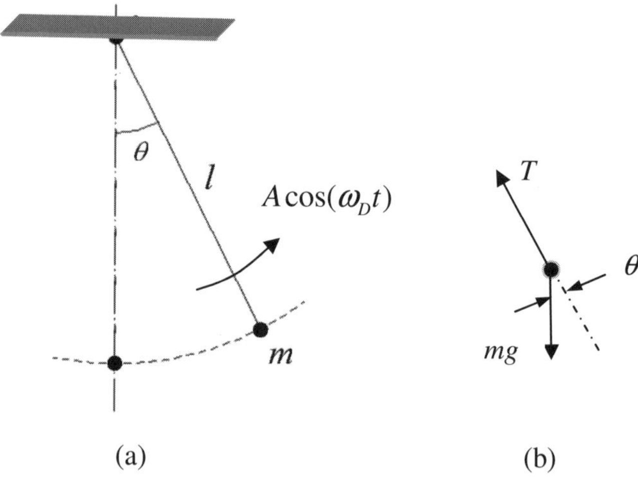

Figure 1.3.   Layout of a driven pendulum and the free body diagram of the mass.

In the mathematical model, the masses of the sleeve and the slim connecting rod are ignored; the connecting rod is assumed to be rigid; the mass is considered as concentrated at the bottom end of the rod with a constant distance l. from the pivot; and the resistance force at the pivot is assumed as negligible. It is reasonable to consider that the driving force acting on the pendulum is a function of time, $f(t)$. With the mathematical model such established, the governing equation or the equations for analysis purpose can be derived.

In many cases the mathematic models need to be refined or gradually improved for more accurate results. The pendulum, for example, can be refined to consider that the material of the pendulum system is elastic and may have deformation during its oscillatory motion. If this is considered, the connecting rod together with the mass of the pendulum can be replaced by a spring-mass system, and the pendulum system then becomes a spring-mass system oscillating about the pivot.

Derivation of Governing Equations

To model the variations or describe the behaviors of the phenomena, the mathematical tools such as linear and nonlinear differential equations are commonly used, and the differential equations are critical for modeling and quantitatively and qualitatively analyzing the phenomena. The differential equations used for modeling the phenomena in nature, such as those in the fields of physics, chemistry, engineering, biology, astronomy etc., are usually established by the basic scientific principles or the laws of nature that govern the behavior of the phenomena. Once the governing equations corresponding to the mathematical model for a physical system is established with implementation of the differential equations or systems of differential equations, theoretically, the solutions of the system can be approached and the nature of the system can be quantitatively studied.

As an example, let us use the pendulum model described above. By implementing Newton's second law of motion, the governing equation for the pendulum system can be conveniently derived with the free-body-diagram shown in Figure 1.3(b), which exhibits the mass separated from the pendulum system with all the forces acting on it. With the free-body-diagram, along the tangential direction of the motion (perpendicular to the rod), we may create the following differential equation as the governing equation for the pendulum system.

$$ml^2 \frac{d^2\theta}{dt^2} + mgl\sin\theta = 0 \tag{1.2}$$

where m is the mass and θ designates the angular displacement of the pendulum as shown in Figure 1.3. This equation represents a pendulum in free oscillation and Figure 1.3(a) exhibits the mathematical model. For a real pendulum system used in a clock, resistance force or damping is inevitable. Also, maintenance of the oscillation of the pendulum needs an external driving force. If we assume that a linear damping is applied onto the pendulum due to the friction at the pivot and a periodic driving force also known as external excitation in the form of $A\sin(\omega_d t)$ is applied at the pendulum, where ω_d is the frequency of the external excitation, a more accurate and complete governing equation can be obtained and expressed as

$$\frac{d^2\theta}{dt^2} + c\frac{d\theta}{dt} + \omega^2 \sin\theta = F\cos(\omega_d t) \tag{1.3}$$

where c is the linear damping coefficient, $\omega = \sqrt{g/l}$ and $F = A/ml$.

  The readers may notice that the governing equations (1.2) and (1.3) may also be derived by utilizing the other approaches such as d'Alembert's principle, principle of virtual work, Lagrange's equations, and principle of conversation of energy, with the identical mathematical model shown in Figure 1.3.

  Equation (1.3) is a nonlinear differential equation as $\sin\theta$ is involved. If the angular displacement θ is small such that $\sin\theta \approx \theta$, the following linear equation can be obtained.

$$\frac{d^2\theta}{dt^2} + c\frac{d\theta}{dt} + \omega^2\theta = F\cos(\omega_d t) \tag{1.4}$$

  Obviously, the application of the linear system governed by equation (1.4) is much restrictive and the system governed by nonlinear differential equation (1.3) is much closer to the reality pendulum system but more involved and much harder to solve in comparing with that of the linear system.

Solution Development

When the governing equation of a system becomes available, in general, the response of the system can be studied. In fact, much of our comprehension of the nature for a system that we are interested comes from our abilities to solve for the differential equations of the system. As can be seen from the example discussed previously, for an equation to be a differential equation, at least one derivative of a function of a variable or variables must appear. The objective of "solving" the differential equation is actually to find the function. Basically, there are two categories of approaches that are employed for solving the differential equations: analytical approach and numerical approach. For solving the governing differential equations with the two approaches, it is also necessary to classify the differential equations into two groups of linear and nonlinear differential equations. The linear differential equations are the ones contain no square or higher powers of variables or their derivatives. For this reason, the differential equations and the corresponding systems are called linear. Otherwise, the differential equations and the corresponding systems are known as nonlinear.

The linear systems and the associated linear differential equations have been well studied and the mathematical techniques for solving the linear systems are well developed. The exact solutions or the analytical solutions of closed form for the linear differential equations can be developed by many methods, such as the standard methods of solving differential equations, Laplace transformation, and calculus of variations. They may also be solved by numerical methods. To the linear systems such as that governed by equation (1.4), for example, the solutions in closed form corresponding to the systems with or without damping and the systems with various types of external excitations can be analytically obtained (Weaver *et al.* 1990). The general analytical solution for equation (1.4) can be given by

$$\theta(t) = \Theta_0 e^{-\xi \omega t} \cos(\omega_n t - \phi_0) + \Theta \cos(\omega_d t - \phi) \qquad (1.5)$$

where

$$\Theta = \frac{F}{\sqrt{(\omega^2 - \omega_d^2)^2 + c^2 \omega_d^2}}, \qquad \omega_n = \omega \sqrt{1 - \xi^2}$$

$$\phi = \tan^{-1}\left(\frac{c\omega_d}{\omega^2 - \omega_d^2}\right), \qquad \xi = \frac{c}{2\omega}$$

and Θ_0 and ϕ_0 are determined by initial conditions.

The advantages of the analytical solutions are the explicit description of the response of the system with respect to time and accurate expectation of the entire possible ranges of the solutions. By the solution in equation (1.5), for instance, the first term in the right side of the equal sign vanishes with increase of time. The analytical methods may also provide the characteristics of the response of the system in relating to systems parameters. Therefore, the analytical solutions are usually desirable in solving for linear or nonlinear systems.

The problems in the real world are often more nonlinear than linear (Thompson and Stewart 1986, Lakshmanan and Rajasekar 2003). Nonlinearity of the systems may be introduced through the systems' fundamental components such as geometry, resistence actions and material properties. In many cases, the linear systems are not sufficient for representing the actual behaviors of the systems, like the linear system governed by equation (1.4) for the pendulums system which is actually nonlinear. Analytical solutions for the nonlinear systems are much difficult to develop, though the advantages of analytical solutions are obvious. There are only very few nonlinear systems have closed form solutions, and most of the nonlinear differential equations can hardly be analytically solved for closed-form solutions, with utilization of the existing methods for solving differential equations. The principle of superposition, which holds for linear systems, is no longer valid for nonlinear systems. The nonlinear systems are therefore solved or analyzed by either approximate and semi-analytical methods or numerical methods for most of the cases. The approximate and semi-analytical methods commonly used in solving nonlinear systems are the perturbation method, iterative method, Ritz-Garlerkin method, and graphical methods (Nayfeh and Mook 1979, Lakshmanan and Rajasekar 2003, Stoker 1950). Nevertheless, these methods may not provide exact solutions for nonlinear systems. In utilizing these methods, one may have to pay attention to the simplifications, linearizations and restrictions of application that may have applied in developing for the solutions.

With the dramatically progressing development in computer hardware and software, numerically solving for the differential equations, which are difficult or impossible for solving with the existing theoretical or analytical approaches, becomes available. For many cases in solving nonlinear dynamic systems, numerical analysis is not only efficient but also necessary. The following numerical methods are found widely used in applications: the Runge-Kutta method, finite difference method, the Houbolt method, the Wilson method, the Newmark method and finite element methods (Nakamura 1977, Nayfeh 1979). It should be noticed that the numerical solutions are approximate solutions. Many factors such as the numerical methods used, the hardware and software employed for obtaining the numerical solutions and even the duration of the numerical calculations may affect the accuracy of the numerical solutions. The characteristics of the numerical approaches for solving linear and nonlinear systems are discussed in details in Chapter 2 and the applications of numerical analyses may also be found in the other subsequent chapters.

Result Analysis and Interpretation

A solution of the governing equation corresponding to a linear or nonlinear system concerned describes a function of a variable or variables. The solution may therefore be utilized to explain the behaviors of the system such as the stability, periodicity, variation of the solution with respect to the variable or variables, and the nonlinear behaviors like bifurcation and chaos. In order to analyze or interpret the results generated from the governing equations, therefore, one must have a clear view on the purpose of the analysis and the desired application of the results. It should be noticed, however, the interpretations of the solutions are merely related to the associated mathematical model and may only reflect the nature behavior imbedded in the mathematical model. The solution of the mathematical model described by equation (1.4) may only express the periodic or other linear behavior of the system. If nonlinear behavior of the pendulum system is desired, equation (1.3) or a more involved mathematical model needs to be used. In interpreting the results, additionally, one may have to attend to the simplifications and hypotheses used in the mathematical modeling.

If analytical solutions are available to the systems considered, the behaviors of the systems can be explicitly described and analyzed with the exact solution obtained. The mathematical manipulations such as derivation can be applied to the analytical solutions. General conclusions about the behavior of the systems concerned can therefore be developed. With the numerical results generated by computers, though, it is difficult to draw the general conclusions about the behavior of the systems.

It may also worth to mention that the solutions obtained with implementation of the mathematical models can only be analytical, semi-analytical or numerical. In many cases, especially the cases found in practices, experimental analyses or the experimental analyses jointed with analytical or numerical analyses are necessary for a better understanding of the behavior of the systems considered.

1.4. Fundamentals of Dynamic System Modeling in Science and Engineering

With the discussions on the procedures of modeling and analyzing the systems governed by differential equations presented in the previous section, the methodology for modeling conventional continuous systems can be demonstrated with examples that are commonly seen in the fields of science and engineering. Concentration will be given to the typical dynamic systems governed by the differential equations that describe the phenomena varying with time, as the dynamic systems and their behavior are the main concerns of this book.

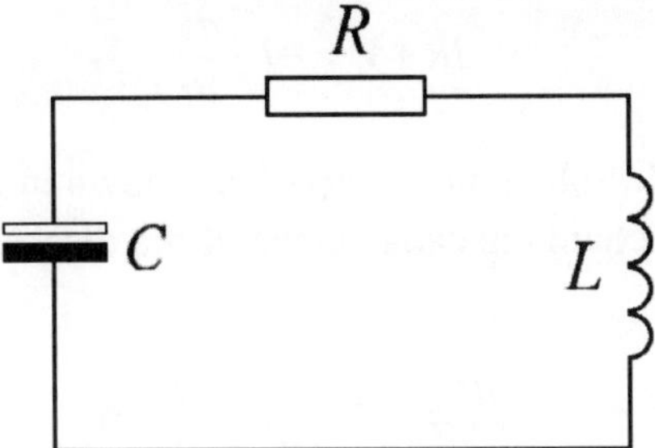

Figure 1.4.   A conductor-resistor-inductor circuit.

Discharging Capacitor

Conductors, resistors and inductances are widely used for electrical circuits in electric engineering practice. A typical conductor-resistor-inductor circuit is shown in Figure 1.4.

In the figure, C represents a conductor, R a resistor and L designates an inductance, as the convention and the three electric parts are connected in series. C, R and L also represent the quantities of capacitance, resistance and inductance respectively.

Suppose that the electric parts are all perfectly connected and the fundamental electric laws such as Ohm's law and Kichhoff's Law (Toro 1986) can be applied. Let Q be the quantity of the electric charge on the capacitor and Q is varying with time. Assume that the electric current strength I in the circuit equals to the changing rate of the quantity of the electric charge per unit time, such that the following linear relation holds.

$$I = \frac{dQ}{dt} \tag{1.6}$$

Assume that the electric potential difference between the two polar panels of the capacitor is a function of time, $V(t)$. Thus, the total electric potential drop of the circuit can be expressed by $IR+V$ per Ohm's law. According to Kichhoff's Law, the total potential drop should equal to the electromotive force in the circuit. As can be seen from Figure 1.4, the only electromotive force is due to the self-inductance, $-L \cdot (dI/dt)$, the following relation exists.

$$IR + V = -L \cdot \frac{dI}{dt} \tag{1.7}$$

Since, $V = Q/C$, substitute $I = dQ/dt$ as shown in equation (1.6) into equation (1.7), the changing capacitance of the discharging capacitor can be expressed as

$$L \cdot \frac{d^2Q}{dt^2} + R \cdot \frac{dQ}{dt} + \frac{Q}{C} = 0 \tag{1.8}$$

Thus, the mathematical model and the corresponding differential equation governing the discharge of the capacitor are developed. One may notice that this is a second-order linear ordinary differential equation to which an analytical solution is available.

Driven Froude Pendulum

As the systems in real world are more often nonlinear than linear, knowledge on modeling of nonlinear systems is practically significant.

Consider a driven Froude pendulum illustrated by a simplified physical model shown in Figure 1.5.

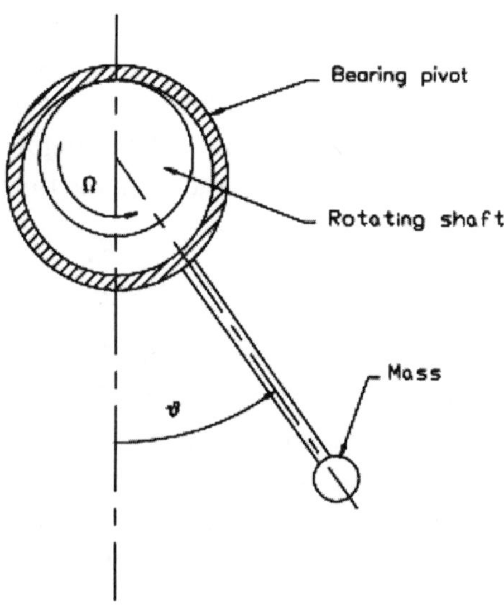

Figure 1.5. Sketch of a Froude pendulum.

This model is a simplification of a pendulum system driven by the friction generated by a rotating shaft which is connected to an engine. Assume that the engine rotates at a angular velocity Ω and the pendulum is swinging under the frictional torque M generated by the friction between the shaft and the bearing pivot. Such a pendulum system is known as a Froude pendulum (Butenin 1965). The frictional torque M can be considered as related to the slipping angular velocity $\dot{\theta}$ such that $M\,(\Omega-\dot{\theta})$ (Butenin 1965). The Froude pendulum considered here is in addition subjected to an external sinusoidal torque as shown in the figure. The governing equation for the Froude pendulum can thus be developed with the nonlinear differential equation in the following form:

$$I\ddot{\theta} + c\dot{\theta} + mgl\sin\theta = M\left(\Omega - \dot{\theta}\right) + A\cos\omega t \qquad (1.9)$$

where m designates the mass of the pendulum, I is the total moment of inertia of all rotating components of the pendulum, c represents the viscous coefficient of the system, l is the distance from the axis of rotation to the center of gravity of the pendulum, and $A\cos\omega t$ is the external sinusoidal torque acting on the pendulum with amplitude A and frequency ω. Assume that Ω can be measured, we expand the frictional torque $M\,(\Omega - \theta)$ in a power series by Taylor series expansion about a given angular velocity Ω, such that

$$M\left(\Omega-\dot{\theta}\right) = M\left(\Omega\right) - M'\left(\Omega\right)\dot{\theta} + \frac{1}{2}M''\left(\Omega\right)\dot{\theta}^2$$
$$- \frac{1}{6}M'''\left(\Omega\right)\dot{\theta}^3 + \cdots \qquad (1.10)$$

If we simplify the function for the torque by considering only the first four terms shown on the right-hand side of equation (1.10) and choose Ω as a point of inflexion of $M\,(\Omega)$ such that $M''(\Omega) = 0$, equation (1.9) can be expressed as

$$I\ddot{\theta} + c\dot{\theta} + mgl\sin\theta = M\left(\Omega\right) - M'\left(\Omega\right)\dot{\theta}$$
$$- \frac{1}{6}M'''\left(\Omega\right)\dot{\theta}^3 + A\cos\omega t \qquad (1.11)$$

This is a nonlinear dynamic system with nonlinear terms of $\sin\theta$ and $\dot{\theta}^3$. Numerical method with high accuracy may have to be used for solving it.

Taylor series expansion is a powerful tool due to its advantage of approximating any function to any desired degree of accuracy. Taylor series expansion is therefore commonly used in simplifying nonlinear systems. The nonlinear pendulum governed by equation (1.3), for example, can be rewritten by the following equation with simplified geometric nonlinearity.

$$\frac{d^2\theta}{dt^2} + c\frac{d\theta}{dt} + \omega^2\left(\theta - \frac{1}{6}\theta^3\right) = F\cos(\omega_d t) \qquad (1.12)$$

This is actually another form of Duffing's equation (Duffing 1918) which will be discussed later.

The Froude pendulum system has been studied by the author in details. The readers interested in this topic may refer to the reference (Dai and Singh 1998).

Workpiece-Cutter System

The modeling examples discussed previously in this section are all linear and nonlinear single-degree-of-freedom systems. Most of the systems in science and engineering are actually multi-degree-of-freedom and continuous systems counting geometry and/or deformation of the bodies involved. Partial differential equations are usually used for modeling the continuous systems and they are much difficult to handle. As a common practice, the continuous systems are usually simplified to multi-degree-of-freedom systems which are relatively easier to solve. Implementation of finite element analysis is also common in solving for continuous systems. Let us now model a dynamic system which can be considered as the combination of a multi-degree-of-freedom system and a continuous system.

Turning operation with a lathe is common in manufacturing. Both the workpiece and the cutter vibrate simultaneously during the turning operation. Consider that the workpiece as a rotating beam and the cutter

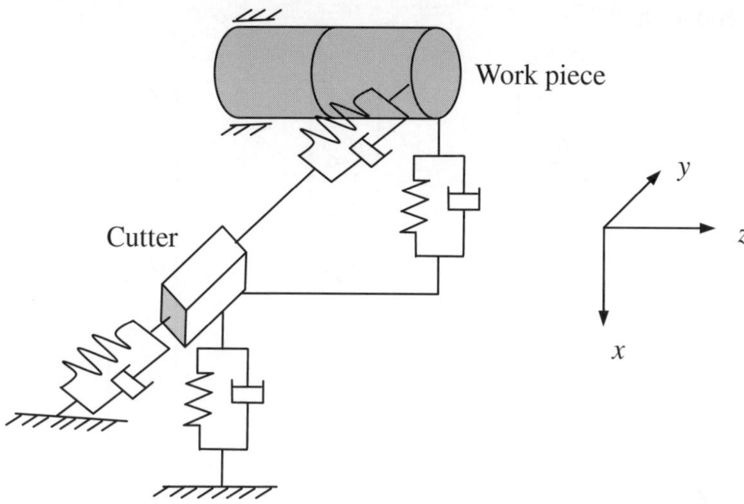

Figure 1.6.   Mathematical model of a coupled workpiece-cutter system.

is coupled with the beam by a spring-damper system to include the deformation of the materials between the cutter and workpiece. The cutter itself may be considered as a spring-mass system as its vibration is concerned. Both deformation and vibration of the workpiece are important for the analysis of this system. The workpiece is then discretized into two identical elements for implementing finite element method. With these considerations, the cutting system in the turning operation is structured as the coupling of the cutting-tool and of the workpiece, as shown in Figure 1.6, as a mathematical model for the system (Dai and Wang 2007).

Assume that all the materials involved in the system are perfect elastic. We also expect that the cutting process is continuous and the vibrations are mainly in the *x-y* plane. Based on the mathematical model shown in Figure 1.6, the governing equation for this system can be given in the following matrix form with fourteen degree of freedom.

$$[M_X][\ddot{X}]+[C_X][\dot{X}]+[K_X][X]=[F_X]$$

$$[M_Y][\ddot{Y}]+[C_Y][\dot{Y}]+[K_Y][Y]=[F_Y]$$

(1.13)

In the equation,

$$\left[\ddot{X}\right]=\begin{bmatrix} \ddot{x}_{w1} \\ \ddot{\theta}_{wx1} \\ \ddot{x}_{w2} \\ \ddot{\theta}_{wx2} \\ \ddot{x}_{w3} \\ \ddot{\theta}_{wx3} \\ \ddot{x}_{u} \end{bmatrix}, \quad \left[\dot{X}\right]=\begin{bmatrix} \dot{x}_{w1} \\ \dot{\theta}_{wx1} \\ \dot{x}_{w2} \\ \dot{\theta}_{wx2} \\ \dot{x}_{w3} \\ \dot{\theta}_{wx3} \\ \dot{x}_{u} \end{bmatrix}, \quad \left[X\right]=\begin{bmatrix} x_{w1} \\ \theta_{wx1} \\ x_{w2} \\ \theta_{wx2} \\ x_{w3} \\ \theta_{wx3} \\ x_{u} \end{bmatrix} \quad (1.14)$$

where x designates displacement, θ is the angular displacement, the subscript w represents the workpiece and the subscript u denotes the cutter.

The excitations along the x direction can be expressed as

$$\left[F_X\right]=\begin{bmatrix} F_{wx1} \\ M_{wx1} \\ F_{wx2} \\ M_{wx2} \\ F_{wx3} \\ M_{wx3} \\ F_{tx} \end{bmatrix} \quad (1.15)$$

where F represents the force and M the bending moment acting on the system.

In equation (1.13), the global mass matrix

$$[M_X] = \begin{bmatrix} m_{w11} & m_{w12} & m_{w13} & m_{w14} & 0 & 0 & 0 \\ m_{w21} & m_{w22} & m_{w23} & m_{w24} & 0 & 0 & 0 \\ m_{w31} & m_{w32} & m_{w33} & m_{w34} & m_{w35} & m_{w36} & 0 \\ m_{w41} & m_{w42} & m_{w43} & m_{w44} & m_{w45} & m_{w46} & 0 \\ 0 & 0 & m_{w53} & m_{w54} & m_{w55} & m_{w56} & 0 \\ 0 & 0 & m_{w63} & m_{w64} & m_{w65} & m_{w66} & 0 \\ 0 & 0 & 0 & 0 & 0 & 0 & m_u \end{bmatrix}$$

(1.16)

in which the elements m_{wij} designate the equivalent masses at each node of the workpiece. The stiffness matrix in the equation can be expressed in the following equation in which k_{wij} represent the corresponding stiffness coefficients at each node of the workpiece.

$$[K_X] = \begin{bmatrix} k_{w11} & k_{w12} & k_{w13} & k_{w14} & 0 & 0 & 0 \\ k_{w21} & k_{w22} & k_{w23} & k_{w24} & 0 & 0 & 0 \\ k_{w31} & k_{w32} & k_{w33} & k_{w34} & k_{w35} & k_{w36} & 0 \\ k_{w41} & k_{w42} & k_{w43} & k_{w44} & k_{w45} & k_{w46} & 0 \\ 0 & 0 & k_{w53} & k_{w54} & k_{w55}+k_{cx} & k_{w56} & -2k_{cx} \\ 0 & 0 & k_{w63} & k_{w64} & k_{w65} & k_{w66} & 0 \\ 0 & 0 & 0 & 0 & -2k_{cx} & 0 & k_{ux}+k_{cx} \end{bmatrix}$$

(1.17)

The two matrices $[M_X]$ and $[K_X]$ are generated per the construction rule of finite element method (Ross 1990) and combination of the workpiece and the cutter. The subscripts 1, 2, 3 designate node numbers, and the subscript c denotes the connecting structure between the work piece and the cutting tool. The matrix $[C_X]$ is the damping matrix which can be determined by the Rayleigh approach (Schmitz and Donaldson 2000) with the mass and stiffness matrices defined.

The matrices in the y direction are in a similar form as that shown in the above equations. The readers may easily derive them based on the equations provided above.

With implementation of the model established, the dynamic response of the system coupled the workpiece and the cutting-tool can be studied with a given cutting force. It is significant that effects of the vibration of the workpiece and the cutter on the surface quality of the machined product can be quantitatively determined by the model established (Dai and Wang 2007).

1.5. Piecewise Constant Systems and Their Modeling

In describing the fundamental principles of the differential equations and the modeling of the physical systems in previous sections, the differential equations considered together with the variables and functions involved in the differential equations are all continuous. However, the actual phenomena in nature are much complicated. In modeling the phenomena or physical systems in the real world with implementation of differential equations, it is difficult or impossible to take into account all the factors involved, as demonstrated previously. Moreover, in many cases, the variables or functions used for the modeling with differential equations may have to be discontinuous such as piecewise constant or other types like piecewise linear, stepwise, impulsive ... and so on, to reflect the actual characteristics of the systems to be modeled.

Piecewise constantly varying phenomena can be found in physics, biology, engineering and many other fields in science. Busenberg and Cooke (1982) first established a mathematical model with a piecewise constant argument for analyzing vertically transmitted diseases. In actual physics and engineering systems, motions under stepwise or piecewise constant forces are common, and many of such systems can be described mathematically by the second-order differential equations with piecewise constant arguments. Examples in practice include vertically transmitted diseases, machinery driven by servo units, charged particles moving in a piecewise constantly varying electric field, and elastic systems impelled by a Geneva wheel. Piecewise constant systems are usually discontinuous systems as they involve with piecewise constantly varying variables or functions. The behaviors of these systems are in general different from that of the continuous systems. To mathematically model a piecewise constant system for analytical or numerical investigation purpose, some special tools such as the greatest integer functions and unique conditions may have to be employed. The present research in current literature on the piecewise constant systems, the concepts of greatest integer functions, modeling of the piecewise constant systems with examples, and the considerations in modeling the systems are presented in this section, for the readers new in this field.

1.5.1. *Greatest Integer Functions*

For modeling and analyzing a piecewise constant system, a special function know as the greatest integer function is needed. The greatest integer functions are seen in many areas such as mechanics, biology, and engineering. The greatest integer function is represented by the symbol [•]. The function value of a greatest integer function is an integer corresponding to the variable of the function. As the greatest integer function always rounds down the variable values to nearest integer, it is also know as floor function. The variable values are usually positive and negative real values. Consider a simple greatest integer function $[ax]$ where a is a constant and x is the variable of the function. Without loss of generality, let $a = 1$, the function values of this function can be given as the following for the corresponding values of the variables.

$$[ax] = \begin{cases} 0, & x = 0; \\ 1, & x = 1; \\ -4, & x = -3.3; \\ 11, & x = 11.99; \\ -2, & x = -\sqrt{3}; \\ 0, & x = \sin(\pi/3) \end{cases}$$

Obviously, the integer function is not continuous, but piecewise constant, i.e., the function value keeps a constant on each of the time intervals. To demonstrate this clearly, consider an equation in the following form with greatest integer functions.

$$f(t) = f_1(t) + f_2(t) \tag{1.18}$$

where

$$f_1(t) = 2[2t] \qquad \text{and} \qquad f_2(t) = 5[3\cos t]$$

To visualize the characteristic of the piecewise constant function, the function $f(t)$ superposed with $f_1(t)$ and $f_2(t)$ is plotted in the following figure. As can be seen from the figure, the function values change as the greatest integer of the variable varies.

Take $f_1(t)$ as an example, $f_1(t)$ keeps the initial value zero from $t = 0$ till $t = 0.5$ at which $f_1(t)$ jumps to 2. $f_1(t)$ then maintain this constant value until $t = 1, \ldots.$

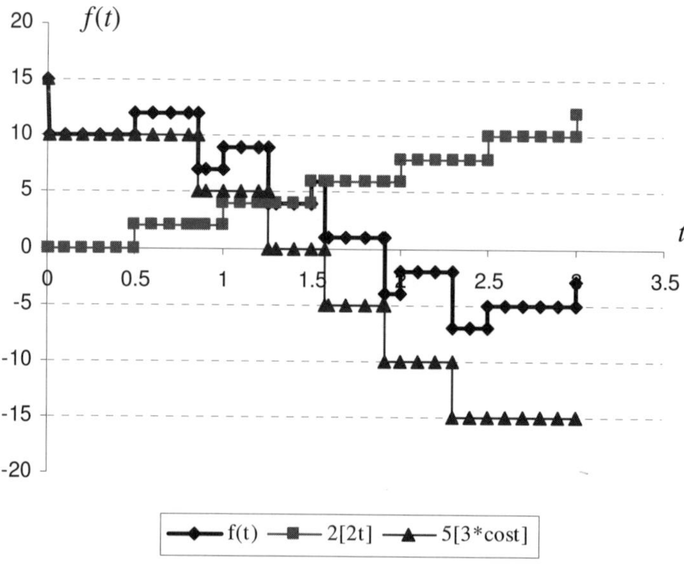

Figure 1.7. Plots of piecewise constant functions $f_1(t) = 2[2t]$ and $f_2(t) = 5[3\cos t]$ and their combination $f(t) = f_1(t) + f_2(t)$.

We may also notice that $f_1(t)$ remains as a constant in each of the constant time interval with a time span of 0.5. However, the time intervals in which the function values are constants may not necessarily be constant. This can be seen from the plot for $f_2(t)$, to which the interval length varies from interval to interval. It is also worth to notice that $f_2(t)$ may not necessarily vary its value at the integer point of time. Combining the two functions $f_1(t)$ and $f_2(t)$, the piecewise constant function $f(t)$ varies piecewise-constantly with the varying time intervals in which the function values are constant. The feature of the combined function $f(t)$ is more complex than either $f_1(t)$ or $f_2(t)$ as shown in the above figure.

Generally speaking, a dynamic system governed by the differential equations containing greatest integer functions are piecewise constant systems. The functions consisting of greatest integer functions are piecewise constant functions. With the characteristics of greatest integer functions, differential equations consisting of piecewise constant functions may show different and unique behaviors in comparing with that of regular differential equations containing only continuous variables and functions.

1.5.2. *Piecewise Constant System Modeling in Science and Engineering*

As mentioned previously, many phenomena in physics, biology, engineering and other fields involve piecewise changes. The analytical or numerical models for the systems involving piecewise constant variables can be constructed by employing the conventional modeling techniques and the greatest integer functions described above. For the sake of clarity, the methodology for modeling the phenomena involving piecewise constant variables is demonstrated in the following simple examples.

Vertically Transmitted Diseases Among American Dog Ticks

In modeling the diseases propagating among vertebrate and invertebrate vectors, numerous factors such as disease transmission type, environmental effects and proportion of females must be taken into consideration. For many diseases, the disease propagation can be divides into two different mechanisms of horizontal and vertical transmission. In vertical transmission, the disease is passed on to a proportion of the offspring of infected parentage, whereas the horizontal transmission refers to the individuals in a population that pick up the disease through direct or indirect contact with infected individuals.

A detailed study was made by Garvie *et al.* (1978) on a type of arthropod, named Dermacentor variabilis or American dog tick in Nova Scotia, Canada. The female tick adults laid the eggs from which the ticks hatch into their larval form. The larvae engorge with a blood meal before molting and emerging as adults. In quantitatively estimate the relative influence and importance of the diseases which are propagated by the vectors of the American dog ticks studied by Garvie *et al.*, Busenberg and Cooke (1982) recognized the significance of the disease transmission among the tick with discrete generations and established a model with a simple piecewise constant argument in the form of $[t]$. The females were classified into the susceptible group and infectious group respectively. A system of equations describing the dynamics of the disease for generation $n = 1,2,3,\ldots$, was developed in the following form for the susceptible and infected proportion of the female population of generation n, named $I^{(n)}$ and $S^{(n)}$ respectively.

$$\begin{cases} \dfrac{dI^{(n)}}{dt} = -c(t)I^{(n)}(t) + k(t)S^{(n)}(t)I^{(n)}(t), & n < t \leq n+1 \\[3mm] \dfrac{dS^{(n)}}{dt} = -c(t)S^{(n)}(t) - k(t)S^{(n)}(t)I^{(n)}(t), & n < t \leq n+1 \end{cases} \tag{1.19}$$

For $n = 2,3,4,\ldots$, the functional relations for $I^{(n)}$ and $S^{(n)}$ are given by

$$\begin{cases} I^{(n)}(n) = \dfrac{1-p}{m_2 - m_1} \displaystyle\int_{m_1+n-1}^{m_2+n-1} b_I(t)I^{(n-1)}(t)[1 - I^{(n-1)}(t) - S^{(n-1)}(t)]dt \\[4mm] S^{(n)}(n) = \dfrac{1}{m_2 - m_1} \displaystyle\int_{m_1+n-1}^{m_2+n-1} [b_S(t)S^{(n-1)}(t) + pb_I(t)I^{(n-1)}(t)] \\[4mm] \qquad\qquad \times [1 - I^{(n-1)}(t) - S^{(n-1)}(t)]dt \end{cases} \tag{1.20}$$

$$0 \leq m_1 \leq m_2 \leq 1 \qquad\qquad \text{and} \qquad\qquad 0 \leq p \leq 1$$

with initial conditions:

$$I^{(1)} = I_0 \qquad\text{and}\qquad S^{(1)} = S_0 \qquad\qquad (1.21)$$

In the equation, c is the death rate, b_I and b_S are the birth rates, m_1 and m_2 are the maturation window limits, and k the horizontal transmission factor. The total population of generation n at time t is given by

$$P^{(n)}(t) = S^{(n)}(t) + I^{(n)}(t) \qquad\qquad (1.22)$$

The general solutions for the model described by equations (1.19), (1.20) and (1.21) are $S(t)$ and $I(t)$, the infected and susceptible vectors at any time desired. In epidemiological practice, however, the infected and susceptible vectors in the current generation are what one may really concern. With this consideration, the infected and susceptible vectors for a given generation with respected to time t, i.e., the solutions, can be expressed as

$$S(t) = S^{([t])} \qquad\text{and}\qquad I(t) = I^{([t])}, \qquad t \geq 1 \qquad\qquad (1.23)$$

where $[t]$ denotes the greatest integer function, and the solutions are valid for the time interval of $[t] < t \leq [t] + 1$.

The model for the vertically transmitted disease such developed have to base on some assumptions and simplifications as those used in developing most of the fundamental dynamic systems. The disease progress was assumed to follow directly the track of the female portion of the tick population. Once a tick is infected by the disease, it was assumed that the tick remains as infected for the balance of its life. The infection was assumed to be transmitted from one generation to the next by vertical path only.... One of the important assumptions made was that the generations were discrete and each generation was involved in the disease dynamics for one unit of time. This provides the foundation for the solutions to be developed with the piecewise constant argument.

With the study of the model established about the vertically transmitted disease, Busenberg and Cooke (1982) gave a piecewise constant system governed by the differential equation in the following general form.

$$\frac{dx(t)}{dt} = F(t, x_t), \quad [t] < t \le [t]+1, \quad x_{[t]} = \phi_{[t]},$$

$$\phi_{[t]} = G([t], x_{[t]}), \quad [t] \ge 2, \quad \phi_1 = H \tag{1.24}$$

The solution of this equation $x(t)$ is thought on the interval $[0, \infty]$ and x_t is the past history of x defined as

$$x_t(s) = \begin{cases} x(t+s), & s \in [-t, 0] \\ 0, & s < -t \end{cases} \tag{1.25}$$

F and G in equation (1.24) are used as piecewise continuous functions while $\phi_{[t]}$ can be considered as a local initial condition corresponding to the governing equation on the time interval of $[t] < t \le [t]+1$.

There are a few characteristics need to be emphasized for such a piecewise constant system established.

1.   The solution of the system needs may not necessarily be continuous, unless some specific conditions such conditions of continuity are applied.
2.   The solution of the system is continuous on each of the time intervals of $[t] < t \le [t]+1$ for $t \ge 0$.
3.   With the greatest integer function in the simplest form $[t]$, the piecewise constant function is related to a time interval of a unit length. This implies that the solution of the system is given with respect to each of the integer points along the time axis.
4.   The behavior of the system reflects both the continuous and discrete nature of their dynamics.

Geneva Wheel

Geneva wheel is a mechanical system that is widely used in watches and instruments. The motion of the system is piecewise continuous. A sketch of a Geneva wheel is shown in following figure.

In the sketch for the Geneva wheel, the small wheel rotating about the axis O' is mounted with a crank equipped with a pin A that can be engaged with a slot of the bigger wheel that may rotate about an axis at O.

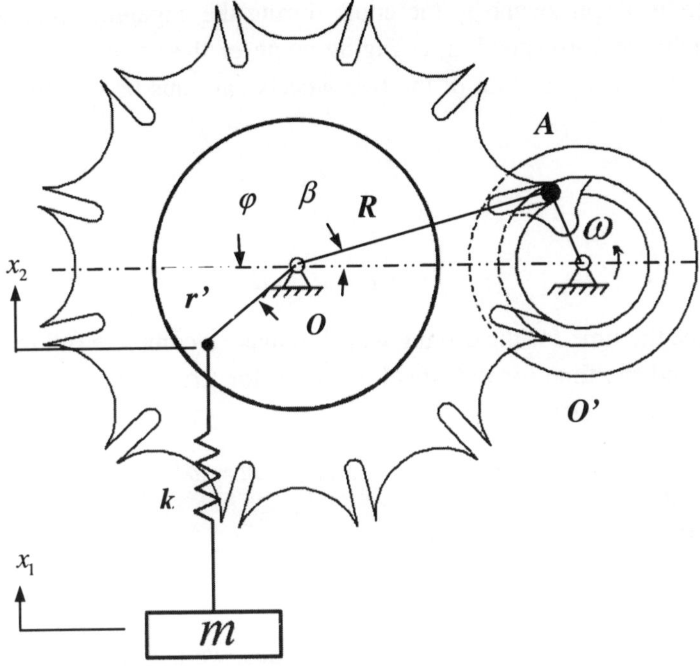

Figure 1.8.   Sketch of a Geneva wheel system.

The small wheel is usually rotating at a constant speed, say ω, and is the driving wheel of the system. The bigger wheel is the driven wheel; it rotates when pin A is engaged with one of its slots and this slot-wheel remains at rest when the pin and the slot are not engaged while the driving wheel keeps rotating. As such, the bigger wheel's revolution is discontinuous. Also, the revolution of the bigger wheel with slots is not constant when the pin of the small wheel and a slot is engaged. The distance between the two axes of the wheels OO' is fixed. The crank length $O'A = r$, and it is a constant. The distance between the engaging point A and the axis of the slot-wheel is R which is changing with time therefore the revolution of the slot-wheel is also varying with time, when the two wheels are engaged. The spring of a spring-mass system is fixed on the slot-wheel with a distance r' from the center of the wheel, as shown in the figure. The angle $\angle AOO'$ is β and assume the angel $\angle AO'O = \alpha$. Therefore, the engaging angle of the driving wheel (the maximum angel swept by the crank during the engaging time) is $2\alpha_0$. As such, the corresponding engaging angle of the driven wheel is $2\beta_0$. The kinematical relation of the two wheels can thus be described as the following.

$$\omega = \frac{d\alpha}{dt} \tag{1.26}$$

$$r\sin\alpha = R\sin\beta \tag{1.27}$$

Assume the time for the driving wheel to make a complete full revolution is T_0 and the time for the wheel to sweep for $2\alpha_0$ is t_0, we hence have the following relations.

$$2\alpha_0 = \omega t_0 \quad \text{and} \quad 2\pi = \omega T_0 \tag{1.28}$$

From the figure, we may define

$$x_2 = r'\sin\varphi \tag{1.29}$$

With the above formulas and definitions, we may now construct the governing equation in the following form for the spring-mass system mounted on the Geneva system, assume that the transversal motion of the. spring-mass system is negligible.

$$m\ddot{x} + k(x_1 - x_2) = 0 \tag{1.30}$$

This system and its solution should be continuous for $t \geq 0$. However, x_2 in the equation is discontinuous, as the driven wheel rotates for a short period of time and then remains at rest. The numerical calculation of the system could be more efficient, if the governing equation can be expressed explicitly in a continuous form. Notice that the engaging time t_0 is smaller than that of the time required for a complete revolution of the driving wheel T_0, and t_0 is much smaller than the time needed for the driven wheel to complete a revolution. Moreover, x_2 actually jumps vertically with an amount of $r'\sin(2\beta)$ for every revolution of the driving wheel. With this consideration, we may rearrange equation (1.29) and approximately describe that

$$x_2 = r'\sin\left(2\beta\left[\frac{t}{T_0}\right]\right) \tag{1.31}$$

In the above equation, the greatest integer function would change to be $[t]$ if T_0 is a unit time, i.e., the driven wheel would quickly rotate 2β radians for every unit time. Substituting equation (1.31) into (1.30), the single governing equation corresponding to the continuous time t is then expressible in the following form.

$$m\ddot{x} + kx_1 = kr'\sin\left(2\beta\left[\frac{t}{T_0}\right]\right) \tag{1.32}$$

The spring-mass system mounted on a Geneva wheel is thus equivalent to a free-vibration system subjected to an external piecewise constant excitation. Though the greatest integer function is a function of time in general, the external excitation is a constant in a time interval of length T_0. Corresponding to this time interval, the system governed by equation (1.32) is a linear vibration system under a constant force, therefore, the solution of the system can be easily obtained for this time interval. The complete solution of the piecewise constant system can be gained with the conditions that the displacement and velocity of the mass are both continuous. The behavior of this system under the piecewise constant force is different from that of the same system subjected to a continuous sinusoidal excitation. Detailed analyses on the behaviors of piecewise constant systems and procedures for developing for the solutions of the systems will be presented in the following chapters.

Flexible Support of Electrodynamic Shaker

An electrodynamic shaker or exciter is an electromagnetic device that may generate various types of forces with adjustable magnitude and frequency. Electrodynamic shakers therefore are widely used in vibration and other experiments. A schematic illustration of a typical electro-dynamic shaker is shown in the following figure.

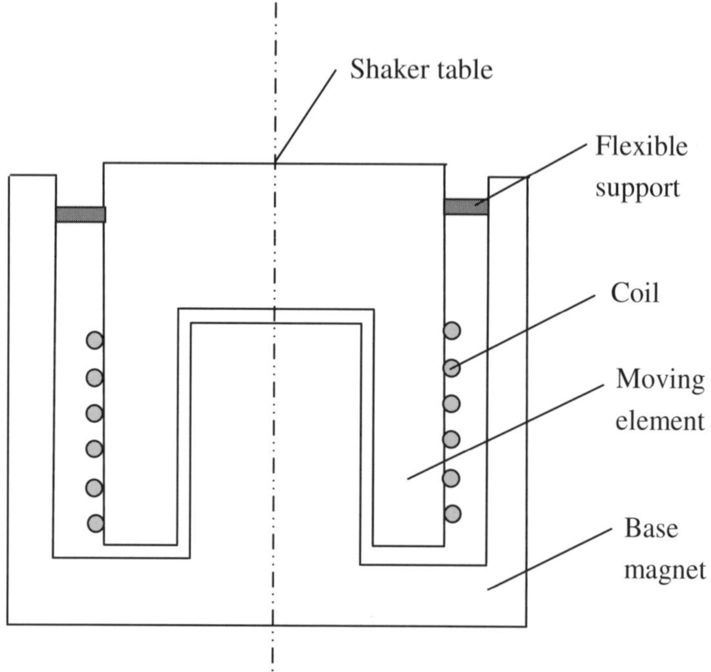

Figure 1.9.   Schematic illustration of an electrodynamic shaker.

During the experiment implementing the shaker, a test component is usually conjunct with the shaker table of the moving element. The force is then passed on to the test component with desired magnitude and frequency. The flexible support of the shaker is to ensure that the moving element may move linearly. A force F will be generated when there is an electric current passing through the coil which is placed in the magnetic field produced by the base magnet. The force generated can be quantified theoretically by the following formula (Buzdugan *et al.* 1986)

$$F = DIL \tag{1.33}$$

where F is the force in Newtons, D is the magnetic flux intensity in Telsas, I is the current in amperes, and L is the length of the coil in meters. The force may therefore vary with the current. When a.c current is flowing through the coil, the force generated is sinusoidal or harmonious; if a d.c current is used, the force will then be constant. A piecewise constant force may also be applied onto the shaker provided that the current flowing through the coil is piecewise constant.

The flexible support of an electrodynamic shaker is an important part; its properties, such as natural frequency and viscoelasticity, appreciably affect the quality and operation of the shaker. The behavior of the flexible support subjected to various excitations including piecewise constant forces is therefore significant to be evaluated for the quality and operation concerns. The flexible support can be simplified as a nonlinear spring-mass system of single degree of freedom. The governing equation for such a flexible support subjected to a piecewise constant force can then be given by the following equation, with considerations that the flexible support is viscoelastic and the stiffness of the support is nonlinear.

$$m\ddot{x} + (c + C_v)\dot{x} + kx - \mu x^3 = A\sin\left(\Omega \frac{[at]}{b}\right) \tag{1.34}$$

In the equation, m is the mass of the flexible support, c the damping coefficient, C_v is the viscoelastic damping constant that is usually a function of velocity or time, k and μ are spring constants. This implies that the support is considered as a soft spring and the viscoelastic material of the support provides a viscoelastic damping. A piecewise constant excitation in the form shown in the right side of the equal sign is assumed, though the other types of excitation can also be considered. A is the amplitude of the excitation acted by the moving element onto the support, Ω is the frequency of the excitation, a and b are parameters controlling the magnitude and duration of the piecewise constant excitation in each of the time interval in which the excitation is a constant. The size of the time interval may also vary with time when b is set as a function of time, if so desired.

Equation (1.34) governs a nonlinear piecewise constant system. The analytical solution of the equation is not available, not even for the time interval in which the excitation of the system is a constant, per the conventional mathematical approaches. The procedure for solving this type of system will be discussed in Chapters 4 and 5, in which the semi-analytical and numerical solutions for this type of nonlinear piecewise constant dynamic system will be developed.

1.6. Implementing Piecewise Constant Arguments in Dynamic Problem Solving

The analytical solutions for nonlinear differential equations are difficult, if not impossible, to obtain. Approximate or numerical approaches are usually inevitable to be employed for solving the nonlinear differential equations that governing the nonlinear dynamic problems. As indicated previously, differential equations with piecewise constant arguments exhibit the joined properties of differential equations and difference equations. They can therefore be considered as hybrid dynamical systems as piecewise constant differential equations combine continuous and discrete systems. It would be practically acceptable if an approximate or numerical solution can be developed, for a given nonlinear differential equation or a system of nonlinear differential equations, via some procedures of simplification or linearization with implementation of piecewise constant arguments. Moreover, it would be desirable if the piecewise constant system, converted from the continuous system with piecewise constant arguments, can be explicitly solved by a direct integration or through an existing method over each of the intervals of the piecewise constant system.

Some researchers (Liu and Gopalsamy 1999, Fan and Wang 2002, Elabbasy and Saker 2005) have recognized such advantages of implementing the piecewise constant arguments in solving for nonlinear differential equations. A typical application of an approach with the implementation of piecewise constant arguments in solving differential equations is a research reported by Sun and Saker (2006) in their recent study on the solutions of a predator–prey system initiated by Holling on an investigation of .predation of small mammals on pine sawflies(1959, 1963). The differential equations that Sun and Saker considered are in the following form. In the equation for the differential equations, y_1 and y_2 are the functions representing the non-interacting two-preys and y_3 is their common natural enemy (predator), a_i are the natural growth rate of y_1 and y_2, b_i and c_i are the system coefficients. Due to the environmental variation, a_i and the coefficients are all functions of time.

$$\begin{cases} \dfrac{1}{y_1(t)}\dfrac{dy_1}{dt} = a_1(t) - b_1(t)y_1(t) - \dfrac{c_1(t)y_2(t)}{y_1(t)+1} \\[3ex] \dfrac{1}{y_2(t)}\dfrac{dy_2}{dt} = -a_2(t) + \dfrac{c_2(t)y_1(t)}{y_1(t)+1} - \dfrac{c_3(t)y_3(t)}{y_2(t)+1} \\[3ex] \dfrac{1}{y_3(t)}\dfrac{dy_3}{dt} = -a_3(t) + \dfrac{c_4(t)y_2(t)}{y_2(t)+1} \end{cases} \tag{1.35}$$

For such a system of nonlinear differential equations, the solution of closed form is difficult to develop. With the assumption that the variables and functions are changes only at regular intervals of time, the following equation is constructed through the conversion from equation (1.35) with implementing a greatest integer function $[t]$.

$$\begin{cases} \dfrac{1}{y_1(t)}\dfrac{dy_1}{dt} = a_1([t]) - b_1([t])y_1([t]) - \dfrac{c_1([t])y_2([t])}{y_1([t])+1} \\[3ex] \dfrac{1}{y_2(t)}\dfrac{dy_2}{dt} = -a_2([t]) + \dfrac{c_2([t])y_1([t])}{y_1([t])+1} - \dfrac{c_3([t])y_3([t])}{y_2([t])+1} \\[3ex] \dfrac{1}{y_3(t)}\dfrac{dy_3}{dt} = -a_3([t]) + \dfrac{c_4([t])y_2([t])}{y_2([t])+1} \end{cases} \tag{1.36}$$

With this conversion making used of the piecewise constant argument $[t]$, all the differential equations in (1.36) can be easily solved by directly integration over a time interval of unit length, as all the functions and variables on the right-side of the equal sign of the equation (1.36) are constants in the interval.

The approach of solving the differential equations in complex form with the direct implementation of piecewise constant arguments, as demonstrated above, is straight forward and efficient to certain extent. However, the accuracy and reliability of the solutions such obtained is low. In applying the approach with direct implementation of piecewise constant arguments, following negative aspects of this approach should be kept in mind.

1. The solution obtained from the piecewise constant system in equation (1.36) can only provide the first-order accuracy in comparing with the numerical approach employing direct Taylor series expansion. (This will be further discussed in Chapter 5).

2. With the conversion from the continuous system in equation (1.35) to the piecewise constant system in equation (1.36), the ecological information embedded in the original continuous system is damaged. The accuracy and reliability of the solution such obtained are in turn hurt significantly.

3. Employment of the greatest integer function $[t]$ implies that all the functions and variables may only allow changing their values after a time duration of one unit length. This further reduces the accuracy of the solutions.

Nevertheless, the approach with the direct implementation of piecewise constant arguments is easy to use in solving for nonlinear dynamic systems and the governing equations with the piecewise constant arguments are convenient to use in numerical calculations on computers. It would be ideal if a methodology can be established in such a way that may keep the advantages of implementing the piecewise constant arguments while maintain high accuracy and reliability for the solutions.

A newly developed semi-analytical and numerical method named P-T method with the implementation of piecewise constant arguments will be presented in Chapter 5. A new piecewise constant argument in the form of $[Nt]/N$ will be introduced, where N can be constant. This allows the control of the time duration for piecewise constant system. With the P-T method, the accuracy of the solutions can actually be desired with Taylor series expansion. The solutions generated by the P-T method are continuous over the time intervals and the entire time domain considered. This method can also be used as a numerical method for solving nonlinear dynamic problems. The numerical results such provided with the P-T method are very accurate in comparing with the existing numerical methods such as Runge-Kutta method.

To demonstrate the main concepts of the P-T method, take the Froude pendulum modeled previously as an example. The Froude pendulum's governing equation (1.11) can be rewritten as the following second-order differential equation with a simplification procedure (Dai and Singh 1998).

$$\frac{d^2\theta_i}{d\tau^2} + a\frac{d\theta_i}{d\tau} = Q(\Omega) - bv_i^3 - h\sin d_i = f(d_i, v_i, \tau) \qquad (1.37)$$

This system is highly nonlinear. For desired accuracy, say the forth-order accuracy, for the solutions to be generated, expand the function $f(d_i, v_i, \tau)$ with Taylor series on an ith time interval $[N\tau]/N \le \tau \le ([N\tau]+1)/N$, we may have the following governing equation on the interval.

$$\frac{d^2\theta_i}{d\tau^2} + a\frac{d\theta_i}{d\tau} = f_{[N\tau]/N} + f'_{[N\tau]/N}\left(\tau - \frac{[N\tau]}{N}\right) + \frac{1}{2!}f''_{[N\tau]/N}\left(\tau - \frac{[N\tau]}{N}\right)^2$$

$$+ \frac{1}{3!}f'''_{[N\tau]/N}\left(\tau - \frac{[N\tau]}{N}\right)^3 \qquad (1.38)$$

As can be observed from equation (1.38), the right-side of the equal sign of the equation is now a function of time τ. Therefore, the solution for this second-order differential equation is readily available per the existing method ordinary differential equations. As such, the complete solution for equation (1.38) can be expressed in the following form on the interval $[N\tau]/N \le \tau \le ([N\tau]+1)/N$.

$$\theta_i = C_1 + C_2 e^{-a(\tau - [N\tau]/N)} + B_1\left(\tau - \frac{[N\tau]}{N}\right) + B_2\left(\tau - \frac{[N\tau]}{N}\right)^2$$

$$+ B_3\left(\tau - \frac{[N\tau]}{N}\right)^3 + B_4\left(\tau - \frac{[N\tau]}{N}\right)^4 \qquad (1.39)$$

In this expression, B_i are the coefficients to be determined with the conditions at one end of the interval. Practically, continuity of the angular displacement and velocity of the pendulum is not broken by the piecewise constant excitation described by the piecewise constant arguments. With the continuity and the solutions for each of the time intervals shown in equation (1.39), the complete solution over the entire time domain can be obtained.

The solution such developed can be very accurate when the procedures and the controlling parameters are properly used. It can be theoretically and numerically demonstrated that the solution generated by the P-T method has a better accuracy in comparing with that of the Runge-Kutta method. Also, the solution obtained is continuous everywhere in the time domain considered, in contrary to that of the other numerical methods that generate the solutions at the discrete points. Therefore, the solution developed is actually a semi-analytical one to the governing equation (1.37) with high accuracy. As will be exhibited later, the time intervals to the solution in equation (1.39) can as small as infinitesimal. Thus, theoretically, the solution may become the exact solution to the differential equation (1.37) when N approaches infinity.

Detailed procedures and techniques for obtaining the solutions of nonlinear dynamic systems with the piecewise constant arguments and the P-T method, together with the corresponding mathematical manipulations, will be presented in the subsequent chapters.

References

Aftabizadeh, A. R. and Wiener, J., "Oscillatory and Periodic Solutions for Systems of Two First Order Linear Differential Equations with Piecewise Constant Argument," Applicable Analysis, Vol. 26, pp. 327-338, 1988.

Busenberg, S. and Cooke, K. L., Models of Vertically Transmitted Disease with Sequential Continuous Dynamics, in "Nonlinear Phenomena in Mathematical Sciences" (V. Lakshmikantham, Ed.), Academic Press, New York, pp. 179-189, 1982.

Butenin, N. V., Elements of the Theory of Nonlinear Oscillations, Blaisdel, New York, 1965.

Buzdugan, G., Mihailescu, E. and Rades, M., Vibration Measurement, Marinus Nijhoff, Dordrecht, The Netherlands, 1986.

Cooke, K. L. and Wiener, J., "Retarded Differential Equations with Piecewise Constant Delays," Journal of Mathematical Analysis and Applications, Vol. 99, No. 1, pp. 256-297, 1984.

Cooke, K. L. and Wiener, J., "An Equation Alternate of Retarded and Advanced Type," Proceedings of the American Mathematical Society, Vol. 99, pp. 726-732, 1987.

Dai, L. and Singh, M. C., "On Oscillatory Motion of Spring-Mass Systems Subjected to Piecewise Constant Forces," Journal of Sound and Vibration, Vol. 173, pp. 217-233, 1994.

Dai, L. and Singh, M. C., "Periodic, Quasiperiodic and Chaotic Behavior of a Driven Froude Pendulum," International Journal of Non-Linear Mechanics, Vol. 33, No. 6, pp. 947-965, 1998.

Dai, L. and Singh, M. C., "A New Approach to Approximate and Numerical Solutions of Oscillatory Problems," Journal of Sound and Vibration, Vol. 263, No. 3, pp. 535-548, 2003.

Dai, L. and Wang, J. "A Study on the Effects of Workpiece Deflection and Motor Features on Nonlinear Vibrations in Machining," Journal of Vibration and Control, Vol. 13, No. 5, pp. 557-582, 2007.

Duffing, G., Erzwungene Schwingungen bei veränderlicher Eigenfrequenz und ihre technische Bedeutung, Braunsweig, F. Vieweg u. Sohn, 1918.

Elabbasy, E. M. and Saker, S. H., "Periodic Solutions and Oscillation of Discrete Nonlinear Delay Population Dynamics Model with External Force," IMA Journal of Applied Mathematics, pp. 1-15, 2005.

Garvie, M. B., McKiel, J. A., Sonenshine, D. E., and Campbell, A., "Seasonal Dynamics of American Dog Tick, Dermacentor Variabilis (say), Population in South Western Nova Scotia," Canadian Journal of Zoology, Vol. 65, pp. 28-39, 1978.

Fan, M. and Wang, K., "Periodic Solutions of a Discrete Time Non-Autonomous Ratio-Dependent Predator–Prey System," Mathematical and Computer Modelling, Vol. 35, pp. 951–961, 2002.

Holling, C. S., "Some Characteristics of Simple Types of Predation and Parasitism," The Canadian Entomologist, Vol. 91, pp. 385–398, 1959.

Holling, C. S., "The Functional Response of Predators to Prey Density and Its Role in Mimiery and Population Regulation," Memoirs of the Entomological Society of Canada, Vol. 45, pp. 3–60, 1963.

Huang, Y. K., "Oscillations and Asymptotic Stability of Solutions of First Order Neutral Differential Equations with Piecewise Constant Argument," Journal of Mathematical Analysis and Applications, Vol. 149, pp. 70-85, 1990.

Jayasree, K. N. and Deo, S. G., "Variation of Parameters Formula for the Equation of Cooke and Wiener," Proceedings of the American Mathematical Society, Vol. 112, No. 1. pp. 75-80, 1991.

Lakshmanan, M. and Rajasekar, S., Nonlinear Dynamics: Integrability, Chaos, and Patterns, Springer, New York, 2003.

Leung, A.Y.T., "Direct Method for the Steady State Response of Structures," Journal of Sound and Vibration, Vol. 124, pp. 135-139, 1988.

Liu, P. and Gopalsamy, K., "Global Stability and Chaos in a Population Model with Piecewise Constant Arguments," Applied Mathematics and Computation, Vol. 101, pp. 63-88, 1999.

Nayfeh, A. H. and Mook, D. T., Nonlinear Oscillations, New York, Wiley-Interscience, 1979.

Papaschinopoulos, G., Stefanidou, G., and Efraimidis, P., "Existence, Uniqueness and Asymptotic Behavior of the Solutions of a Fuzzy Differential Equation with Piecewise Constant Argument," Information Sciences, Vol. 177, pp. 3855-3870, 2007.

Ross, C. T. F., Finite Element Analysis in Engineering Science, Ellis Horwood, New York, 1990.

Schmitz, T. and Donaldson, R., Predicting High-Speed Machining Dynamics by Substructure Analysis, CIRP Annals, Vol. 49, No. 1, pp. 303-308, 2000.

Shah, S. M. and Wiener, J., "Advanced Differential Equations with Piecewise Constant Argument Deviations," International Journal of Mathematics and Mathematical Science, Vol. 6, pp. 671-703, 1983.

Stoker, J. J., Nonlinear Vibration in Mechanical and Electrical Systems, Interscience Publishers, New York, 1950.

Sun, Y. G, Saker, S. H., "Positive Periodic Solutions of Discrete Three-Level Food-Chain Model of Holling Type II," Applied Mathematics and Computation, Vol. 180, pp. 353-365, 2006.

Thompson, J. M. T. and Stewart, H. B., Nonlinear Dynamics and Chaos, John Wiley and Sons Ltd., New York, 1986.

Toro, V. D., Electrical Engineering Fundamentals (second edition), Prentice-Hall, Englewood Cliffs, New Jersey, 1986.

Wang, Y. and Yan, J., "Oscillation of a Differential Equation with Fractional Delay and Piecewise Constant Arguments," Computers and Mathematics with Applications, Vol. 52, pp. 1099-1106, 2006.

Weaver, W. J., Timoshenko, S. and Young, D. H., Vibration Problems in Engineering, John Wiley & Sons, Inc., New York, 1990.

Wiener, J., Generalized Solutions of Functional Differential Equations, World Scientific, New Jersey, 1993.

Wiener, J. and Aftabizadeh, A. R., "Differential Equations Alternately of Retarded and Advanced Types," Journal of Mathematical Analysis and Applications, Vol. 129, pp. 243-255, 1988.

Yuan, R., "The Existence of Almost Periodic Solutions of Retarded Differential Equations with Piecewise Constant Argument," Nonlinear Analysis, Vol. 48, pp. 1013-1032, 2002.

Zhang, B. G, and Parhi, N. "Oscillatory and Nonoscillatory Properties of First Order Differential Equations with Piecewise Constant Deviating Arguments," Journal of Mathematical Analysis and Applications, Vol. 139, pp. 23-25, 1989.

CHAPTER 2

Preliminary Theorems and Techniques for Analysis of Nonlinear Piecewise Constant Systems

2.1. Introduction

Before a thorough study on the methodology of solving and analyzing nonlinear piecewise constant systems, some fundamental knowledge on regular nonlinear dynamics systems and techniques for solving and analyzing the systems are necessary. These knowledge and techniques together with the others will be used in the consequent chapters for the piecewise constant systems. This chapter is for describing the knowledge on the fundamental behaviors of the nonlinear dynamic systems, such as periodic, quasiperiodic and chaotic responses of the nonlinear dynamic systems. These nonlinear behaviors together with the others are also common to nonlinear piecewise constant systems and even linear piecewise constant systems. The widely used tools of nonlinear mechanics, such as wave form, phase diagram, Lyapunov exponent, bifurcation, and the others are also useful for analyzing the behaviors of the piecewise constant systems. These tools will be briefly described in this chapter for the analysis of the piecewise constant systems.

2.2. Nonlinear Behaviors and Fundamental Analytical and Geometric Tools of Nonlinear Dynamics

Nonlinear dynamic systems show much richer behaviors in comparing with that of linear systems. In order to analyze the behaviors of nonlinear piecewise-constant dynamic systems, the knowledge on fundamental behaviors of nonlinear dynamic systems are necessary. Furthermore, some tools commonly employed in analyzing the behaviors of regular nonlinear dynamic systems are also needed in analyzing nonlinear piecewise-constant dynamic systems. Familiarity of the tools is important for analyzing the piecewise constant systems.

In comparing with linear dynamic systems, the behaviors of nonlinear dynamic systems are much involved and complicated (Weaver *et al.* 1990, Strogatz 2000, Lakshmanan 2003). The typical features of the nonlinear dynamic systems can be described as the following:

1. The technique of superposition used in studying linear dynamic systems is no longer valid for nonlinear dynamic systems.
2. The nonlinear dynamic system may have stable periodic solutions even for the case in which no harmonic excitations are applied.
3. The nonlinear dynamic system may provide multi-solutions.
4. The responses of nonlinear dynamic systems may be closely related to the stability of the system and the initial conditions.
5. The responses of nonlinear dynamic systems may not only show the behaviors with the same frequency as that of the external excitations, but also demonstrate the behaviors with multiple or fractional frequencies to that of the external excitations.
6. The frequencies, amplitudes and the other behaviors of nonlinear vibration or oscillatory systems may be sensitive to initial conditions.
7. The natural frequency of a nonlinear vibration system may correlate to the amplitude of the vibration of the system.
8. "Jumping phenomena" may occur to nonlinear dynamic systems.
9. The complex behaviors such as bifurcation, quasiperiodic oscillations and chaos may occur to nonlinear dynamic systems.

2.2.1. *Periodic Responses of Linear and Nonlinear Dynamic Systems*

A simplest linear dynamic system is probably an undamped free vibration system that can be represented by a one dimensional spring-mass system governed by the following homogeneous equation.

$$m\ddot{x} + kx = 0 \tag{2.1}$$

where m is the mass of the system, and x represents the displacement of the system from its equilibrium position. As the solution of the above equation must be a function that cancels its derivatives, the function should be similar in form to its derivatives. Conventionally, the solution for this vibration system is taken in the following form (Weaver *et al.* 1990).

$$x = e^{rt} \tag{2.2}$$

This leads to a characteristic equation corresponding to equation (2.1), such that

$$mr^2 + k = 0 \tag{2.3}$$

The roots of the characteristic equation are known as eigenvalues of the system. The general solution of the system can then be given by

$$x(t) = A_0 \sin(\omega t + \varphi_0) \tag{2.4}$$

where

$$A_0 = \left[x_0^2 + \left(\frac{v_0}{\omega} \right) \right]^{21/2} \quad \text{and} \quad \varphi_0 = \tan^{-1} \left(\frac{x_0 \omega}{v_0} \right)$$

in which $\omega = \sqrt{k/m}$ and x_0 and v_0 are the initial displacement and velocity respectively.

The velocity of the response of the system is then

$$\dot{x}(t) = A_0 \omega \cos(\omega t + \varphi_0) \tag{2.5}$$

The response of the system can be illustrated by a useful tool, x-t curve of the system, known as a wave form diagram shown in Figure 2.1. This solution of the system is obviously representing a periodic motion with period $2\pi/\omega$.

The diamonds ($\blacklozenge$) in the figure is a "Poincare Point" which will be defined and discussed in detail in the following section.

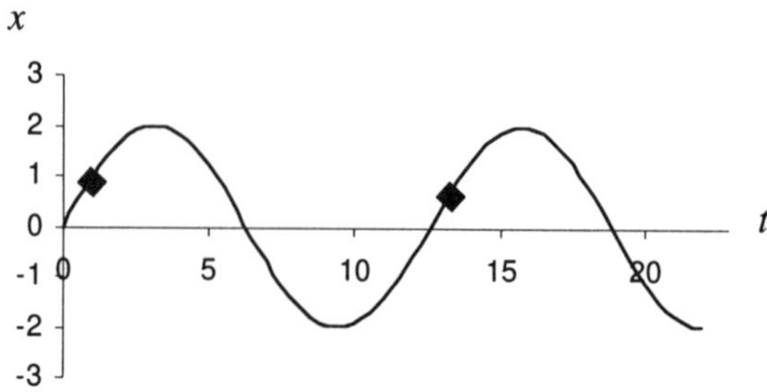

Figure 2.1.   Wave form (*x-t* curve) of an undamped free vibration. $\omega = 0.5$, $x_0 = 0$, $v_0 = 1$, amplitude is 2 and period is 4π.

A diagram shown the displacement x versus velocity $\dot{x}$ is known as phase diagram, which is also a very important tool for analyzing the behavior of a nonlinear dynamic system. It can be proved from equations (2.4) and (2.5) that the phase diagram of this system is an ellipse with one half-axis equals to A_0 and the other half-axis equals to $A_0\omega$ as shown in Figure 2.2. Such a motion of the spring-mass system is known as simple harmonic motion.

The harmonic and periodic responses may also be found in nonlinear dynamic systems. Taking a nonlinear Duffing's equation as an example, with the governing equation given as follow.

$$\ddot{x} + k\dot{x} + x^3 = B\cos t \qquad (2.6)$$

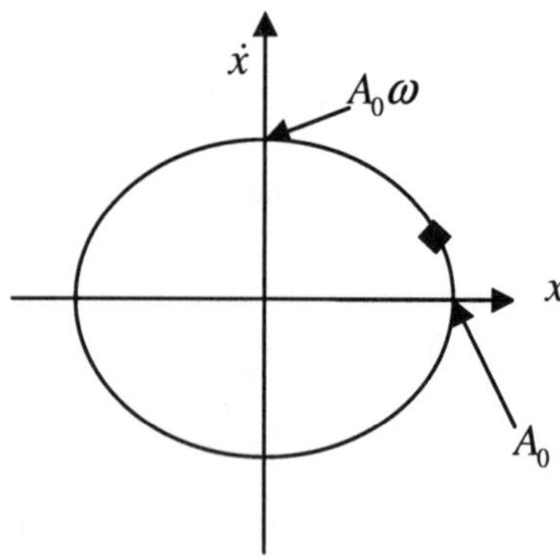

Figure 2.2.   Phase diagram of a simple harmonic motion.

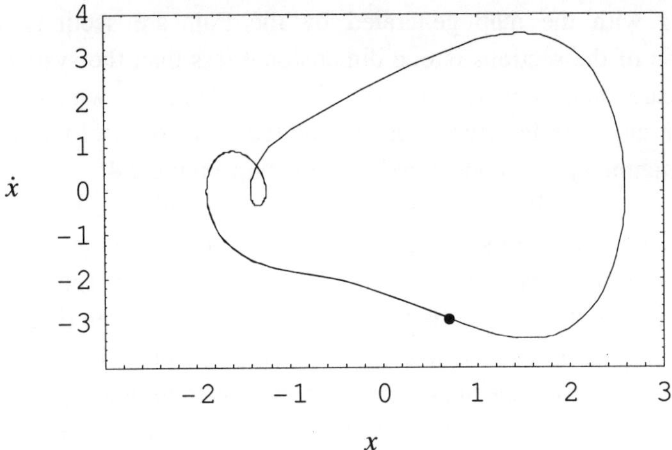

Figure 2.3. Phase diagram superposed with corresponding Poincare map for Duffing's equation, $k = 0.2$, and $B = 4$.

where x designates displacement, k represents damping coefficient, and B can be considered as the amplitude of the periodic external excitation. A phase diagram showing a periodic case of the system is given in Figure 2.3 above.

2.2.2. *Poincare Map*

More complex responses such as quasiperiodic and chaotic cases may happen to a nonlinear dynamic system, rather than periodic responses. Another tool named Poincare map after Henri Poincare (1946) shows great advantages in geometrically distinguishing between different types of responses of a nonlinear system. Poincare map is therefore widely used in analyzing the behavior of nonlinear dynamic systems.

A Poincare map is generated by Poincare sections in a phase space of the solution of a nonlinear system. Consider an autonomous system of n-dimension in matrix form.

$$\dot{x} = f(x,t) \tag{2.7}$$

One may generate a map with Poincare sections of n-1 dimension. It is expected that the analysis on the behavior of a dynamic systems will be easier with the map generated by the Poincare sections, as the dimension of the sections is one dimensional less than the system itself. A Poincare map consists of the Poincare points, which are the intersections of the Poincare sections and the trajectories of the response of the dynamic system considered, as shown in Figure 2.4.

The curves in Figure 2.4 represent the trajectories in a phase space and the dots on the Poincare section are the Poincare points. A Poincare map is therefore consists of discrete points on a Poincare section, which can be a phase plane. Assume that x^* in Figure 2.4 is a starting point of the trajectory. If in the case that x_k, x_{k+1} and all the other Poincare points overlap with x^*, the trajectory repeatedly returns to its starting point after a period T. The corresponding response of the dynamic system must be periodic. This can be seen from Figure 2.2, in which the diamond represents the Poincare points overlapped. Therefore, the Poincare map of a harmonious case consists only one single point and the trajectory in the phase diagram is closed. In fact, if a Poincare map shows merely finite number of points, the corresponding trajectory in phase diagram must be closed. Accordingly, the response of the dynamic system can only be periodic. To illustrate this geometrically, the phase diagram and Poincare map for a periodic case of a Duffing's equation are shown in Figures 2.5 and 2.6 respectively. It should be noticed that the Poincare map shown in Figure 2.5 is plotted for the response at the steady state and each of the points (Poincare points) shown in the figure actually

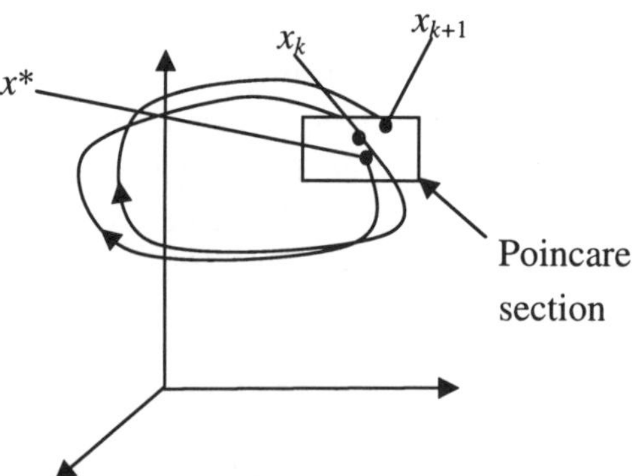

Figure 2.4.   Poincare section and phase trajectories.

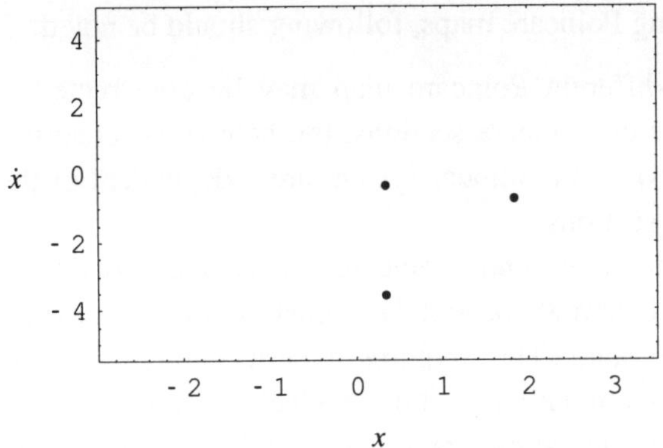

Figure 2.5.   Poincare map for Duffing's equation, $k = 0.2$, and $B = 8$.

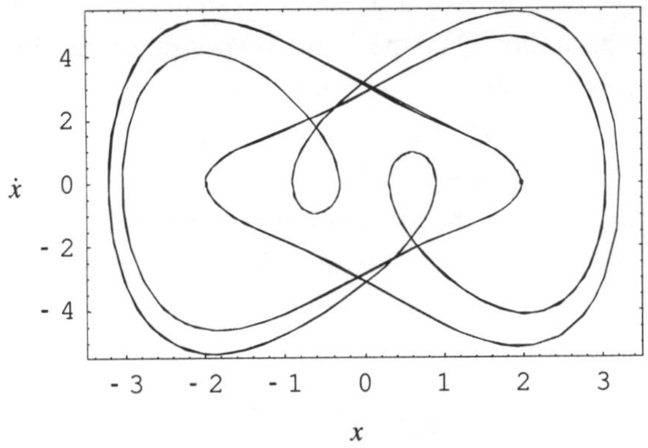

Figure 2.6.   Phase diagram corresponding to the Poincare map in Figure 2.5 for Duffing's equation, $k = 0.2$, and $B = 8$.

represent many points overlapped with each other. This implies that the curve shown in Figure 2.6 represents that many identical curves along the time axis (imagine an axis of time t perpendicular to the phase diagram) are overlapped in the phase diagram.

Obviously, with implementing Poincare maps, diagnosing the behaviors of a dynamic system becomes much efficient in comparing with the wave forms and phase diagrams.

In plotting Poincare maps, following should be noted:

1. Though different Poincare map may be constructed with different selections of Poincare sections, the behaviors (chaotic, periodic and the others) of a nonlinear system are independent of the selection of Poincare sections.
2. Phase diagrams and Poincare maps are usually generated by numerical simulations and they illustrate the responses of a system at its steady state. This requires a large enough time domain in the numerical simulations and the neglect of the starting iterations of the numerical calculation that are affected by the initial conditions.

2.2.3. *Quasiperiodic Response of Nonlinear Systems*

This is a type of response of a dynamic system that is close to periodic but not quite periodic. Quasiperiodic response of nonlinear dynamic systems can also be shown geometrically. Again, take the Duffing's equation shown in equation (2.6) as an example. The phase diagram of a quasiperiodic case is shown in Figure 2.7. As can be seen from the figure, the trajectories of the system are not perfectly overlapped as that of the perfect periodic case as shown in Figure 2.4, but the patterns of the trajectories show high similarity. Figure 2.8 shows the Poincare map of

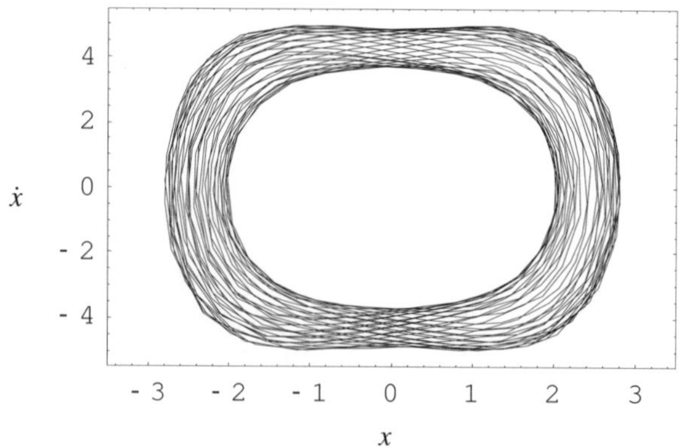

Figure 2.7. Phase diagram of a quasiperiodic case for a quasiperiodic motion governed by $\ddot{x} + k\dot{x} + x^3 = B\cos t$, where $k = 0.00001$, B = 1.8, initial conditions: $x_0 = -2$ and $\dot{x}_0 = 0$.

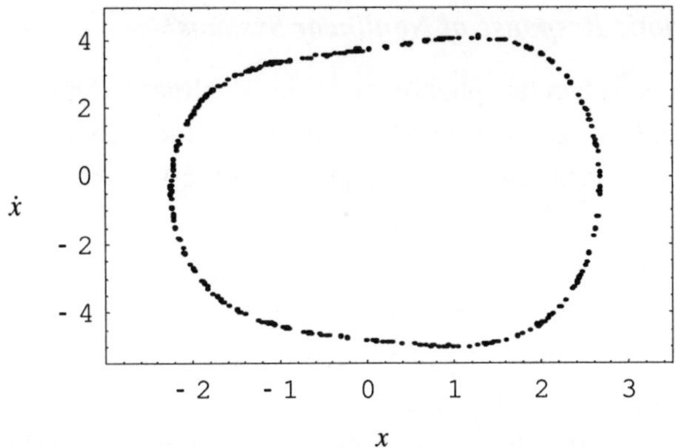

Figure 2.8.   Poincare map of a quasiperiodic motion corresponding to Figure 2.7.

this case. Connection of the points in the Poincare map may form a smooth curve.

The Poincare map of a quasiperiodic system can form a smooth curve as shown in Figure 2.8.

A good example of quasiperiodic case can also be found from a van der Pol system with harmonic external excitation (Lakshmanan 2003) governed by

$$\ddot{x} + \alpha(x^2 - 1)\dot{x} + \omega^2 x = F \cos \Omega t \qquad (2.8)$$

in which ω is the natural frequency of the system, Ω is the frequency of the excitation. It is found that if the ratio ω/Ω is rational, the system is periodic with period of Ω or $b\Omega$, where b is a rational number. When the ratio of ω/Ω becomes irrational, the response of the system is quasiperiodic (Lakshmanan 2003).

Periodic and quasiperiodic responses of a nonlinear system are not necessarily sensitive to initial conditions and the responses of the system for these cases can be predicted (periodic cases) or fairly predicted (quasiperiodic cases) as the responses either have the identical patterns (periodic cases) or similar patterns (quasiperiodic cases). Therefore, periodic and quasiperiodic responses are usually considered as regular responses.

2.2.4. *Chaotic Response of Nonlinear Systems*

Chaos is a spectacular phenomenon in nonlinear dynamics. Though precise definition for chaotic response of nonlinear dynamic systems is not availed yet, following are acceptable to the researchers in this field (Strogatz 2000, Nayfeh and Mook 1979).

1. Chaos is a phenomenon of nonlinear dynamic systems that is highly sensitive to initial conditions in comparing with linear dynamic system. The neighboring two trajectories of a chaotic system separate exponentially fast with time.
2. Though the trajectories of a chaotic system are random like, the range of the trajectories is approximately known and limited.
3. The dynamic system itself is deterministic. The random like responses of a chaotic system are due to the nonlinearity of the system.
4. Long-term prediction of the behavior of a chaotic system is impossible, though short-term prediction can be roughly made.

   A typical chaotic system has a Poincare map consisting of lines or clouds. Chaotic response of a nonlinear dynamic system can be geometrically identified by Poincare maps or phase diagrams. Figures 2.9 and 2.10 exhibit a phase diagram and its corresponding Poincare map for a chaotic case.

2.2.5. *Bifurcation of Nonlinear Systems*

In nonlinear dynamics, the behavior of a nonlinear system may change quantitatively as some control parameters of the system are changed. This quantitative change is called bifurcation in nonlinear dynamics and the parameter value at which the change occurs is knows as bifurcation point. More precisely, if we use v to represent a parameter of a nonlinear system, the system can be described with the following governing equation in matrix form.

$$\dot{x} = f(x, v) \tag{2.9}$$

   If the variation of the parameter from v to a critical value v_c causes the change of the behavior or the topological behavior change of

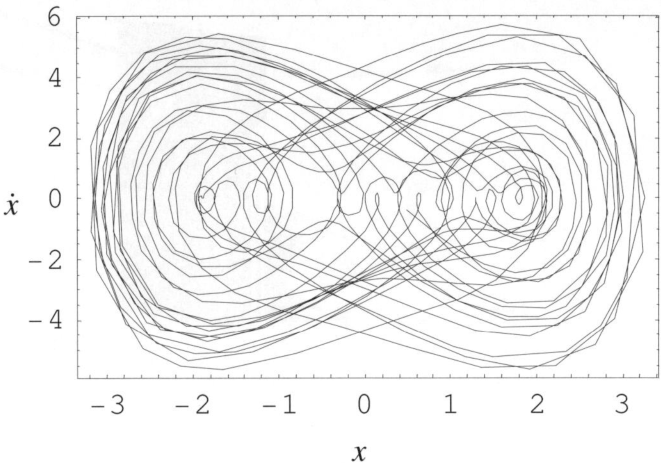

Figure 2.9. Phase diagram for a chaotic case governed by the Duffing's equation $\ddot{x} + k\dot{x} + x^3 = B\cos t$, where $k = 0.05$, B = 7, initial conditions: $x_0 = -2$ and $\dot{x}_0 = 0$.

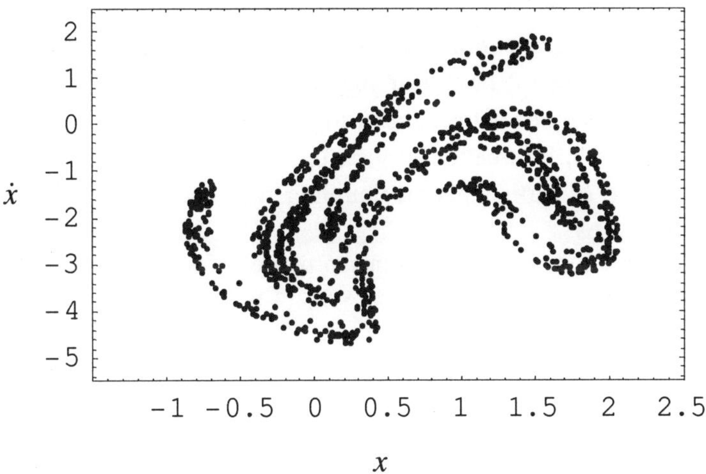

Figure 2.10. Poincare map corresponding to Figure 2.9 for a chaotic case governed by the Duffing's equation $\ddot{x} + k\dot{x} + x^3 = B\cos t$, where $k = 0.05$, B = 7, initial conditions: $x_0 = -2$ and $\dot{x}_0 = 0$.

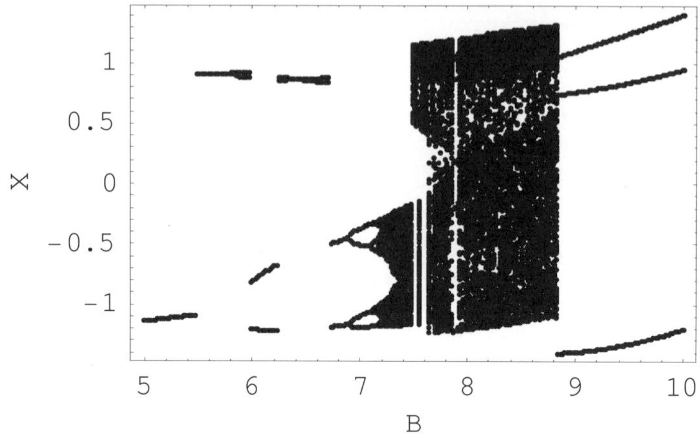

Figure 2.11.   Bifurcation diagram of governed by Duffing's equation $\ddot{x} + k\dot{x} + x^3 = B\cos t$, where $k = 0.25$, initial conditions: $x_0 = -2$ and $\dot{x}_0 = 0$.

the system, the change of the behavior is defined as bifurcation and v_c is called bifurcation point (Strogatz 2000, Nayfeh and Mook 1979).

Bifurcation can be graphically demonstrated in many ways. One of the bifurcation diagram commonly used is shown in Figure 2.11.

This is to show the nonlinear characteristics of a Duffing's equation. The vertical axis is the displacement. It can be clearly visualized when B reaches about 5.5, bifurcation occurs, this is known as period doubling. Further reduce b, further period doubling will occur till appearance of chaotic responses.

Therefore, bifurcation diagrams provide information regarding the sates of the behavior of a nonlinear system with respect to the variation of a control parameter of the system.

2.3.   Lyapunov Exponent

Behaviors of nonlinear dynamic systems are much more complex and hard to solve in comparing with that of linear dynamic systems. Periodic, quasiperiodic and chaotic responses may occur for a given single nonlinear dynamic system. In analyzing the characteristics of nonlinear dynamic systems, it is therefore crucial for developing criteria to distinguish the nonlinear behaviors from the regular ones and to determine under what conditions chaos or the other types of responses may occur. Many criteria and techniques for diagnosing chaos as well as regular and the other types of nonlinear responses of nonlinear systems have been developed by the researchers in this field. Among these, the Power Spectral, Fractals, Poincare Maps, Entropy, and Lyapunov-Exponent are commonly seen in using in nonlinear analyses (Strogatz 2000, Baker and Gollub 1990, Solari *et al.* 1996). Lyapunov Exponent is probably a most quantitatively and theoretically sound criterion among all the other criteria and therefore widely used in nonlinear analysis practice. It measures the sensitivity of a nonlinear system its initial conditions, and the sensitive response to initial condition is one of the important behavior of chaotic systems (Lyapunov 1935). Lyapunov-Exponent is hence used as a criterion for determining whether a response of a dynamic system is chaotic or periodic.

With Lyapunov-Exponent, the sensitivity of a dynamic system to initial conditions is determined by measuring the average exponential rates of divergence or convergence of close orbits of the system in the phase space. Assume there are two identical nonlinear dynamic systems evolving from two slightly different initial conditions, x and $x + \varepsilon_0$, where ε_0 can be a very small number. After n iterations of evolution in a numerical simulation of the dynamic systems, the divergence of the two systems can be described approximately as:

$$\varepsilon(n) \approx \varepsilon_0 e^{\lambda n} \tag{2.10}$$

λ in the equation is defined as Lyapunov-Exponent, which actually gives an average rate of divergence of two nearby trajectories starting from slightly different initial states (Lakshmanan and Rajasekar 2003).

Consider a nonlinear dynamic system; say a system which is governed by a general equation that can be numerically determined by the following mapping relation.

$$x_{n+1} = f(x_n) \qquad (n = 0, 1, 2, \ldots) \tag{2.11}$$

Assuming Δx is a small value indicating the difference between the two identical systems at the initial states, after n iterations, the divergence happens and can be expressed as

$$f^n(x+\Delta x) - f^n(x) \approx \Delta x e^{n\lambda} \tag{2.12}$$

By taking the natural logarithm to the equation above, one may obtain

$$\text{Ln}\left[\frac{f^n(x+\Delta x) - f^n(x)}{\Delta x}\right] \approx n\lambda. \tag{2.13}$$

in which

$$f^n(x) = f[f[\cdots f[f(x)]\cdots]] \tag{2.14}$$

Because Δx is a small number, equation (2.13) may also take the following form:

$$\lambda = \frac{1}{n}\text{Ln}\left|\frac{df^n}{dx}\right| \tag{2.15}$$

This is a general expression of Lyapunov-Exponent (Baker and Gollub 1990).

Making use of equation (2.14) and taking the chain rule for the derivative of the nth iterate, Lyapunov-Exponent can be determined. Obviously, the description of the Lyapunov-Exponent can be true only if infinite number of n can be taken. As such, Lyapunov-Exponent is defined as the following.

$$\lambda = \lim_{n\to\infty}\frac{1}{n}\sum_{i=0}^{n-1}\text{Ln}\left|f'(x_i)\right| \tag{2.16}$$

This may also be given by

$$\lambda = \lim_{n\to\infty}\lim_{\varepsilon_0\to 0}\frac{1}{n}\text{Ln}\left|\frac{\varepsilon(n)}{\varepsilon_0}\right| \tag{2.17}$$

Lyapunov-Exponent such defined actually gives the stretching rate per iteration, averaged over the trajectory of the system. With this definition, if λ is negative, slightly separated trajectories of the system converge and the evolution is not chaotic. If λ is positive, nearby trajectories diverge, this implies that the evolution of the systems considered is sensitive to initial conditions and therefore the system is chaotic.

Numerically determining Lyapunov-Exponent requires a small deviation $\varepsilon(n)$. When the deviation or distance between the trajectories of the two identical systems are getting large, equation (2.16) or (2.17) is hard to use. Practically, the following equation is used in numerically calculating for the Lyapunov-Exponent (Bennetin *et al.* 1978).

$$\lambda = \lim_{m \to \infty} \frac{1}{m\Delta t} \sum_{i=0}^{m} \mathrm{Ln}\,\frac{d_i}{d_0} \tag{2.18}$$

Obviously, m in this equation must be a large enough finite number in the numerical calculation. Δt is a constant time span and increment. d_i in the equation represents the distance between the two trajectories at the beginning of the ith time span, $d_i = |y_i - x_i|$, $i = 1,2,\dots m$, letting y and x be the solutions of the two trajectories respectively and x be the solution of the base trajectory. In performing the calculation corresponding to a time span, y is always started from a constant distance d_0 from the base trajectory.

It should be noted however that one Lyapunov-Exponent calculated may represent the divergence of the corresponding two trajectories in one direction only. This implies that the number of Lyapunov-Exponents for a given system is equal to the dimension of the system, i.e., the number of first order differential equations of the governing equations of the system. It is therefore the largest Lyapunov-Exponent to be used for diagnosing whether a system is chaotic or regular.

It should also be noted that Lyapunov-Exponents are determined numerically in nonlinear analysis practice. Therefore, the accuracy and reliability of the Lyapunov-Exponent calculated highly depend on the accuracy and reliability of the numerical calculations, m in equation (2.18) used, accuracy of the solutions of the system, and the numerical method used.

In quantitatively diagnosing chaos, Lyapunov-Exponent is probably the most popular criterion. However, this criterion may not necessarily provide reliable results for all the responses of nonlinear systems. Also,

Lyapunov-Exponent may determine whether a response is chaotic or periodic, but the responses other than chaotic and periodic can not be diagnosed by this criterion. The author recently developed a new quantitative criterion named Periodicity-Ratio which may diagnose chaotic responses that Lyapunov-Exponent has difficulties with. Periodicity-Ratio quantitatively describes the periodicity of the responses of nonlinear systems. It may be employed to predict chaos, periodic and the other responses in between pure chaos and completely periodic ones. Periodicity-Ratio will be described in details in Chapter 7 together with the comparisons with Lyapunov-Exponent.

Though many criteria and techniques are available for diagnosing nonlinear behavior of a dynamic system, as recognized in the field of nonlinear dynamics (Baker and Gollub 1990, Strogatz 2000), there is still lack of a theoretically and practically sound technique which can completely diagnose all the behavior of nonlinear systems with high accuracy and reliability.

2.4.   Characteristics of Numerical Solutions and Runge-Kutta Method

Closed-form or analytical solutions are hard if not impossible to obtain for the differential equations governing nonlinear dynamic systems as indicated in Chapter 1. Only a very few nonlinear systems are found having analytical solutions with the existing analytical methods and the analytical solutions of nonlinear systems are usually associated with the special functions (not the fundamental functions for linear systems) such as elliptic functions. This implies that the analytical methods existing in the field of differential equations are not satisfied for solving all the nonlinear systems. Numerical analysis is therefore not only important but also necessary for solving many of nonlinear problems in engineering and other applications. The availability of high-speed digital computers has made the numerical analyses feasible, effective and reliable.

Numerous numerical methods have been created for numerical calculations, such as the Euler method, finite difference method, the

Taylor-series method, the Runge-Kutta method, the Houbolt method, the Wilson method, the Milne method and the Newmark method (Nakamura 1977, Nayfeh 1979). Among these methods, Runge-Kutta method especially the 4th order Runge-Kutta method is probably the most popular numerical method used in physics and engineering fields due to its high accuracy and reliability. The 4th order Runge-Kutta method is related to Taylor series expansion and its accuracy is equivalent to 4th order accuracy of Taylor expansion. For obtaining numerical solutions with the Runge-Kutta method to higher-order differential equations, one may have to reduce the differential equation to a system of first-order differential equations. A general one-dimensional dynamic system can be described by

$$m\ddot{x} + c\dot{x} + kx = F \qquad (2.19)$$

where m, c, and k are system parameters and F can be a function of time, x and $\dot{x}$. Rewrite the governing equation (2.19) in two first-order differential equations:

$$\begin{cases} \dot{x} = y \\ \dot{y} = \dfrac{1}{m}(F - c\dot{x} - kx) \end{cases} \qquad (2.20)$$

Define the matrices

$$X = \begin{Bmatrix} x(t) \\ y(t) \end{Bmatrix} \quad \text{and} \quad f = \begin{Bmatrix} y(t) \\ \dfrac{1}{m}(F - c\dot{x} - kx) \end{Bmatrix} \qquad (2.21)$$

With the formulas of 4th order Runge-Kutta method, taking a time step h, we may find the numerical solution of equation (2.21) by the following recurrence relations.

$$X_{i+1} = X_i + \frac{1}{6}(k_1 + 2k_2 + 2k_3 + k_4) \qquad (2.22)$$

where

$$k_1 = hf(X_i, t_i); \qquad k_2 = hf\left(X_i + \frac{1}{2}k_1, t_i + \frac{1}{2}h\right)$$

$$k_3 = hf\left(X_i + \frac{1}{2}k_2, t_i + \frac{1}{2}h\right); \qquad k_4 = hf(X_i + k_3, t_i + h)$$

Implementing the formulas and the initial conditions $x_{t=0} = x_0$ and $y_{t=0} = \dot{x}_{t=0} = v_0$, the solution $x(t)$ as well as $\dot{x}(t)$ can be numerically determined for $t \geq 0$ as desired.

Some characteristics of the Runge-Kutta method can be listed here as the reference for the comparison with a new numerical method named P-T method to be discussed in Chapter 5.

1. Runge-Kutta method is a single-step method and it is also known as an explicit integration method (Nakamura 1977). This characteristic shows advantage in numerical calculations with high efficiency.
2. The numerical solution generated by Runge-Kutta method is discrete. The solution does not satisfy the governing equation at all time t but only at the discrete times separated by time intervals.
3. A dynamic system governed by a second-order differential equation are usually reduced into two first-order differential equations before applying the Runge-Kutta method.
4. The variations of the displacement x and its first and second derivatives are based on the formulas of the Runge-Kutta method with the order of accuracy desired.
5. The local truncation error for the 4th order Runge-Kutta method is $x^{(5)}(\xi)h^5/180$ where ξ is a point between x_0 and x_n (Simmons and Kantz 2007).

As indicated above, numerical solution is an approximate solution to the governing differential equations of dynamic systems. Errors of numerical calculations are inevitable. For numerical solutions of high accuracy and reliability, a great attention needs to be paid to the errors and the sources of the errors. The major error sources can be listed as the following:

- Truncation error. Some numerical methods such as Runge-Kutta method use only the first few terms of Taylor series expansion, and the rest of terms are truncated to raise truncation errors. This type of error is due to the numerical method itself. It is also known as discretization error.

- Round-off error. Any numerical calculation with computers may calculate only to a finite number of decimal places, or the numerical values calculated are rounded off to the decimal place limited by the accuracy of the computer used. This error may accumulate as the number of computations is increases.

- Propagation error. For most of the numerical methods, the solution x_i is based on the previous solution x_{i-1}. Therefore, x_i inherits error from x_{i-1}, and the error may propagate with the performance of the numerical calculations.

The knowledge and techniques described in this chapter are mainly for the use in the development of the concepts and methodologies of nonlinear piecewise constant systems to be discussed in the following chapters. The responses of regular continuous nonlinear dynamic systems can be more complex and the analyses for them can be more involved. Some of them will be discussed at the places where they are needed. Further information regarding the regular continuous nonlinear dynamic systems can be found accordingly from the literature in this field.

References

Baker, G. L. and Gollub, J. P., Chaotic Dynamics an Introduction, Cambridge University Press, Cambridge, 1990.

Bennetin, G., Galgani, L., Giorgilli, A., and Strelcyn, J. M., "Lyapunov Characteristic Exponents for Smooth Dynamical Systems and for Hamiltonian Systems: A Method for Computing All of Them," Meccanica, Vol. 15, pp. 9-30, 1980.

Lakshmanan, M. and Rajasekar, S., Nonlinear Dynamics: Integrability, Chaos, and Patterns, Springer, New York, 2003.

Lyapunov A. M., The General Problem of the Stability of Motion, United Scientific and Technical Publishing House, 1935.

Nakamura, S., Computational Methods in Engineering and Science, Wiley, New York, 1977.

Nayfeh, A. H. and Mook, D. T., Nonlinear Oscillations, New York, Wiley-Interscience, 1979.

Poincare, H., The Foundation of Science: Science and Method (English Translation), The Science Press, Lancaster, PA, 1946.

Simmons, G. F. and Kantz, S. G., Differential Equations: Theory, Technique, and Practice, McGraw-Hill, Boston, 2007.

Solari, H., Natiello, M. and Mindlin, G., Nonlinear Dynamics, A Two Way Trip from Physics to Math, Institute of Physics Publishing, London, 1996.

Strogatz, S. H., Nonlinear Dynamics and Chaos: With Applications to Physics, Biology, Chemistry, and Engineering, Westview Press, Cambridge, MA, 2000.

Weaver, W. Jr., Timoshenko, S. and Young, D. H., Vibration Problems in Engineering, John Wiley & Sons Inc., New York, 1990.

CHAPTER 3

Piecewise Constant Dynamical Systems and Their Behavior

3.1. Introduction

With the knowledge and techniques on the characteristics of the piecewise constant variables and the functions containing piecewise constant valuables as described in the previous chapters, the investigation on the differential equations and linear and nonlinear dynamic systems containing piecewise constant variables is readily available. In this chapter, a systematic study of the motions of dynamical systems subject to displacement related piecewise constant external forces of the form $f(x([t]))$ is undertaken. Detailed methodology on the development of the governing equations for the dynamic systems and their corresponding solutions are to be presented and the results will be interpreted both on a theoretically and physically sound basis.

Unlike the classical equations of motion for linear vibration in which the external forcing function $f(t)$ is known and continuous in the range of $t \in [0, \infty)$, the governing equations investigated in this chapter involve a forcing function $f(x([t]))$ in which $[t]$ is a greatest integer function with the definition and characteristics discussed in Chapter 1. The function $f(x([t]))$ is discontinuous and unknown until the complete solution, $x(t)$, is obtained. As the convention (Inglis 1944, Weaver, Timoshenko, and Young 1990, Den Hartog 1951), the displacement x is usually measured from the equilibrium position of any vibration or oscillatory systems considered and the time t is sought for the range $t \in [0, \infty)$. The forcing function $f(x([t]))$ is displacement

61

related with a constant value in each unit duration interval in between two integer points of time. A time plot of a typical forcing function $f(x([t]))$ is shown graphically in Figure 3.1. As can be seen in Figure 3.1, the piecewise constant force $f(x([t]))$ is discontinuous with a different amplitude from interval to interval.

Within a time interval of unit duration, the motion considered has the structure of a continuous dynamic system of linear vibration. Continuity of displacement and velocity of the motion produces a recurrence relation which provides initial conditions for the motion in the adjacent unit time intervals at each integral point of time. Initial value problems so constructed are discussed in detail in the following sections for undamped and damped vibration systems subjected to piecewise constant forces. Complete solutions of the linear and nonlinear problems under

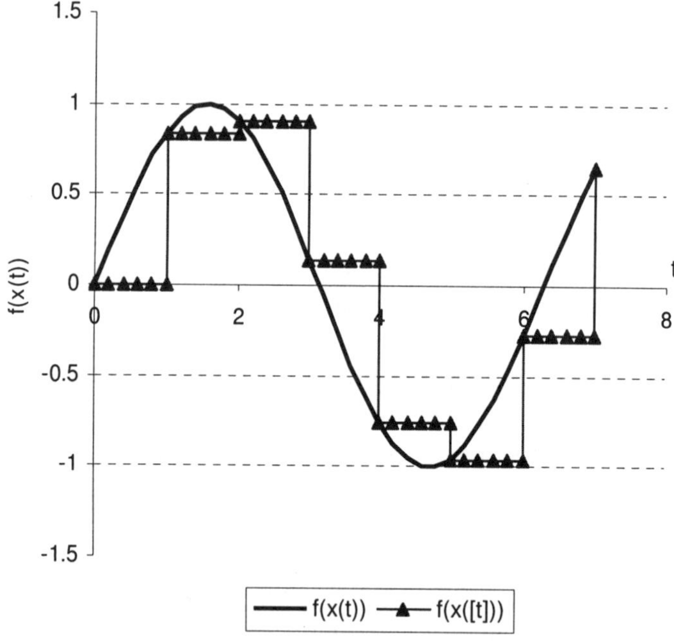

Figure 3.1.   Time plot of a typical piecewise constant force $f(x([t]))$ with comparison to a continuous force $f(x(t))$.

the piecewise constant forces of the form $f(x([t]))$ or $f([t])$ are derived with motions entirely different from those of the classical continuous linear or nonlinear vibrations. However, it is significant to note that the solution and its first derivative for a given dynamic system subjected to the discontinuous forces are continuous on the whole time range.

As indicated in Chapter 1, a mathematical model consisting of a piecewise constant variable was first constructed by Busenberg and Cooke (1982). Since then, the linear first-order differential equations involving piecewise constant variables of retarded and advanced types have been studied by various research scholars in this area. In the last few years, there has been increasing interest in the study of the oscillatory behavior of the solution for differential equations with piecewise constant variables. The important contributions in this area are found in the following works.

A scalar differential equation of first-order with piecewise constant variables was first analyzed by Cooke and Wiener (1984). The oscillatory properties of its solution were reported with detailed analysis later by Cooke and Wiener (1987) and Zhang and Parhi (1989). Since then, many academic works on theories of the differential equations with piecewise constant variables and behavior of the solutions of the equations with different types of piecewise constant variables have been published (Jayasree and Deo 1991, Zhang and Parhi 1989, Aftabizadeh and Wiener 1988, Wiener and Aftabizadeh 1988, Wang and Yan 2006). However, most of the investigations in this field are pure mathematical approaches concentrating on stability, oscillation and existence of the solutions, as discussed in the Chapter 1. Therefore, a systematic study of the modeling and properties of the practically sound dynamic systems governed by second-order differential equations with piecewise constant arguments is not only significant for the applications in science and engineering fields but also necessary for better comprehension of the linear and nonlinear dynamic systems.

Motion of a dynamic system disturbed by a piecewise constant force shows an entirely different behavior from that of the corresponding continuous system (Dai and Singh 1994). The motion's characteristics of all the systems considered will be graphically presented for various combinations of coefficients in the equations of motion and different initial conditions. The solutions obtained will also be compared with those corresponding to the conventional continuous systems. For analysis of motion, the numerical results will be presented graphically.

Due to the characteristics of the discontinuous disturbance, dynamic systems subjected to piecewise constant forces may show some peculiar behavior under certain conditions. The typical peculiar behavior will be demonstrated for several dynamic systems under piecewise constant excitations.

It should be noted that some exponential matrices may be derived for the solutions of piecewise constant systems in many cases and the matrices characterize the oscillatory behavior of a system subjected to piecewise constant forces. Analysis of the properties of motion for the systems may therefore be based on the attributes of the exponential matrices. The oscillatory condition for the motion of the systems can therefore be given with respect to the eigenvalues of the corresponding exponential matrices. Oscillatory and asymptotic properties of motion of an undamped spring-mass system subjected to piecewise constant forces are thus ready to be analyzed both theoretically and numerically on the basis of the exponential matrix of the system.

3.2. Governing Equations of Dynamic Systems with Piecewise Constant Variables

Analysis of the behavior of a piecewise constant system can be started with a damped spring-mass system subjected to an external piecewise constant force expressible in the form $f(x([t]))$. A linear spring-mass system would seem to be one of the simplest physical systems. However, the behavior of the spring-mass systems may be rich and complex when it is subjected to a piecewise constant force. The study of the motion of a spring-mass system subjected to a piecewise constant force can be facilitated by mathematical models formulated as second-order differential equations with piecewise constant arguments. A piecewise constant force acting on the spring-mass system has intervals of constancy and may vary its magnitude or its magnitude and direction simultaneously at certain points of time. Continuity of displacement and velocity of the system at a point joining any two consecutive intervals then implies recurrence relation for the solution at such points.

The equation of motion for such piecewise constant dynamic system can be expressed in general as the following.

$$m\ddot{x} + c\dot{x} + kx = f(x([t])) \tag{3.1}$$

where a dot denotes differentiation with respect to time t, $x(t)$ is the displacement of the mass from the equilibrium position of the system, m is the mass, c is the damping coefficient, and k is the spring constant. The exerting force $f(x([t]))$ in equation (3.1) is piecewise continuous over the entire range $t \in [0,\infty)$, and varies its magnitude or magnitude and direction simultaneously at the moment $t = [t]$ and keeps a constant value in the interval between the two integer points of time, $[t] \le t < [t]+1$.

Therefore, the equation of motion of the system represents a continuous dynamic system of linear vibrations within a unit time interval. In general, the force $f(x([t]))$ so exerted on the mass will produce a motion with entirely different behavior from the simple harmonic vibration.

Equation (3.1) will be solved for each unit-duration interval with the conditions at the corresponding integer points of time. Solution $x(t)$ on the range of $t \in [0,\infty)$ satisfies the following conditions:

(1) the displacement $x(t)$ and its derivative $\dot{x}(t)$ are continuous over $t \in [0,\infty)$;

(2) in $t \in [0,\infty)$, the derivative $\ddot{x}(t)$ exists at each point, with the possible exception at the integer points $t = [t]$ at which left-hand side derivatives exist;

(3) the solution $x(t)$ satisfies equation (3.1) on each time interval $n \le t < n+1$, where n is an integer point on $t \in [0,\infty)$;

(4) there is a continuous linear dynamic system on each of the time interval $n \le t < n+1$, corresponding to the piecewise constant system governed by equation (3.1).

3.3.   Solution Development of Simple Dynamic Systems Subjected to Piecewise Constant Excitations

There is no direct solutions for piecewise constant systems, which are fundamentally discrete systems. Development of the solutions for piecewise constant dynamic systems is therefore unique in comparing with that of conventional continuous systems. For purpose of clarity and simplification, the solution for the following equation of motion of an undamped spring-mass system is first considered for demonstrating the basic procedures of solution development of piecewise constant dynamic systems.

$$m\ddot{x}(t) + kx(t) = Ax([t]) \tag{3.2}$$

where, A is a parameter which refers to the magnitude of the exerting force. The governing equation models a spring-mass system subjected to a piecewise constant force related to the displacement.

The equation of motion can also be expressed as

$$\ddot{x}(t) + \omega^2 x(t) = \beta x([t]) . \tag{3.3}$$

where $\omega^2 = k/m$ and $\beta^2 = A/m$. The initial conditions for the system may be assumed as follows:

$$x(0) = d_0 \quad \text{and} \quad \dot{x}(0) = v_0 \tag{3.4}$$

In an arbitrary time interval $n \leq t < n+1$, designate $x_n(t)$ as a displacement, equation (3.3) can be written as

$$\ddot{x}_n(t) + \omega^2 x_n(t) = \beta x_n([t]) \tag{3.5}$$

The initial conditions on the basis of equation (3.4) are expressible at an instant $t = [t] = n$ as

$$x_n(n) = d_n \quad \text{and} \quad \dot{x}_n(n) = v_n \tag{3.6}$$

As the excitation force $f = \beta x[t]$ employed in equation (3.3) is a constant on the time segment $n \leq t < n+1$, the solution to the linear ordinary differential equation (3.5) with the conditions in equation (3.6) is readily available. The equation (3.5) has a general solution in $n \le t < n+1$ as below

$$x_n(t) = d_n \left\{ \left(1 - \frac{\beta}{\omega^2}\right) \cos[\omega(t-n)] + \frac{\beta}{\omega^2} \right\} + v_n \frac{\sin[\omega(t-n)]}{\omega} \quad (3.7)$$

Similarly, the corresponding solution on the interval $n-1 \le t < n$ is

$$x_{n-1}(t) = d_{n-1} \left\{ \left(1 - \frac{\beta}{\omega^2}\right) \cos[\omega(t-n+1)] + \frac{\beta}{\omega^2} \right\}$$
$$+ v_{n-1} \frac{\sin[\omega(t-n+1)]}{\omega} \quad (3.8)$$

where d_{n-1} and v_{n-1} are displacement and velocity of the system at the very time of $t = n-1$.

As mentioned previously, the displacement $x(t)$ and velocity $\dot{x}(t)$ are continuous in $t \in [0,\infty)$. Physically, this implies that there is no jump (break) or discontinuity of $x(t)$ and $\dot{x}(t)$ on $t \in [0,\infty)$, this is rational for most of the physical problems. It can be assumed that at the right moment of $t = [t]$, a change of the force F from $Ax([t])$ to $Ax([t]+1)$ affects the acceleration $\ddot{x}(t)$, however, the change of F itself will not physically disrupt the continuity of displacement $x(t)$ and velocity $\dot{x}(t)$. Therefore, the displacement $x(t)$ and velocity $\dot{x}(t)$ must be continuous for all $t > 0$ satisfying the following conditions of continuity.

$$x_n(n) = x_{n-1}(n) \quad \text{and} \quad \dot{x}_n(n) = \dot{x}_{n-1}(n) \quad (3.9)$$

at the points joining the consecutive unit duration intervals. x_n and $\dot{x}_n$ in the above equations are the end conditions for the system in an interval of unit duration, i.e., $n-1 \le t < n$. They are obviously also the local initial conditions for the succeeding interval of unit duration, $n \le t < n+1$. A recurrence relation between d_n, v_n and d_{n-1}, v_{n-1} can thus be determined from the conditions of continuity (3.9) and the solutions given by equations (3.7) and (3.8) in a matrix form

$$\begin{bmatrix} d_n \\ v_n \end{bmatrix} = \begin{bmatrix} (1-\beta/\omega^2)\cos\omega + \beta/\omega^2 & \sin\omega/\omega \\ -\omega(1-\beta/\omega^2)\sin\omega & \cos\omega \end{bmatrix}\begin{bmatrix} d_{n-1} \\ v_{n-1} \end{bmatrix} \tag{3.10}$$

Through an iterative procedure, final values of d_n and v_n can be associated with the initial displacement and velocity d_0 and v_0 by multiply the n square matrices, such that

$$\begin{bmatrix} d_n \\ v_n \end{bmatrix} = \begin{bmatrix} (1-\beta/\omega^2)\cos\omega + \beta/\omega^2 & \sin\omega/\omega \\ -\omega(1-\beta/\omega^2)\sin\omega & \cos\omega \end{bmatrix}^n\begin{bmatrix} d_0 \\ v_0 \end{bmatrix} \tag{3.11}$$

which describes the displacements and velocities of the system at the integral values of time. It can be seen from this equation that the displacement d_n and velocity v_n at an integer point of t are independent of the antecedent values of d_{n-1}, v_{n-1}, d_{n-2}, $v_{n-2}, \ldots$, at the integer points of time, except for the initial values of d_0 and v_0 which are known to us. Therefore, the instantaneous values of displacement and velocity of the system at each integer point $t=[t]$ for all the time $t>0$ can be obtained by equation (3.11) as long as the constants ω and β, and the initial conditions are given.

Since x_n represents the displacement with respect to time on an arbitrary interval $n \leq t < n+1$, and the motion of the system is continuous, the complete solution of equation (3.3) can therefore be expressed on the entire time range of $t \in [0,\infty)$ as follows.

$$x(t) = \left[\left(1 - \frac{\beta}{\omega^2}\right)\cos[\omega(t-[t])] + \frac{\beta}{\omega^2}\frac{\sin[\omega(t-[t])]}{\omega} \right]\begin{bmatrix} d_{[t]} \\ v_{[t]} \end{bmatrix} \tag{3.12}$$

where the displacement and velocity at $t=n$ are related to the given initial displacement d_0 and initial velocity v_0 by the following equation.

$$\begin{bmatrix} d_{[t]} \\ v_{[t]} \end{bmatrix} = \begin{bmatrix} (1-\beta/\omega^2)\cos\omega + \beta/\omega^2 & \sin\omega/\omega \\ -\omega(1-\beta/\omega^2)\sin\omega & \cos\omega \end{bmatrix}^{[t]}\begin{bmatrix} d_0 \\ v_0 \end{bmatrix} \tag{3.13}$$

With equations (3.12) and (3.13), in other words, the displacement of the piecewise constant system governed by equation (3.3) can be completely determined at a given time t so long as the initial conditions are known.

3.4. Development of Analytical Solutions via Piecewise Constant Variables

In certain cases, analytical solutions can be developed for some linear dynamic systems with implementation of piecewise constant variables. Let us take the piecewise constant system considered in the previous section as an example. One may have noted that there is no restriction whatsoever on the values or ranges of the parameters ω and β in the derivation of the solution of equation (3.12). The equation therefore indeed represents a general solution to the system governed by equation (3.3). We may now consider two extreme cases as β and ω approach zero respectively. These limiting cases may lead to some innovative approaches for solving differential equations. The technique and results obtained in this section will be implemented for the studies in the subsequent sections.

When β approaches zero, the spring-mass system tends to be a free undamped continuous system and its solution on the basis of equation (3.12) assumes the form

$$x(t) = \begin{bmatrix} \cos[\omega(t-[t])] & \dfrac{\sin[\omega(t-[t])]}{\omega} \end{bmatrix} \begin{bmatrix} d_{[t]} \\ v_{[t]} \end{bmatrix} \tag{3.14}$$

where

$$\begin{bmatrix} d_{[t]} \\ v_{[t]} \end{bmatrix} = \begin{bmatrix} \cos\omega & \sin\omega/\omega \\ -\omega\sin\omega & \cos\omega \end{bmatrix}^{[t]} \begin{bmatrix} d_0 \\ v_0 \end{bmatrix} \tag{3.15}$$

Significantly, as can be proved (see Appendix A), the exponential matrix in the equation above can be rewritten as a single matrix, such that

$$\begin{bmatrix} \cos\omega & \dfrac{\sin\omega}{\omega} \\ -\omega\sin\omega & \cos\omega \end{bmatrix}^{[t]} = \begin{bmatrix} \cos(\omega[t]) & \dfrac{\sin(\omega[t])}{\omega} \\ -\omega\sin(\omega[t]) & \cos(\omega[t]) \end{bmatrix} \tag{3.16}$$

Substituting this result into equation (3.14) and rearranging it, one may rewrite equation (3.14) in the following form.

$$x(t) = \begin{bmatrix} \cos(\omega t) & \dfrac{1}{\omega}\sin(\omega t) \end{bmatrix} \begin{bmatrix} d_0 \\ v_0 \end{bmatrix} \tag{3.17}$$

which is identical to the motion for a freely vibrating spring-mass system governed by the equation of motion:

$$\ddot{x}(t) + \omega^2 x(t) = 0 \tag{3.18}$$

as expected (Weaver *et al.* 1990).

The conventional continuous solution for a free vibrating system is thus obtained by an equation in matrix form involving the discontinuous variable $[t]$. It should be noted that the solution showing in (3.17) is obtained through a totally different approach from the conventional method in solving for the dynamic equations. With the integer $[t]$, a solution in the form of equation (3.14) can be seen to have an advantage over the classical approach to the analysis and numerical calculation of a free vibratory system.

There is another case worth analyzing in which the parameter ω vanishes. As ω in equation (3.3) approaches zero, the equation of motion becomes

$$\ddot{x}(t) = \beta x([t]) \tag{3.19}$$

satisfying the identical initial conditions as shown in equation (3.4).

This governing equation can be considered as the one to describe a physical case in which the force $F = Ax([t])$ exerts on a body of mass m moving on a horizontal frictionless plane. As one may expect, the solution of equation (3.19) can also be obtained directly from the solution of equation (3.12) by taking a limiting case as ω approaches zero, such that

$$x(t) = \lim_{\omega \to 0} \begin{bmatrix} \left(1 - \dfrac{\beta}{\omega^2}\right)\cos[\omega(t-[t])] + \dfrac{\beta}{\omega^2}\dfrac{\sin[\omega(t-[t])]}{\omega} \end{bmatrix} \begin{bmatrix} d_{[t]} \\ v_{[t]} \end{bmatrix} \tag{3.20}$$

where

$$\begin{bmatrix} d_{[t]} \\ v_{[t]} \end{bmatrix} = \begin{bmatrix} (1-\beta/\omega^2)\cos\omega + \beta/\omega^2 & \sin\omega/\omega \\ -\omega(1-\beta/\omega^2)\sin\omega & \cos\omega \end{bmatrix} \begin{bmatrix} d_0 \\ v_0 \end{bmatrix} \tag{3.21}$$

In this equation, the following conclusions can be derived for the limiting cases as ω approaches zero (see Appendix A).

$$\lim_{\omega \to 0}\left\{\left(1-\frac{\beta}{\omega^2}\right)\cos[\omega(t-[t])]+\frac{\beta}{\omega^2}\right\}=1+\frac{\beta}{2}(t-[t])^2 \qquad (3.22)$$

$$\lim_{\omega \to 0}\left\{\left(1-\frac{\beta}{\omega^2}\right)\cos\omega+\frac{\beta}{\omega^2}\right\}=1+\frac{\beta}{2} \qquad (3.23)$$

and

$$\lim_{\omega \to 0}\left\{\frac{\beta-\omega^2}{\omega}\sin\omega\right\}=\beta \qquad (3.24)$$

These are useful conclusions for seeking the solutions involving limits of matrices as shown in equation (3.21). In performing the above limits, L'Hopital's rule has been used. These results show great importance in carrying out the limit of equation (3.20) to yield the following solution for equation (3.19).

$$x(t)=\left[1+\frac{\beta}{2}(t-[t]) \quad t-[t]\right]\left[\begin{array}{c} d_{[t]} \\ v_{[t]} \end{array}\right] \qquad (3.25)$$

where

$$\left[\begin{array}{c} d_{[t]} \\ v_{[t]} \end{array}\right]=\left[\begin{array}{cc} 1+\beta/2 & 1 \\ \beta & 1 \end{array}\right]^{[t]}\left[\begin{array}{c} d_0 \\ v_0 \end{array}\right] \qquad (3.26)$$

This analytical solution, obtained directly from the solution of equation (3.12), is in an identical form as the one obtained from equation (3.19) by the other conventional approaches, as it should be (Weaver *et al.* 1990). Again, the analytical solution is obtained via an approach employing the piecewise constant argument, which is completely different from that of the conventional approaches.

It may also be seen from the discussion above that the solution obtained in the proceeding section is indeed general and covers a broad range of values of the parameters ω and β. Therefore, the solutions derived previously are readily available for analyzing the properties of motion of the systems represented by equation (3.3) with varying parameters and initial conditions. This will be further discussed in the following sections.

3.5. General Vibration Systems under Piecewise Constant Excitations

In general, damping and resistance exertions against motion may be involved in a piecewise constant system. For solving for the motion of a damped spring-mass system governed by the general equation (3.1), the equation of motion in the following specific form is taken into consideration.

$$m\ddot{x} + c\dot{x} + kx = Ax([t]) \tag{3.27}$$

with the initial conditions the same as that shown in equation (3.4). The solution for this problem can be developed through a procedure similar to that used in the previous section. In an arbitrary interval $n \leq t < n+1$, therefore, the general solution to the above equation is expressible in the following form:

$$x_n(t) = e^{-\theta(t-[t])} \left\{ \begin{array}{l} d_n\left(1-\dfrac{A}{k}\right)\cos(\xi t - \xi[t]) \\[2ex] + \dfrac{1}{\xi}\left[v_n + \theta d_n\left(1-\dfrac{A}{k}\right)\right]\sin(\xi t - \xi[t]) \end{array} \right\} + \dfrac{A}{k}d_n \tag{3.28}$$

in which

$$\theta = \frac{c}{2m}$$

$$\xi = \left(\frac{k}{m} - \frac{c^2}{4m^2}\right)^{\frac{1}{2}} \tag{3.29}$$

and it is considered here that

$$\frac{k}{m} > \frac{c^2}{4m^2} \tag{3.30}$$

Again the continuity of displacement and velocity must be preserved through out the motion. Making use of the continuity conditions as shown in equation (3.9), the recurrence relation corresponding to equation (3.27) can be derived as follows.

$$
\begin{bmatrix} d_n \\ v_n \end{bmatrix} = e^{-\theta t} \times
$$

$$
\begin{bmatrix} (1-A/k)[\cos\xi+(\theta\sin\xi)/\xi]+(Ae^{\theta})/k & \sin\xi/\xi \\ -(1-A/k)[(\theta^2/\xi+\xi)\sin\xi] & \cos\xi-(\theta\sin\xi)/\xi \end{bmatrix} \begin{bmatrix} d_{n-1} \\ v_{n-1} \end{bmatrix}
$$

$$(3.31)$$

The complete solution to the damped piecewise constant spring-mass system can then be given by

$$
x(t) = \left\{ \begin{array}{l} \left(1-\dfrac{A}{k}\right)[\cos(\xi t-\xi[t]) \\ \\ +\dfrac{\theta}{\xi}\sin(\xi t-\xi[t])]+\dfrac{A}{k}e^{\theta(t-[t])} \end{array} \quad \dfrac{1}{\xi}\sin(\xi t-\xi[t]) \right\} D_1 \qquad (3.32)
$$

where the matrix D_1 is

$$
D_1 = e^{-\theta t} \begin{bmatrix} (1-A/k)[\cos\xi+(\theta\sin\xi)/\xi] \\ \quad +(Ae^{\theta})/k & \sin\omega/\omega \\ -(1-A/k)[(\theta^2/\xi+\xi)\sin\xi & \cos\xi-(\theta\sin\xi)/\xi \end{bmatrix}^{[t]} \begin{bmatrix} d_0 \\ v_0 \end{bmatrix}
$$

$$(3.33)$$

Now consider a spring-mass system subjected to a sinusoidal piecewise constant force $f([t]) = A\cos(\Omega[t])$ with the equation of motion as

$$
m\ddot{x} + c\dot{x} + kx = A\cos(\Omega[t]) \qquad (3.34)
$$

A procedure similar to that discussed previously produces the following recurrence relations corresponding to the above equation.

$$
d_n = e^{-\theta}\left[(d_{n-1}-\gamma_{n-1})\cos\xi + \frac{1}{\xi}(v_{n-1}+\theta d_{n-1}-\theta\gamma_{n-1})\sin\xi \right] + \gamma_{n-1} \qquad (3.35)
$$

where

$$v_n = e^{-\theta} \left\{ v_{n-1}\cos\xi - \left[(v_{n-1} + \theta d_{n-1} - \theta\gamma_{n-1})\frac{\theta}{\xi} + \xi(d_{n-1} - \gamma n - 1) \right]\sin\xi \right\} \tag{3.36}$$

in which

$$\gamma_{n-1} = \frac{F}{\omega^2}\cos(\Omega[t] - \Omega) \tag{3.37}$$

$$\xi = (\omega^2 - \theta^2)^{\frac{1}{2}} \tag{3.38}$$

and $\theta = c/2m$. $\omega^\gamma = k/m$. The motion governed by equation (3.34) is then expressible in the following form:

$$x(t) = e^{-\theta(t-[t])}\left[\cos(\xi t - \xi[t]) + \frac{\theta}{\xi}\sin(\xi t - \xi[t]) \quad \frac{1}{\xi}\sin(\xi t - \xi[t]) \right] D_2$$

$$+ \frac{A}{k}\cos(\Omega[t])e^{-\theta(t-[t])}\left[e^{\theta(t-[t])} - \cos(\xi t - \xi[t]) - \frac{\theta}{\xi}\sin(\xi t - \xi[t]) \right]$$

$$\tag{3.39}$$

where D_2 represents

$$D_2 = e^{-\theta[t]}S^{[t]}\begin{bmatrix} d_0 \\ v_0 \end{bmatrix}$$

$$+ \sum_{j=1}^{[t]} e^{-\theta j}S^{j-1}\begin{bmatrix} e^\theta - \cos\xi - \dfrac{\theta}{\xi}\sin\xi \\[2mm] \left(\dfrac{\theta^2}{\xi} + \xi\right)\sin\xi \end{bmatrix}\frac{A}{k}\cos[\Omega([t] - j)] \tag{3.40}$$

· in which the square matrix S has the form

$$S = \begin{bmatrix} \cos\xi + \dfrac{\theta}{\xi}\sin\xi & \dfrac{1}{\xi}\sin\xi \\[3mm] -\left(\dfrac{\theta^2}{\xi} + \xi\right)\sin\xi & \cos\xi - \dfrac{\theta}{\xi}\sin\xi \end{bmatrix} \tag{3.41}$$

The complete solution of the differential equation (3.34) consists of a transient portion, which gradually subsides. The complete solution of the differential equation (3.34) also consists of a steady state portion that will dominate the motion of the system after the transient subsides. For the governing equation (3.34), the corresponding transient solution is obtained from equation (3.41) by letting A equal to zero, such that

$$x_t = e^{-\theta(t-[t])} \begin{bmatrix} \cos(\xi t - \xi[t]) \\ +\dfrac{\theta}{\xi}\sin(\xi t - \xi[t]) & \dfrac{1}{\xi}\sin(\xi t - \xi[t]) \end{bmatrix} E_2 \qquad (3.42)$$

the matrix E_2 is

$$E_2 = \begin{bmatrix} \cos\xi + \dfrac{\theta}{\xi}\sin\xi & \dfrac{1}{\xi}\sin\xi \\ -\left(\dfrac{\theta^2}{\xi}+\xi\right)\sin\xi & \cos\xi - \dfrac{\theta}{\xi}\sin\xi \end{bmatrix}^{[t]} \begin{bmatrix} d_0 \\ v_0 \end{bmatrix} \qquad (3.43)$$

The transient solution to equation (3.34) is in fact the free damped vibration governed by

$$m\ddot{x} + c\dot{x} + kx = 0 \qquad (3.44)$$

It is verifiable that the exponential matrix in equation (3.43) can also be expressed as a single matrix (see Appendix A), such that

$$\begin{bmatrix} \cos\xi + \dfrac{\theta}{\xi}\sin\xi & \dfrac{1}{\xi}\sin\xi \\ -\left(\dfrac{\theta^2}{\xi}+\xi\right)\sin\xi & \cos\xi - \dfrac{\theta}{\xi}\sin\xi \end{bmatrix}^{[t]}$$

$$= \begin{bmatrix} \cos(\xi[t]) + \dfrac{\theta}{\xi}\sin(\xi[t]) & \dfrac{1}{\xi}\sin(\xi[t]) \\ -\left(\dfrac{\theta^2}{\xi}+\xi\right)\sin(\xi[t]) & \cos(\xi[t]) - \dfrac{\theta}{\xi}\sin(\xi[t]) \end{bmatrix} \qquad (3.45)$$

Substituting this result into equation (3.42) with rearrangement, the transient solution to equation (3.34) can be rewritten in a continuous form as follows.

$$x_t = e^{-\theta t}\left[\cos(\xi t) + \frac{\theta}{\xi}\sin(\xi t) \quad \frac{1}{\xi}\sin(\xi t)\right]\begin{bmatrix} d_0 \\ v_0 \end{bmatrix} \tag{3.46}$$

As expected, this solution is indeed in the form identical to the analytical solution of equation (3.44) corresponding to a free damped spring-mass system (Weaver *et al.* 1990).

It is interesting to note that an expression with piecewise constant variable, such as equation (3.42) or (3.14), may be represented by a continuous form, such as equation (3.46) or (3.17), and vice versa. Although the variable $[t]$ in equations (3.14) and (3.42) varies discontinuously as time increases, the functions $x(t)$ and x_t are continuous everywhere for $t \geq 0$ and may vary continuously and smoothly as time increases. This can be seen from equations (3.17) and (3.46).

Due to the fact that the variable $[t]$ assumes only integer values, the discrete form solution with integer variables shows great advantage in applications of numerical simulation on computers, such as computational time saving with implementation of integers.

The steady state terms in equation (3.41) form the perennial portion of the complete solution of equation (3.34). The steady state part of the solution represents the long-standing dynamical motion of the system excited by a sinusoidally varying piecewise constant force. Therefore, the steady state portion of the solution is truly the response of the piecewise constant system to the excitation of the discontinuous piecewise constant exertion. In understanding this, the dynamic system studied above actually represents a hybrid of continuous and discrete dynamic systems. The steady state motion of such system is complex and will be discussed in the subsequent chapters.

A more complicated nonlinear oscillatory system can also be analyzed via the same procedure as discussed above. For a damped spring-mass system subjected to a cubic piecewise constant force $Ax^3([t])$, with the equation of motion

$$m\ddot{x} + c\dot{x} + kx = Ax^3([t]) \tag{3.47}$$

for example, the nonlinear oscillatory motion on $n \leq t < n+1$ can be written as

$$x_n(t) = e^{-\theta(t-[t])} \left\{ \begin{array}{l} d_n\left(1-\dfrac{Ad_n^2}{k}\right)\cos[\xi(t-[t])] \\[2ex] +\dfrac{1}{\xi}\left[v_n+\theta d_n\left(1-\dfrac{Ad_n^2}{k}\right)\right]\sin[\xi(t-[t])] \end{array} \right\} + \dfrac{Ad_n^3}{k}$$

$$(3.48)$$

The recurrence relations corresponding to this problem are derived as follows through the procedures as discussed previously, with preservation of the displacement and velocity continuity.

$$d_n = e^{-\theta}\left\{ \begin{array}{l} d_{n-1}\left(1-\dfrac{Ad_{n-1}^2}{k}\right)\cos\xi \\[2ex] +\dfrac{1}{\xi}\left[v_{n-1}+\theta d_{n-1}\left(1-\dfrac{Ad_{n-1}^2}{k}\right)\right]\sin\xi \end{array} \right\} + \dfrac{Ad_{n-1}^3}{k} \qquad (3.49)$$

$$v_n = e^{-\theta}\left\{ \begin{array}{l} v_{n-1}\cos\xi - \sin\xi\left[\dfrac{\theta}{\xi}\left(v_{n-1}+\theta d_{n-1}-\dfrac{Ad_{n-1}^2}{k}\right)\right. \\[2ex] \left. +\xi\left(d_{n-1}-\dfrac{Ad_{n-1}^3}{k}\right)\right] \end{array} \right\} \qquad (3.50)$$

Should initial values of displacement and velocity be given, the displacement d_n and velocity v_n at any integer point can be calculated by these recurrence relations through an iterative procedure. Once the displacement d_n and velocity v_n are known, motion of the system in a unit duration a time segment $n \le t < n+1$ is readily available by equation (3.48). The complete solution of equation (3.47) is then procured by connecting all the segments with the conditions of continuity. It is evident that the transient solution to the governing equation (3.47) can be derived by eliminating A in equation (3.48), through a procedure similar to that described previously for determining the transient solutions of the linear systems.

The recurrence relations above may be expressed in a matrix form as that shown in the previous sections. However, due to the nonlinearity of the forcing function in this case, the displacement and velocity d_n and v_n at an arbitrary integer point can not be directly related to the initial values of d_0 and v_0 as those expressed in the previous sections. d_n and v_n have to be calculated through equations (3.49) and (3.50) with the antecedently calculated values of d_{n-1} and v_{n-1} which are related to d_{n-2}, v_{n-2}, d_{n-3}, v_{n-3},..., d_0 and v_0. As a result, the complete solution to equation (3.47) cannot be expressed in a form as neat as that shown in equations (3.12), (3.32) and (3.39), and the recurrence relations for this problem are not expressible in the form of multiplication of the square matrices as those in the previous sections. Consequently, some convenient manipulations, such as taking the limiting case as one of the system parameters approaches a limit, are difficult to carry out.

Nevertheless, the expressions of the above recurrence relation are straightforward and succinct in style involving integer values of n. Therefore, it is practical and manageable for using the recurrence relations in a computer program to obtain numerical solutions for the nonlinear problem. It should be noticed however that the solution indicated by equations (3.48), (3.49) and (3.50) is continuous and derivable everywhere in the entire time range. The solution describes the motion not only at the discrete points of $[t]$ but also within a unit duration interval and in the time segments over the range $t > 0$.

The solutions of the linear and nonlinear systems investigated above are all developed with the consideration of the intervals of unit duration. Due to the properties of the variable $[t]$, the motion within such an interval has the structure of a continuous dynamic system of linear vibration. In fact, without losing the linearity within a time interval, solutions of the systems involving piecewise constant variables of smaller or larger duration can be derived by the same procedure as presented in the previous sections. Furthermore, the dynamic systems containing piecewise constant variables of unequal durations may also be handled without much difficulty, provided that the pattern of the unequal durations is known. The systems involving piecewise constant variables of fractional durations will be considered later.

3.6. Derivation and Characteristics of Approximate and Numerical Solutions of Dynamic Systems with Piecewise Constant Variables

Among the dynamic systems discussed in the previous section, the sinusoidally varying piecewise constant force shown in the governing equation (3.34) has a given form $A\cos(\Omega[t])$. Obviously, the value of this force is independent of the corresponding solutions of $x(t)$ and can be calculated for any give time or time interval. It is therefore convenient to start the analysis of the properties of motion of a simple undamped spring-mass system subjected to the sinusoidally varying piecewise constant force.

$$\ddot{x} + x = \cos([t]) \tag{3.51}$$

with initial conditions

$$x(0) = 0 \quad \text{and} \quad \dot{x}(0) = 1 \tag{3.52}$$

The solution of this system can be found from equation (3.39) in the form:

$$x(t) = \{\cos(t-[t]) + \sin(t-[t]) \quad \sin(t-[t])\} D_2' \\ + [1 - \cos(t-[t])]\cos([t]) \tag{3.53}$$

where the matrix D_2' is

$$D_2' = \begin{bmatrix} \cos(1) & \sin(1) \\ -\sin(1) & \cos(1) \end{bmatrix}^{[t]} \begin{bmatrix} x(0) \\ \dot{x}(0) \end{bmatrix}$$

$$+ \sum_{j=1}^{[t]} \begin{bmatrix} \cos(1) & \sin(1) \\ -\sin(1) & \cos(1) \end{bmatrix}^{j-1} \begin{bmatrix} 1-\cos(1) \\ \sin(1) \end{bmatrix} \cos([t]-j) \tag{3.54}$$

The motion of this system starts at $t = 0$, and, by equation (3.53) and its derivative, the following expressions can be obtained to represent the motion in the first time interval of unit duration.

$$x_1(t) = -\cos t + \sin t + 1 \tag{3.55}$$

$$\dot{x}_1(t) = \sin t + \cos t \tag{3.56}$$

The diagram of displacement versus time is plotted in Figure 3.2 in which the corresponding time plots of velocity and piecewise constant force are superposed. The motion in this first interval, as can be seen from the governing equation, represents a portion of a linear vibration. At the end point of this interval, the displacement and velocity are given by $x_1(1)$ and $\dot{x}_1(1)$ respectively. These specific results in turn become the starting conditions or the local initial conditions for the motion in the second time interval, $1 \le t \le 2$. With these local initial conditions, the displacement and velocity of the system in the second interval can be given by

$$x_2(t) = [x_1(1) - \cos(1)]\cos(t - [t]) + \dot{x}_1(1)\sin(t - [t]) + \cos(1) \quad (3.57)$$

$$\dot{x}_2(t) = -[x_1(1) - \cos(1)]\sin(t - [t]) + \dot{x}_1(1)\cos(t - [t]) \quad (3.58)$$

which are graphically shown for the interval $1 \le t \le 2$ in Figure 3.2.

The motion in the third time interval is consequently determined from equation (3.53) with the new local initial conditions calculated by

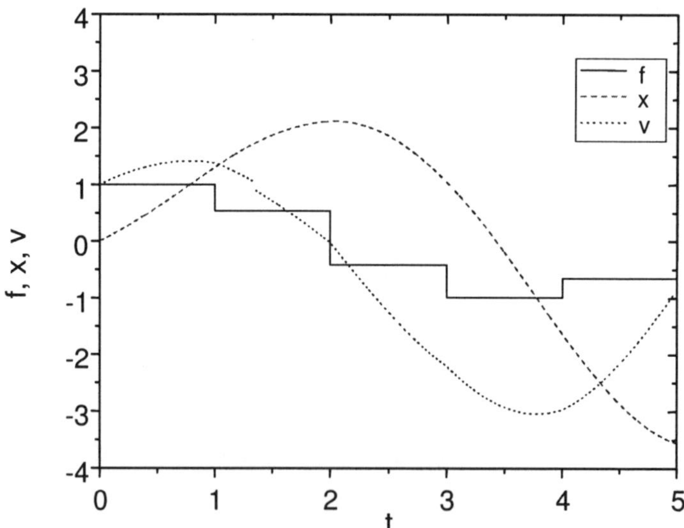

Figure 3.2.   Time plots of displacement, velocity and piecewise constant force acting on a spring-mass system governed by $\ddot{x} + x = \cos([t])$. 'f': piecewise constant force, $\cos[t]$; 'x': displacement; 'v': velocity or first derivative of $x(t)$.

equations (3.57) and (3.58) for $t = 2$. The motion in the succeeding time intervals may be found by the same iterative procedure as for the first three time intervals. With the help of the recurrence relations in equations (3.35) and (3.36), the complete solution to the system, together with the corresponding velocity and piecewise constant forces acting on it, can be determined analytically through an interaction process or numerically on a computer. The calculated results of the system are shown in Figure 3.2.

The solid line in Figure 3.2 indicates the exerting piecewise constant force and the dashed and dotted lines form the wave-like time plots of displacement and velocity respectively. It should be noted in the diagram that the displacement curve and its slope are continuous though the external excitation is discontinuous. The velocity however may show some slope sharp changes at the integer points of time. The sharp changes of the slop are a consequence of the discontinuous piecewise constant force acting on the system. The solution in a time range much further in time is plotted in Figure 3.3 shown below, which reveals the asymptotically divergent oscillatory behavior of the system.

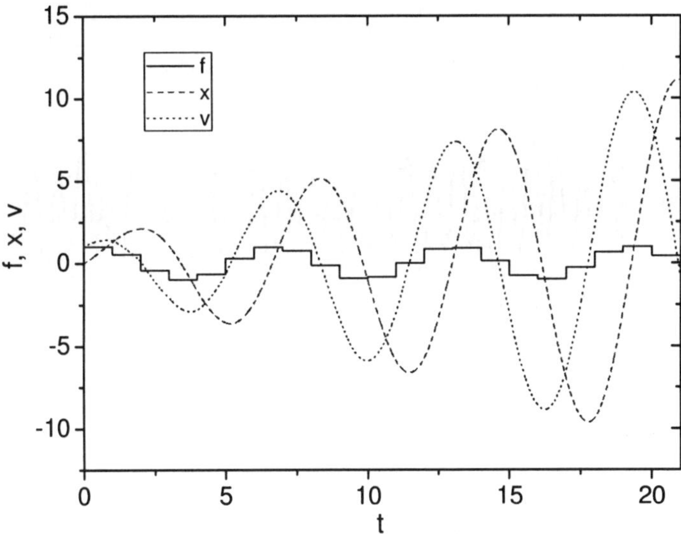

Figure 3.3. Time plots of displacement, velocity and piecewise constant force acting on a spring-mass system governed by $\ddot{x}(t) + x(t) = \cos[t]$. 'f': piecewise constant force $\cos[t]$; 'x': displacement; 'v': velocity.

Consider now a similar system governed by the following equation with a higher stiffness

$$\ddot{x}(t) + 9x(t) = \cos[t] \tag{3.59}$$

Under the same conditions and after a relatively long time from $t = 0$, this system with a higher stiffness spring shows a stable oscillation with constant amplitude as exhibited in Figure 3.4. From Figure 3.4, as one of the interesting characteristics of piecewise constant systems, it can be seen from the figure that the wave forms of displacement are repeating precisely with a period of 2π, although the shapes of the stepwise excitations on the system are different from period to period. The above two examples clearly show that the response of a piecewise constant system may not be related to the relationship between the "natural frequency" and the "frequency of external excitations" as that holding for the continuous dynamic systems under continuous excitations (Weaver *et al.* 1990).

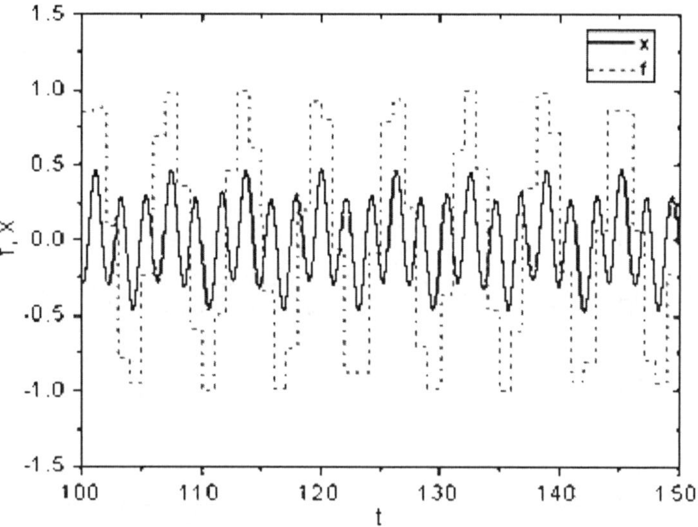

Figure 3.4.   Wave from versus piecewise constant force acting on a spring-mass system governed by $\ddot{x}(t) + 9x(t) = \cos[t]$. 'f': piecewise constant force, $\cos[t]$; 'x': displacement.

Consider now a system in which a piecewise constant force is displacement related, instead of an explicit function of time. Let us start with the following simple linear system first.

$$m\ddot{x}(t) + kx(t) = Ax([t]) \tag{3.60}$$

The complete solution for a time interval of unit duration must first be obtained so that the end conditions for this interval, therefore the starting conditions of the consecutive interval, may be determined. The solution of such a spring-mass system is obtained through the similar procedures as that used for the previous two examples and the motion determined numerically are shown in Figure 3.5. As displayed in the figure, the piecewise constant force exerting on the system has the same value as the displacement at the integer point of time, and the corresponding motion in this case is asymptotically convergent (the system eventually stops at rest).

As can be seen from the characteristics of the piecewise constant systems discussed above, behavior of a dynamic system subjected to piecewise constant forces of the type shown in equation (3.60) is in general unique from that of the corresponding continuous system exerted by continuous forces.

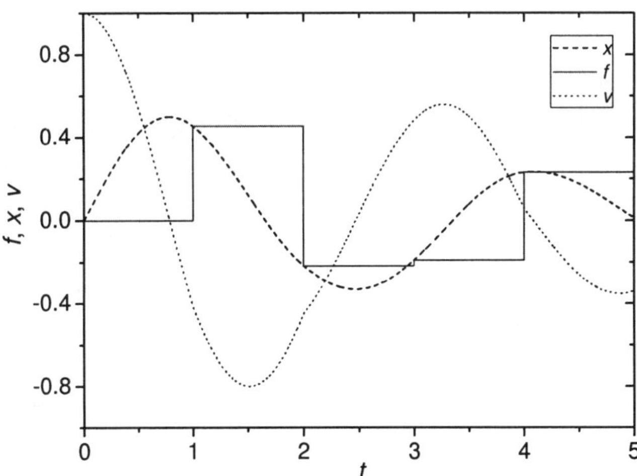

Figure 3.5. Convergence of a spring-mass system under piecewise constant force acting on a spring-mass system governed by $\ddot{x}(t) + 4x(t) = x([t])$; 'f': force, $x[t]$; 'x': displacement. 'v': velocity.

To further describe the uniqueness of the behavior of a piecewise constant system with respect to that of a corresponding continuous system, let us consider a damped spring-mass system subjected to a piecewise constant force governed by the equation of motion

$$\ddot{x}(t) + a_1 \dot{x}(t) + a_2 x(t) = Ax([t]) \tag{3.61}$$

together with a corresponding linear vibration system governed by the following equation of motion (Thomson 1981).

$$\ddot{x}(t) + a_1 \dot{x}(t) + a_2 x(t) = Ax(t) \tag{3.62}$$

where a_1, a_2 and A are constant parameters of the system.

A comparison is illustrated in Figure 3.6 in which the motion with the piecewise constant force $-8x([t])$ diverges asymptotically and, on the contrary, the motion under the continuous force $-8x(t)$ is damped out rapidly with time.

For the spring-mass systems subjected to known forms of piecewise constant and continuous forces respectively, such as $A\cos(\Omega[t])$ and

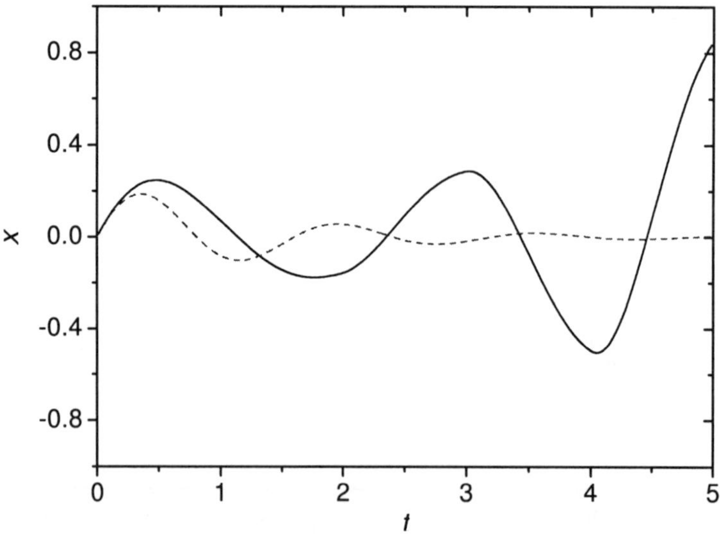

Figure 3.6. Comparison of the responses of a piecewise constant system (solid line) governed by $\ddot{x}(t) + 1.5\dot{x}(t) + 8x(t) = -8x([t])$ and a continuous system (dashed line) of $\ddot{x}(t) + 1.5\dot{x}(t) + 8x(t) = -8x(t)$. Initial conditions: $d_0 = 0, v_0 = 1.0$.

$A\cos(\Omega t)$, the contrast is also evident. As an example, the motion of a system subjected to a piecewise constant force,

$$m\ddot{x}(t) + c\dot{x}(t) + kx(t) = Ax(\Omega[t]) \tag{3.63}$$

is compared with a system subjected to a continuous force

$$m\ddot{x}(t) + c\dot{x}(t) + kx(t) = A\cos(\Omega t) \tag{3.64}$$

as shown in Figure 3.7.

It can be observed from the figure that the motion of the system subjected to the continuous force becomes a steady-state vibration with identical shape as time increases, whereas the system under the piecewise constant forces oscillates with a random like wave form of distinct pattern and amplitudes.

Properties of motion for the dynamic systems exerted by piecewise constant forces under various conditions will be studied graphically and analytically in the following sections.

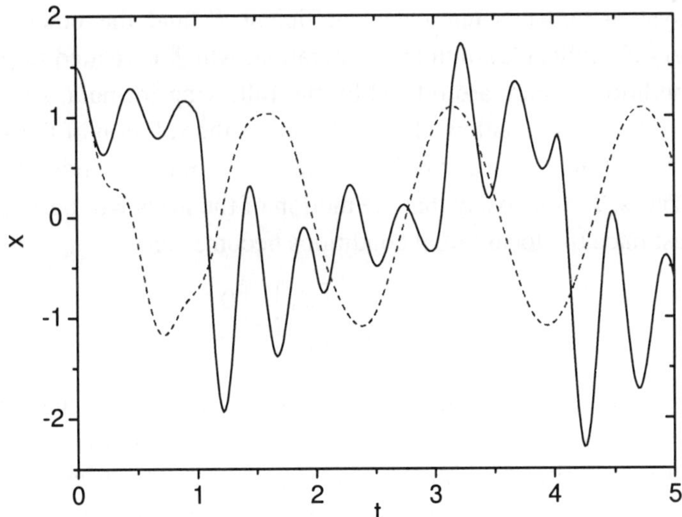

Figure 3.7.   Comparison of motions of a piecewise constant system (solid line) governed by $m\ddot{x}(t) + c\dot{x}(t) + kx(t) = A\cos(\Omega[t])$ and a continuous system (dashed line) of $m\ddot{x}(t) + c\dot{x}(t) + k_2 x(t) = A\cos(\Omega t)$. Initial conditions: $d_0 = 0, v_0 = 1.0$.

3.7.   Extraordinary and Nonlinear Behavior of Linear Piecewise Constant Systems

It has been demonstrated in the previous section that a dynamic system subjected to a piecewise constant force exhibits different oscillatory behavior from that of the corresponding continuous system. In fact, as will be shown in the following sections, the response of a dynamic system may be oscillatory, nonoscillatory, stable or unstable with varying coefficients and initial conditions.

An interesting phenomenon of a dynamic system subjected to a piecewise constant force is that its motion may asymptotically vanish even when the system is free of damping. Because of this, an undamped dynamic system under piecewise constant forces can not be considered as a conservative system. A simple example is shown in Figure 3.8 in which the motion of a mass moving on a frictionless horizontal plate disappears under the piecewise constant force shown. Solution for this problem is presented in equation (3.25). The fading away of the motion of this system depends upon the coefficient β (and the other system parameters if applicable), number of iteration, which is related to $[t]$, and initial conditions which are related by the following formula. As can be seen from the formula below, the condition for the solution of the system and its derivative to be zeros is that the local initial conditions both become zeros. This is to say that the motion of the mass will permanently stay at rest once the local initial conditions become zeros.

$$\begin{bmatrix} 1+\beta/2 & 1 \\ \beta & 1 \end{bmatrix}^{[t]} \begin{bmatrix} d_0 \\ v_0 \end{bmatrix} = \begin{bmatrix} 0 \\ 0 \end{bmatrix} \tag{3.65}$$

In general, when a starting perturbation is preserved, fundamental motions of linear vibrating systems subjected to continuous forces are not sensitive to initial conditions. However, under certain conditions, a dynamic system subjected to piecewise constant forces may show sensitive dependence upon initial conditions. The above system can again be taken as an example for this phenomenon of a piecewise constant system. Based on the results of the numerical simulations of this system, the sensitivity of the system to the initial conditions is illustrated

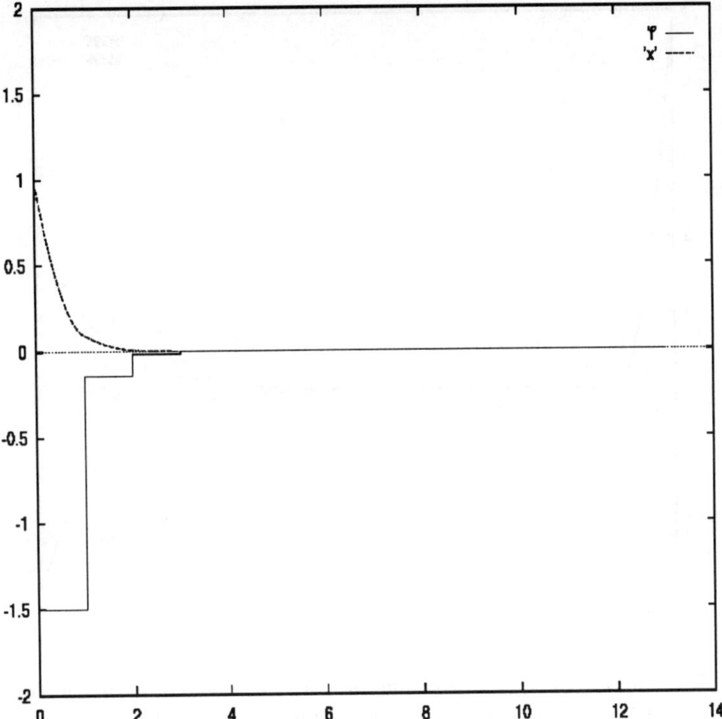

Figure 3.8. Response of a system governed by $\ddot{x}(t) + \beta x([t]) = 0$, where $\beta = -1.5$, initial conditions: $d_0 = 1.0$, $v_0 = 1.656$; 'f' represents the piecewise constant force, $\beta x[t]$; and 'x' is the displacement.

in Figure 3.9 by starting two motions from adjacent states. Under the piecewise constant exertions, the two adjacent starts appear to remain close to each other for a time and then rapidly move apart. This behavior of the linear piecewise constant system is obviously unique in comparing with its corresponding continuous system. It may also be interesting to point out that a small perturbation of a parameter of such a piecewise constant system may also lead to a greater variation of the motion. Such behavior of linear piecewise constant systems are significant as these behavior are usually found in nonlinear continuous systems.

So far, only the behavior with linear piecewise constant external forces are investigated. If nonlinear piecewise constant forces are involved in the dynamic systems, the corresponding motion will become

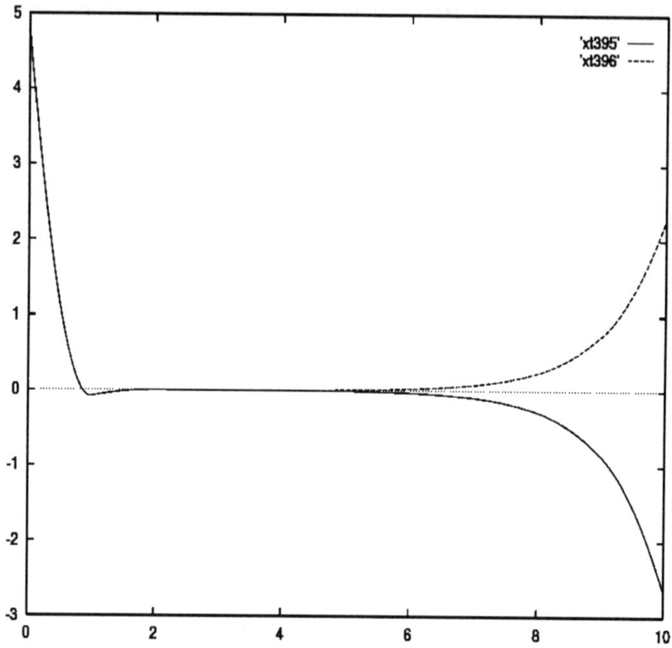

Figure 3.9.   Sensitivity of the solution of $\ddot{x}(t) + \beta x([t]) = 0$ to initial conditions. $\beta = -2.1$; 'xt395': $d_0 = 0.48395$, $v_0 = -1.0$; 'xt396': $d_0 = 0.48396$, $v_0 = -1.0$.

complex and nonlinear properties will be brought into the systems. Replace the nonlinear piecewise constant force $Ax^3([t])$ in equation (3.34) by a continuous force $Ax^3(t)$. The differential equation thus becomes the following famous Duffing's equation (Duffing 1918, Klitter 1951), for which the numerical results are given, among others by Fang and Dowell (1987) and Christopher (1973).

$$m\ddot{x}(t) + c\dot{x}(t) + kx(t) = Ax^3(t) \qquad (3.66)$$

A numerical solution for the following system with the cubic piecewise constant exertion is illustrated in Figure 3.10 in comparison with the numerical solution of the corresponding continuous system presented in equation (3.66).

$$\ddot{x}(t) + a\dot{x}(t) + bx(t) + dx^3([t]) = 0 \qquad (3.67)$$

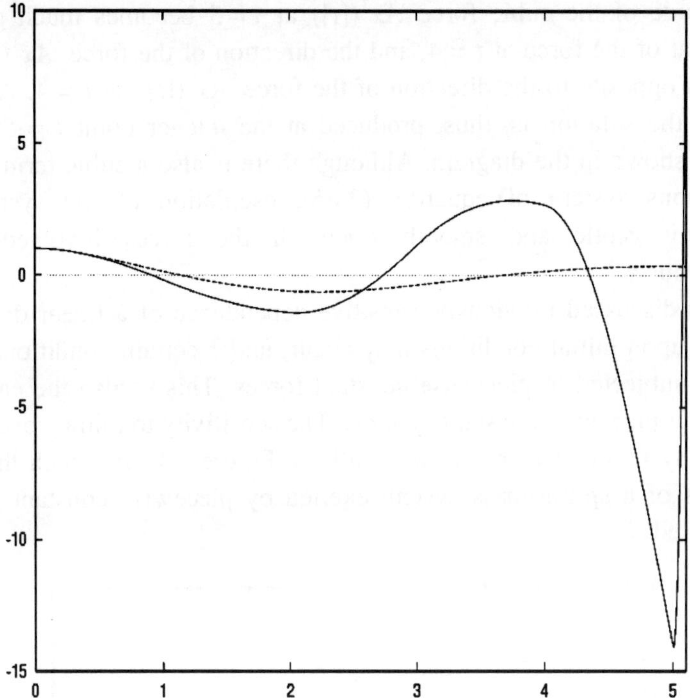

Figure 3.10. Solution of the nonlinear systems described by $\ddot{x}(t) + a\dot{x}(t) + bx(t) + dx^3(t) = 0$ (dashed line) and $\ddot{x}(t) + a\dot{x}(t) + bx(t) + dx^3([t]) = 0$ (solid line). For $a = 0.5$, $b = 1.0$, $d = 2.0$, $d_0 = 1.0$ and $v_0 = 0.0$.

It can be seen from Figure 3.10 that the motion of the system with the piecewise constant exertions are highly distinct from the one under the continuous force. Quantitatively, the motion under the piecewise constant forces varies much conspicuously in terms of amplitude and slope in comparison with the oscillation of the continuous nonlinear system corresponding to the Duffing's equation.

As it may also be observed in Figure 3.10, at $t = 0$ and the integer points $t = 1, 2, 3, 4$, the magnitude of the exertion $Ax^3([t])$ in equation (3.47) does not vary much during the time period $4 \leq t \leq 5$ due to the small displacement x calculated by using equation (3.48). However, the displacement x is rapidly increased in absolute value on $4 \leq t \leq 5$ as shown in the figure. Corresponding to the increased displacement, the magnitude of the cubic force $Ax^3([t])$ at $t=5$ becomes much greater than that of the force at $t = 4$, and the direction of the force $Ax^3([t])$ at $t = 5$ is opposite to the direction of the force $Ax^3([t])$ at $t = 4$. A sharp rise of the solution is, thus, produced at the integer point $t = 5$, as is clearly shown in the diagram. Although there is also a cubic term in the continuous system of equation (3.66), oscillation of the system is relatively gentle and smooth since all the forces involved vary continuously.

As discussed previously, sensitive dependence of a linear dynamic system upon initial conditions may occur, under certain conditions, in a system subjected to piecewise constant forces. This is also the case for nonlinear piecewise constant systems. The sensitivity to initial conditions for the system considered is seen fully in Figure 3.11 in which the two motions of a spring-mass system exerted by piecewise constant forces are plotted.

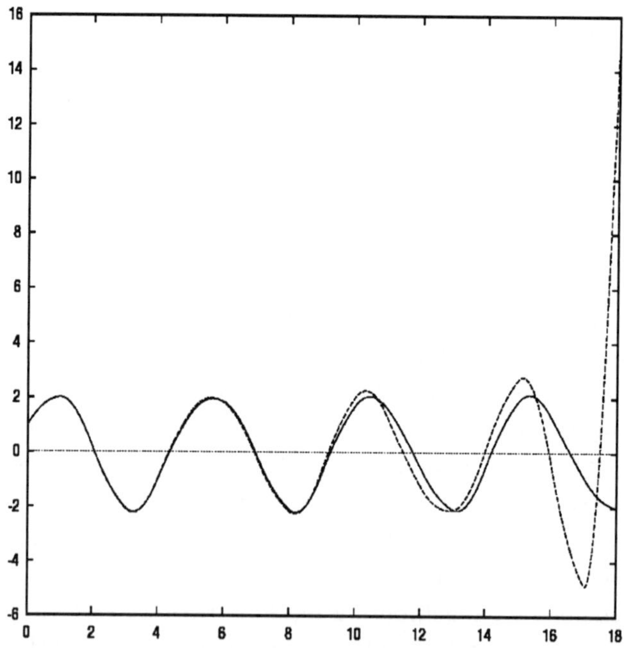

Figure 3.11. Sensitivity of the solution of $\ddot{x}(t) + a\dot{x}(t) + bx(t) + dx^3([t]) = 0$ or $a = 0.6$, $b = 0.5$, $d = 0.46$. Initial conditions: $d_0 = 1.0$, $v_0 = 2.0$ (solid line); and $d_0 = 1.0$, $v_0 = 2.01$ (dashed line).

The two motions from nearby states (slightly different initial velocity and the same initial displacement) remain close for a time, but quickly diverge and become uncorrelated as shown in the figure. Sensitivity of a dynamic system to initial conditions is a significant measure in nonlinear and chaotic dynamics (Thompson and Stewart 1986, Farmer 1982) which will be further discussed in the following chapters.

3.8. Oscillatory Properties of Dynamic Systems with Piecewise Constant Variables

Based on the solutions in the closed form derived in the previous sections, properties of the motion of a system subjected to a piecewise constant force can be analytically studied. For the sake of clarity, consider the motion of an undamped spring-mass system represented by equation (3.3). Regarding the properties of the solution given by equation (3.12), the following points need to be remarked:

(1) The first matrix of equation (3.12) represents a linear vibration in the time range $[t] \leq t < [t]+1$, and the absolute values of $\sin[\omega t - \omega[t]]$ and $\cos[\omega t - \omega[t]]$ in the matrix lie between 0 and unity.

(2) The elements of $d_{[t]}$ and $v_{[t]}$ in the column matrix of equation (3.12) are actually the local initial conditions for the motion between $[t]$ and $[t] + 1$. In this relatively small time interval of unit duration, response of the system is largely affected by the values of $d_{[t]}$ and $v_{[t]}$.

(3) Independent of the linear vibrations between the integer points of time, the values of displacement and velocity at the instants of the integer number of time along $t \geq 0$ are given by the second matrix in equation (3.12).

(4) d_0 and v_0 in equation (3.13) are the initial conditions independent of time. The piecewise constant systems may be sensitive to these values. This behavior is similar to that of a nonlinear dynamic system.

On account of the above points, the properties of the undamped spring-mass system are characterized by the exponential matrix:

$$\Gamma^{[t]} = \left[\begin{array}{cc} \left(1 - \dfrac{\beta}{\omega^2}\right)\cos\omega + \dfrac{\beta}{\omega^2} & \dfrac{\sin\omega}{\omega} \\ -\omega\left(1 - \dfrac{\beta}{\omega^2}\right)\sin\omega & \cos\omega \end{array} \right]^{[t]} \tag{3.68}$$

which appears at the right hand side of equation (3.13).

The properties of the system considered are closely related to the eigenvalues of the matrix Γ. The eigenvalues denoted as $\alpha_{1,2}$ are determined as

$$\alpha_{1,2} = \cos\omega + \frac{\beta}{2\omega^2}(1 - \cos\omega)$$

$$\pm \frac{1}{2}\left\{ (\cos\omega - 1)\left[\frac{\beta^2}{\omega^4}(\cos\omega - 1) - 4(\cos\omega + 1)\left(\frac{\beta}{\omega^2} - 1\right)\right]\right\}^{\frac{1}{2}} \tag{3.69}$$

The square matrix Γ may then be written as a diagonal matrix in terms of the eigenvalues,

$$\Gamma = P \left[\begin{array}{cc} \alpha_1 & 0 \\ 0 & \alpha_2 \end{array} \right] P^{-1} \tag{3.70}$$

In the above expression, the matrix P represents a set of linearly independent eigenvectors for the corresponding eigenvalues obtained and P^{-1} expresses the inverse of P. According to the matrix theory (Gantmacher 1960), for the real non-symmetrical matrix Γ it can be stated that

(1) for $|\alpha| = 1$, the solution $x(t)$ given by equation (3.12) is stable;
(2) for $|\alpha| < 1$, with increase of time t, solution $x(t)$ will be convergent or asymptotically convergent, and therefore, can be considered as another case of a stable solution;
(3) for $|\alpha| > 1$, solution $x(t)$ is monotonically or asymptotically divergent.

It can be shown, through an iterative procedure, the exponential square matrix in equation (3.68) is expressible (see Appendices A and B) in the following form:

$$\Gamma^{[t]} = P \begin{bmatrix} \alpha_1^{[t]} & 0 \\ 0 & \alpha_2^{[t]} \end{bmatrix} P^{-1} \qquad (3.71)$$

Considering that P and P^{-1} are constant matrices, therefore, only the diagonal matrix in equation (3.71) varies with variables $[t]$ and t as time increases. This is actually the advantage to convert the square matrix into the one with the eigenvalue forms. We may thus concentrate on the elements in the diagonal matrix to analyze the properties of the system with respect to time. If α_1 and α_2 are both negative, the signs of $\alpha_{1,2}^{[t]}$ are opposite according to even and odd number $[t]$ and the corresponding displacement $x(t)$ is then alternating and the motion of the system is oscillatory with time t. When the eigenvalues are both positive, on the other hand, the absolute value of the solution $x(t)$ will increase with increasing time t and the corresponding motion is then divergent. If the eigenvalues are complex and expressible as $\alpha = re^{i\theta}$, the corresponding motion is oscillatory since $\alpha^{[t]}$ changes its sign alternately with an increase of time in this case.

As can be observed from equations (3.68) and (3.69), the natural frequency ω is a known value, thence, the parameter β plays an important role in controlling the eigenvalues and, as a consequence, the properties of the motion. According to the conclusions presented above, the properties of motion for various values of β can be discussed as the following.

(a) $\beta \to 0$

As discussed in foregoing sections, the corresponding spring-mass system tends to be a free undamped system and its solution on the basis of equation (3.12) assumes the form

$$x(t) = \begin{bmatrix} \cos(\omega t) & \dfrac{1}{\omega}\sin(\omega t) \end{bmatrix} \begin{bmatrix} d_0 \\ v_0 \end{bmatrix} \qquad (3.72)$$

which, as shown in Figure 3.12, is identical to the harmonic motion of a freely vibrating spring-mass system governed by the equation of motion:

$$\ddot{x}(t) + \omega^2 x(t) = 0 \qquad (3.73)$$

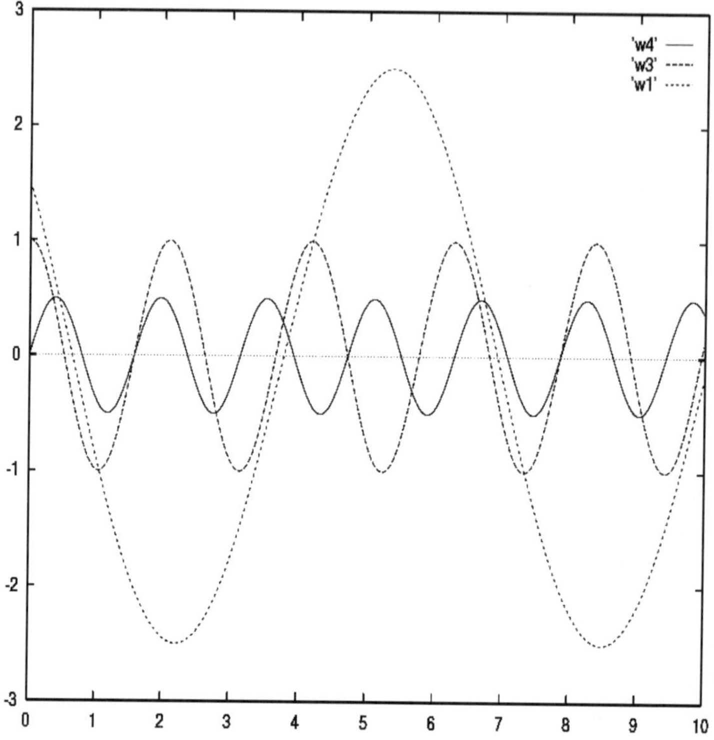

Figure 3.12.   Solution of $\ddot{x}(t) + \omega^2 x = \beta x([t])$ for $\beta = 0$. 'w4': $\omega = 4$, $d_0 = 0$, $v_0 = 2$; 'w3': $\omega = 3$, $d_0 = 1$, $v_0 = 0$; 'w1': $\omega = 1$, $d_0 = 1.5$, $v_0 = -2$.

As presented in the previous sections, the solution to equation (3.73) is derived directly from equation (3.12) where a piecewise constant variable is presented. The numerical solution illustrated in Figure 3.12 is calculated with the integer variables $[t]$.

(b) $\beta = \omega^2$

The eigenvalues in this case become

$$\alpha_{1,2} = \cos\omega + \frac{1}{2}(1 - \cos\omega) \pm \frac{1}{2}[(\cos\omega - 1)(\cos\omega - 1)]^{\frac{1}{2}} \qquad (3.74)$$

which implies that $|\alpha| \leq 1$ on the supposition that $\omega \neq 2n\pi$ or $\omega \neq (2n - 1)\pi$ where $n = 1, 2, 3, \dots$ The corresponding solution $x(t)$ is therefore convergent or asymptotically convergent for different values of ω. As plotted in Figure 3.13, the oscillation eventually vanishes independent of the values of β, d_0 and v_0 as long as $\beta = \omega^2$. In particular as $\dot{x}(0) = 0$ and $x(0) = d_0$, the corresponding solution becomes $x(t) = d_0$ which represents a motionless case. When $\omega = 2n\pi$ or $\omega = (2n - 1)\pi$, the corresponding eigenvalues are $|\alpha_1| = |\alpha_2| = 1$, the motion in this case is stable as shown in Figure 3.14. It is illustrated in Figure 3.14, independent of the number n for the ω above, the mass vibrates about a new equilibrium position in a simple harmonic fashion with different initial conditions.

It can be calculated from equation (3.74), if $\omega = 2n\pi$, the corresponding eigenvalue will always be $\alpha = 1$ independent of β. The response of the system in this case is then a simple harmonic vibration, as shown in Figure 3.14, for various values of parameter β and initial conditions.

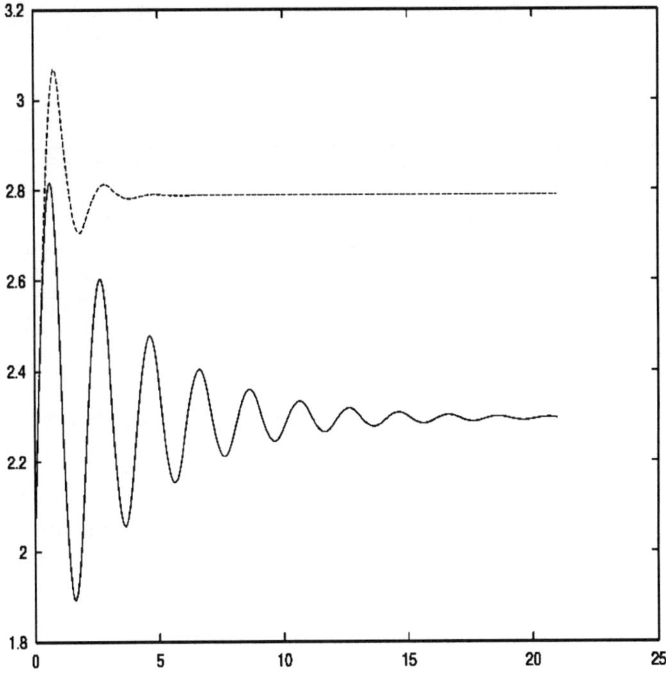

Figure 3.13. Solution of $\ddot{x}(t) + \omega^2 x = \beta x([t])$ as $\beta = \omega^2$, $d_0 = 2$, $v_0 = 2$; $\beta = 6$, $\omega^2 = 6$ for solid line, and $\beta = 3.5$, $\omega^2 = 3.5$ for dashed line.

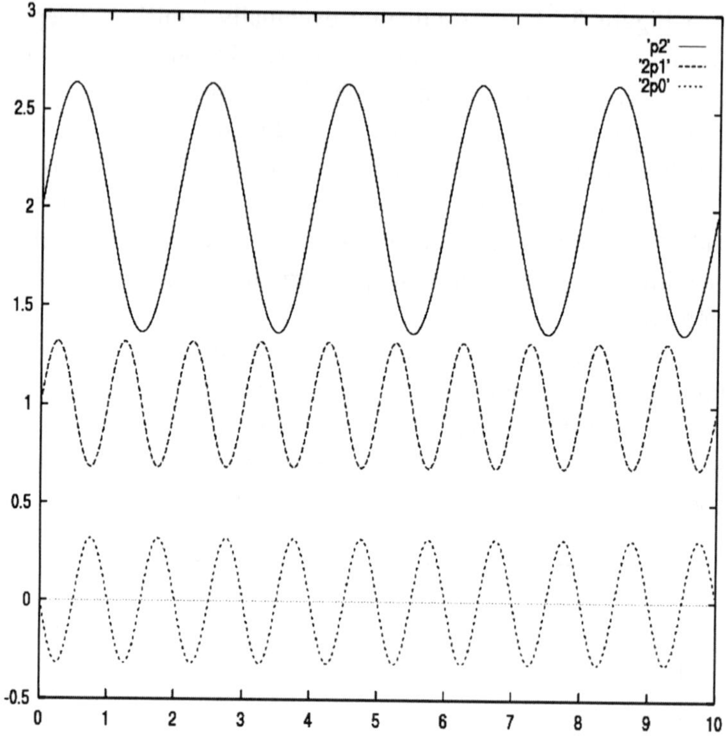

Figure 3.14. Solution of $\ddot{x}(t) + \omega^2 x = \beta x([t])$ as $\beta = \omega^2$, while $\omega = 2n\pi$ or $\omega = (2n-1)\pi$. 'p2': $\omega = \pi$, $d_0 = 2$, $v_0 = 2$; '2p1': $\omega = 2\pi$, $d_0 = 1$, $v_0 = 2$; '2p0': $\omega = 2\pi$, $d_0 = 0$, $v_0 = 2$.

(c) $0 < \beta < \omega^2$

When $\omega = (2n-1)\pi$, $n = 1, 2, 3, ...$, the two eigenvalues satisfy $|\alpha| \leq 1$ in this case.

The motions corresponding to these specific eigenvalues are oscillatory and convergent or asymptotically convergent for different value of n. In fact, the acceleration $\ddot{x}$ in equation (3.3) is entirely dependent upon the resultant of the forces $\beta x([t]) - \omega^2 x$.

Therefore, the absolute value of the resultant force acting on the mass will be smaller than $|\omega^2 x|$, at least at the integral points of time, and the resultant force acting on the mass will keep the sign opposite to that of the displacement x and tend to pull the body of mass m back to the

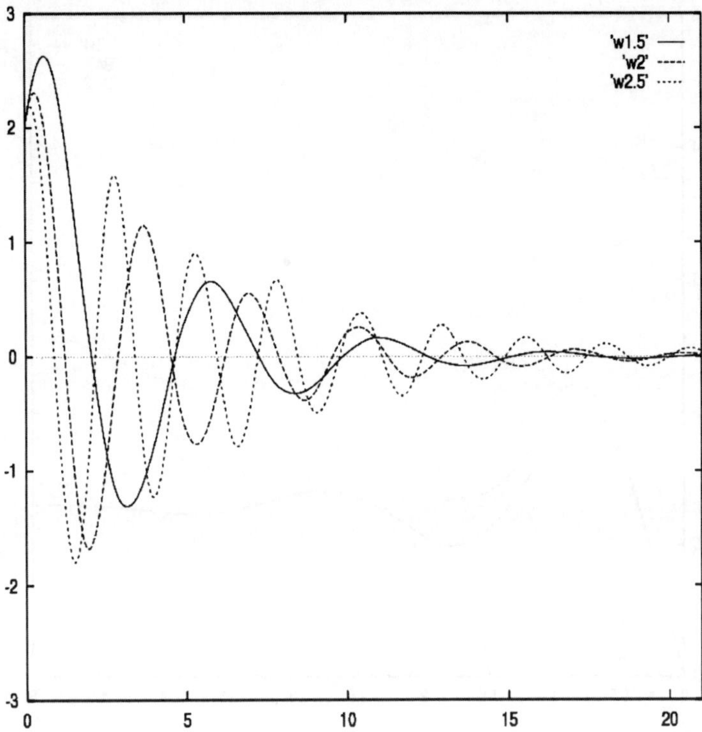

Figure 3.15. Solution of $\ddot{x}(t) + \omega^2 x = \beta x([t])$ as $0 < \beta < \omega^2$. $\beta = 1$, $d_0 = 2$, $v_0 = 2$. 'w1.5': $\omega^2 = 1.5$; 'w2': $\omega^2 = 2$; 'w2.5': $\omega^2 = 2.5$.

equilibrium position of the system. Hence, as shown in Figure 3.15 above, the corresponding solutions in this case are convergent. It may be seen from Figure 3.15 that the period of the oscillations is proportional to the value of ω.

(d) $\beta > \omega^2$

In this case, complex eigenvalues appear and corresponding modules of the eigenvalues are greater than 1. Physically, the resultant force $\beta x([t]) - \omega^2 x$ keeps the same sign as that of the displacement x for $t \in [0, \infty)$ at least at every integral point of time, and tends to push the body away from the equilibrium position, $x = 0$. Therefore, by equation (3.68), the solution in such a case is divergent, as shown in Figure 3.16.

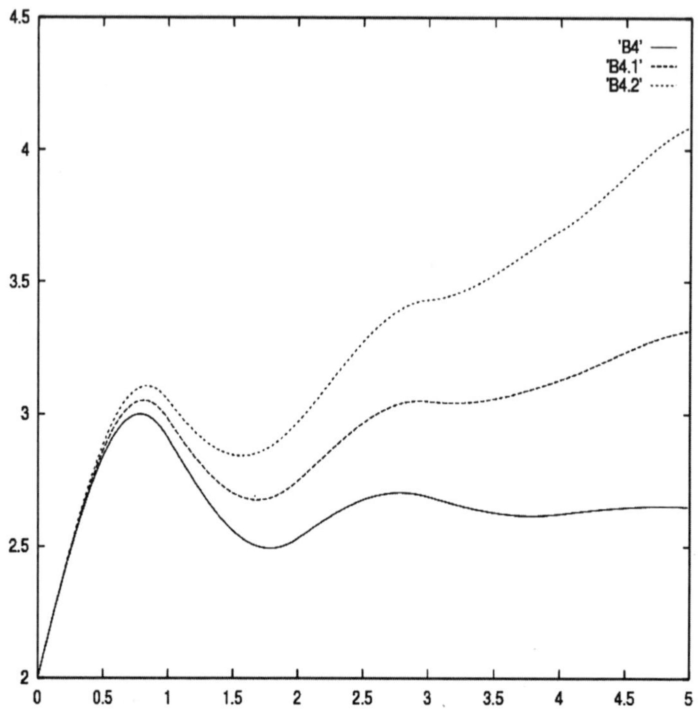

Figure 3.16.   Solution of $\ddot{x}(t) + \omega^2 x = \beta x([t])$ as $\beta > \omega^2$. $\omega^2 = 4$, $d_0 = 2$, $v_0 = 2$. 'B4': $\beta = 4$; 'B4.1': $\beta = 4.1$; 'B4.2': $\beta = 4.2$.

(e) $\beta < 0$

In the case of $\omega = (2n - 1)\pi$, for $\beta < 0$, both eigenvalues are negative and one of them is greater than unity in absolute value. Therefore, the corresponding motions are oscillatory and divergent, as shown in Figure 3.17. The motion in this case is divergent in general (except for some special cases such as $d_0 = v_0 = 0$) since the piecewise constant force acting on the body is in the same direction as that of the spring force. Under the combined action of the piecewise constant force and the spring force, from equation (3.12) the amplitude of vibration will increase with increasing time corresponding to a divergent motion. As exhibited in Figure 3.16, the amplitude of the oscillation in this case is proportional to the absolute value of the amplitude of the piecewise constant excitation acting on the system.

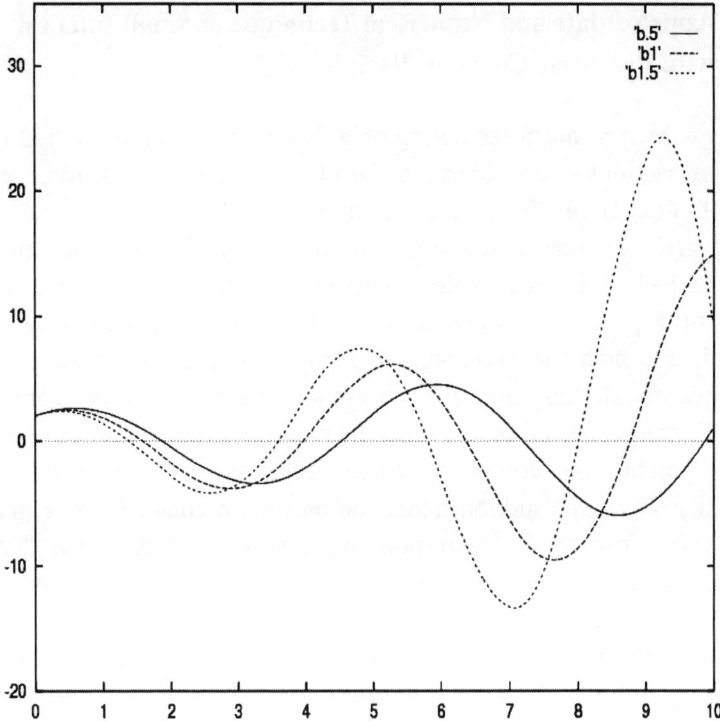

Figure 3.17. Solution of $\ddot{x}(t) + \omega^2 x = \beta x([t])$ as $\beta < 0$. $\omega^2 = 1$, $d_0 = 2$, $v_0 = 2$. 'b.5': $\beta = -0.5$; 'b1': $\beta = -1.0$; 'b1.5': $\beta = -1.5$.

Complete analytical solutions for the other linear dynamic systems discussed previously can also be obtained through the approach with implementation of the piecewise constant variables, as presented in the foregoing sections. It is clear that the properties of motion for these linear and continuous dynamic systems may also be analyzed from their closed form solutions by a similar approach as that for the undamped piecewise constant spring-mass system discussed above. Theoretical analysis for nonlinear systems, such as the continuous system subjected to the cubic excitation, is difficult due to the nonlinear excitations involved. However, a numerical analysis for the nonlinear piecewise constant system can be processed by employing the solution of equations (3.48), (3.49) and (3.50) derived in the prior sections.

3.9.　Approximate and Numerical Technique of Small Interval with Piecewise Constant Variable

A piecewise constant force acting on a dynamic system as studied in the previous sections is considered to be of a unit duration. A force of the form of $f(x[t])$ or $f([t])$ changes its value only at the integer point of time $t = [t]$, whereas it remains constant within the unit time interval $[t] \leq t < [t] + 1$. It is reasonable to expect that the effect of the stepwise action of a piecewise constant force would be weakened if the time interval, in which the exertion remains constant, can be made smaller. Consequently, the response of a system under a piecewise constant force of small interval would be closer to the motion of the system subjected to a corresponding but continuous force. When the time interval is made arbitrarily small, the motion under the piecewise constant force may be made very close to the corresponding continuous force such that the solution such obtained may well approximate the motion under the corresponding continuous force.

According to the inference above, a numerical technique to solve linear and nonlinear dynamic problems can be developed. Visualization of the motions exerted by piecewise constant forces of various time intervals should first be investigated, corresponding to the piecewise constant forces of smaller interval introduced. With a new variable of small time interval, the solution corresponding to a dynamic system subjected to a piecewise constant force of small interval should be easily derived on the basis of the preliminary knowledge presented in the prior chapters. As can be expected, solutions generated through this approach should be comparable with that of the solutions of the corresponding continuous systems, so long as the time intervals can be made properly small. When the time intervals are taken to be sufficiently small, a solution obtained can be a good approximation to the exact solution to a continuous dynamic system, therefore, the approach can be used as a numerical technique for the solutions of linear and nonlinear dynamic problems.

In numerically solving linear and nonlinear problems of dynamics, the commonly used methods, such as the Direct Linear Extrapolation

Methods (Newmark 1959), Iterative Predictor-Corrector Methods (Morino, Leech and Witmer 1974), and Runge-Kutta Method (Runge 1895, Kutta 1901, Burden and Faires 1989, Forsythe, Malcom and Moler 1977), provide the solutions at discrete points of time, say t_n and t_{n+1}, by approximately integrating the function in the interval between t_n and t_{n+1}. Usually, the information in between the adjacent two points t_n and t_{n+1} is simply produced through a linear interpolation. In contrast to these methods, the present technique achieves a continuous solution on the time interval of $t_n \leq t < t_{n+1}$ based on the solution of the equation of motion itself corresponding to a constant force in this interval. It is significant to have a continuous solution describing the motion of the system on the interval $t_n \leq t < t_{n+1}$ and therefore the entire time range. Furthermore, the continuous solution on $t_n \leq t < t_{n+1}$ accomplished by present technique is derived directly from the original differential equation of motion without resorting to numerical integration or interpolation.

As discussed previously, a piecewise constant force of the form of $f(x([t]))$ or $f([t])$ changes its value right at the integral point $t = [t]$, and remains constant within the interval $[t] \leq t < [t]+1$ of unit duration. If the time interval is made different from the unit interval of time, then, the corresponding motion must be different from that of the piecewise constant system with unit time interval. In order to vary and control the time interval, a new variable τ is introduced here, such that:

$$t = \frac{\tau}{p} \tag{3.75}$$

where p is a parameter which controls the scale of time for the new variable τ.

For determining the motion of a system subjected to a piecewise constant force over a small interval, the new variable τ may be substituted into the governing equation to replace the variable t, and a new piecewise constant variable $[\tau]$ may accordingly be employed to replace the variable $[t]$. In this manner, the corresponding piecewise constant force $f(x([\tau]))$ or $f([\tau])$ remains constant in the interval $[\tau] \leq \tau < [\tau]+1$ of changed duration and varies its value only at the integer point of $\tau = [\tau]$. If the parameter p is a value greater than one, the time scale is reduced p times from t to τ, and therefore, the interval in between $[t]$ to $[t]+1$ is correspondingly divided into p segments of unit duration in the new time scale.

It is clear that the change of time scale is actually a change of time unit. For $p > 1$, the unit of τ is p times smaller than that of t. Nevertheless, despite the varied unit of time, there is still a continuous dynamic system corresponding to a linear vibration in between the two integer points of $[\tau]$ and $[\tau]+1$. In other words, the properties of the variable $[\tau]$ are similar to that of the variable $[t]$, except that the unit of τ may be varied by parameter p. Owing to the above characteristics of the variable $[\tau]$, the procedure used for obtaining the solution of a dynamic system discussed previously can be directly applied to derive the solution of a system subjected to a piecewise constant force of a smaller interval of time (under the condition of $p > 1$).

With time τ of small unit, for instance, the governing equation (3.64) can be rewritten in the following form with the use of equation (3.75).

$$m\ddot{x}(\tau) + c\dot{x}(\tau) + kx(\tau) = Ax([\tau]) \tag{3.76}$$

The corresponding conditions of continuity for this system are given by the following expressions.

$$x_{[\tau]}([\tau]) = x_{[\tau]-1}([\tau]), \qquad \dot{x}_{[\tau]}([\tau]) = \dot{x}_{[\tau]-1}([\tau]) \tag{3.77}$$

The solution of equation (3.76) in a range $[\tau] \leq \tau < [\tau]+1$ can thus be obtained.

$$x_{[\tau]}(\tau) = e^{-\theta\frac{\tau-[\tau]}{p}} d_{[\tau]}\left(1 - \frac{A}{k}\right)\cos\left(\frac{\xi\tau - \xi[\tau]}{p}\right)$$

$$+ \frac{e^{-\theta\frac{\tau-[\tau]}{p}}}{\xi}\left[v_{[\tau]} + \theta d_{[\tau]}\left(1 - \frac{A}{k}\right)\right]\sin\left(\frac{\xi\tau - \xi[\tau]}{p}\right) + \frac{A}{k}d_{[\tau]} \tag{3.78}$$

where

$$d_{[\tau]}([\tau]) = x_{[\tau]}([\tau]), \qquad v_{[\tau]} = \dot{x}_{[\tau]}([\tau]) \tag{3.79}$$

and

$$\theta = \frac{c}{2m}, \qquad \xi = \left(\frac{k}{m} - \frac{c^2}{4m^2}\right)^{\frac{1}{2}} \tag{3.80}$$

Based on the solution shown in equation (3.78) and the conditions of continuity, the recurrence relations are derived in the following form.

$$d_{[\tau]} = e^{-\frac{\theta}{p}[\tau]} \left\{ \left[\left(1 - \frac{A}{k}\right)\left(\cos\frac{\xi}{p} + \frac{\theta}{\xi}\sin\frac{\xi}{p}\right) + \frac{A}{k}e^{\frac{\theta}{p}} \right] d_{[\tau]-1} + \frac{\sin\frac{\xi}{p}}{\xi} v_{[\tau]-1} \right\} \tag{3.81}$$

and

$$v_{[\tau]} = e^{-\frac{\theta}{p}[\tau]} \left\{ -\left(1 - \frac{A}{k}\right)\left[\left(\frac{\theta^2}{\xi} + \xi\right)\sin\frac{\xi}{p}\right] d_{[\tau]-1} + \left(\cos\frac{\xi}{p} - \frac{\theta}{\xi}\sin\frac{\xi}{p}\right) v_{[\tau]-1} \right\} \tag{3.82}$$

$d_{[\tau]-1}$ and $v_{[\tau]-1}$ in equations (3.81) and (3.82) can be related to the initial displacement d_0 and initial velocity v_0 through an iterative procedure, such that

$$\begin{bmatrix} d_{[\tau]-1} \\ v_{[\tau]-1} \end{bmatrix} = \begin{bmatrix} \left(1 - \frac{A}{k}\right)\left(\cos\frac{\xi}{p} + \frac{\theta}{\xi}\sin\frac{\xi}{p}\right) + \frac{A}{k}e^{\frac{\theta}{p}} & \frac{1}{\xi}\sin\frac{\xi}{p} \\ -\left(1 - \frac{A}{k}\right)\left[\left(\frac{\theta^2}{\xi} + \xi\right)\sin\frac{\xi}{p}\right] & \cos\frac{\xi}{p} - \frac{\theta}{\xi}\sin\frac{\xi}{p} \end{bmatrix}^{[\tau]} \begin{bmatrix} d_0 \\ v_0 \end{bmatrix} \tag{3.83}$$

The complete solution of the vibration system described by equation (3.76) may thus be obtained as:

$$x(\tau) = \left[\left(1 - \frac{A}{k}\right) \left[\cos\left(\frac{\xi\tau - \xi[\tau]}{p}\right) + \frac{\theta}{\xi}\sin\left(\frac{\xi\tau - \xi[\tau]}{p}\right) \right] \right.$$

$$\left. + \frac{A}{k} e^{\frac{\theta}{p}(\tau - [\tau])} \frac{1}{\xi}\sin\left(\frac{\xi\tau - \xi[\tau]}{p}\right) \right] D_{1p} \tag{3.84}$$

where

$$D_{1p} = e^{-\frac{\theta}{p}\tau} \left[\begin{array}{cc} \left(1 - \frac{A}{k}\right)\left(\cos\frac{\xi}{p} + \frac{\theta}{\xi}\sin\frac{\xi}{p}\right) + \frac{A}{k}e^{\frac{\theta}{p}} & \frac{1}{\xi}\sin\frac{\xi}{p} \\[2ex] -\left(1 - \frac{A}{k}\right)\left[\left(\frac{\theta^2}{\xi} + \xi\right)\sin\frac{\xi}{p}\right] & \cos\frac{\xi}{p} - \frac{\theta}{\xi}\sin\frac{\xi}{p} \end{array} \right]^{[\tau]} \left[\begin{array}{c} d_0 \\[1ex] v_0 \end{array} \right]$$

$$\tag{3.85}$$

Making use of different values of p for this solution, several responses of the system are plotted in Figure 3.18 with time τ for the purpose of visualizing the effects of varied time interval with the parameter p. For purpose of comparison, the analytical solution of the corresponding continuous system governed by equation (3.62) is superposed on the figure.

The continuous system has the same parameters and starts at the same state as that of the system under the piecewise constant force of small interval as shown in equation (3.76). As can be seen in Figure 3.18, the curves corresponding to the system under various piecewise constant forces approach the response of the continuous system as p increases. In comparing the curves to the one corresponding to the system with the variable $[t]$, i.e., $p = 1$, the most significant deviation occurs as the parameter p changes from 1 to 5. When the value of p becomes large enough, as shown in Figure 3.19 (where $p = 60$), the two solutions are almost identical. Therefore, the technique employing a piecewise constant variable $[\tau]$ may be used as a numerical method in approximately solving the dynamic problems, and the parameter p can be considered as a parameter to control the accuracy of the numerical solutions.

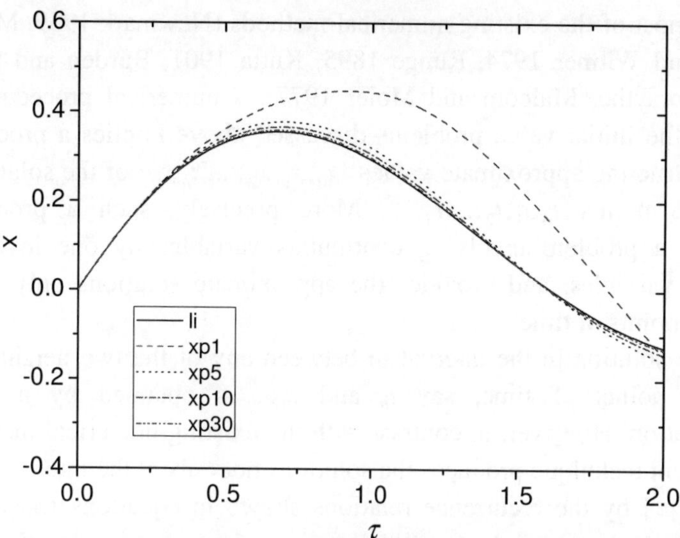

Figure 3.18.   Convergence of a piecewise constant system governed by $\ddot{x} + \dot{x} + 2x = -2x([\tau])$ to a continuous system governed by $\ddot{x} + \dot{x} + 4x = 0$ ('li'). In the figure, 'xp1': $p = 1$; 'xp5': $p = 5$, 'xp10': $p = 10$; and 'xp30': $p = 30$.

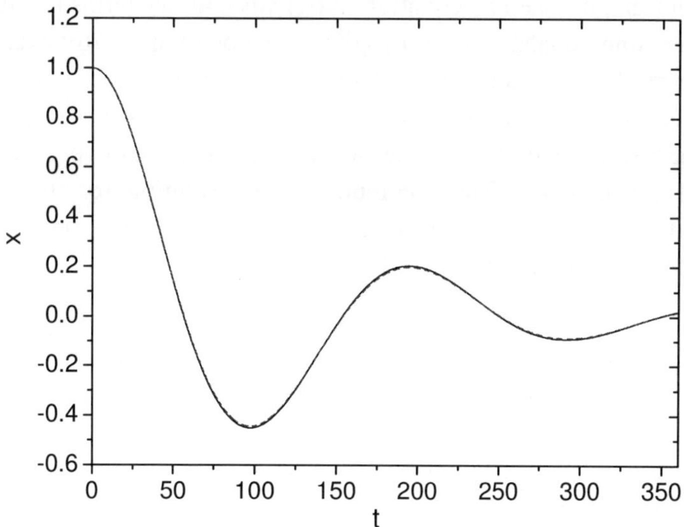

Figure 3.19.   Comparison of solutions of $\ddot{x} + \dot{x} + 2x = -2x([\tau])$ for $p = 60$ (solid line) and $\ddot{x} + \dot{x} + 4x = 0$ (dashed line). Initial conditions: $d_0 = 1$, $v_0 = 0$.

In most of the existing numerical methods (Newmark 1959, Morino Leech and Witmer 1974, Runge 1895, Kutta 1901, Burden and Faires 1989, Forsythe, Malcom and Moler 1977), a numerical procedure for solving the initial value problems discussed above implies a procedure for constructing approximate values $x_0, x_1, x_2, ..., x_n, ...$ of the solution at the time points $t_0, t_1, t_2, ..., t_n, ...$. More precisely, such a procedure replaces a problem involving continuous variables by one involving discrete variables, and provides the approximate solutions only at the discrete points of time.

The solution in the interval in between any of the two neighboring discrete points of time, say t_n and t_{n+1}, is obtained by a linear interpolation. However, in contrast with the existing numerical methods, the present technique produces the solutions not only at the integer points of time $[\tau]$ by the recurrence relations shown in equations (3.81) and (3.82), but also in between the integer points of time as given by equation (3.78). The continuous solution given by equation (3.78) represents a dynamic system of linear vibration in between the two integer points of time. The solution is a good approximation close to the exact solution governed by equation (3.62) since all the terms in equation (3.62) remained unchanged in the governing equation (3.76) except the single term $Ax(t)$ which is considered as a constant $Ax([\tau])$ in a small time interval. With consideration of the conditions of continuity, the present technique actually provides a solution continuous everywhere for the time range $\tau > 0$. The continuity of the solution for the system subjected to the piecewise constant forces as shown in equation (3.76) is clearly described by equation (3.84).

3.10. Characteristics of Approximate Results with Piecewise Constant Variable in Small Intervals

Numerical solutions for the other dynamic systems studied formerly can also be derived by employing the present technique. For a vibratory system subjected to a sinusoidally varying force, such as the system governed by equation (3.63), the corresponding numerical solution may be obtained by solving the following equation of motion.

$$m\ddot{x}(\tau) + c\dot{x}(\tau) + kx(\tau) = A\cos\Omega[\tau] \tag{3.86}$$

Employing the same procedure as the one used for obtaining equation (3.84), the complete solution for equation (3.86) is obtained as shown below.

$$x(\tau) = e^{\frac{\theta}{p}(\tau-[\tau])}\left[\begin{array}{cc}\cos\left(\dfrac{\xi\tau-\xi[\tau]}{p}\right) & \dfrac{1}{\xi}\sin\left(\dfrac{\xi\tau-\xi[\tau]}{p}\right)\\[2ex]+\dfrac{\theta}{\xi}\sin\left(\dfrac{\xi\tau-\xi[\tau]}{p}\right) & \end{array}\right]D_{2p}$$

$$+\frac{A}{k}\cos\left(\frac{\Omega}{p}[\tau]\right)e^{-\frac{\theta}{p}(\tau-[\tau])}\left[e^{\frac{\theta}{p}(\tau-[\tau])}\quad-\cos\left(\frac{\xi\tau-\xi[\tau]}{p}\right)\right.$$

$$\left.-\frac{\theta}{\xi}\sin\left(\frac{\xi\tau-\xi[\tau]}{p}\right)\right] \tag{3.87}$$

where ξ and θ are defined in equation (3.80) and the matrix $\mathbf{D}_{2p}$ is

$$\mathbf{D}_{2p} = e^{\frac{\theta}{p}[\tau]}S_p^{[\tau]}\begin{bmatrix}d_0\\v_0\end{bmatrix} + \sum_{j=1}^{[\tau]}e^{-\frac{\theta j}{p}}S_p^{j-1}\begin{bmatrix}e^{\frac{\theta}{p}}-\cos\dfrac{\xi}{p}-\dfrac{\theta}{\xi}\sin\dfrac{\xi}{p}\\[2ex]\left(\dfrac{\theta^2}{\xi}+\xi\right)\sin\dfrac{\xi}{p}\end{bmatrix} \tag{3.88}$$

$$\frac{A}{k}\cos\left[\frac{\Omega}{p}([\tau]-j)\right]$$

in which the square matrix

$$S_p = \begin{bmatrix}\cos\dfrac{\xi}{p}+\dfrac{\theta}{\xi}\sin\dfrac{\xi}{p} & \dfrac{1}{\xi}\sin\dfrac{\xi}{p}\\[2ex]-\left(\dfrac{\theta^2}{\xi}+\xi\right)\sin\dfrac{\xi}{p} & \cos\dfrac{\xi}{p}-\dfrac{\theta}{\xi}\sin\dfrac{\xi}{p}\end{bmatrix} \tag{3.89}$$

The solution is again continuous everywhere for $\tau>0$.

With exceptions of some special cases, most of the nonlinear initial value problems cannot be solved explicitly (Sanchez, Allen and Kyner 1988, Boyce and DiPrima 1969). Hence, numerical methods are extremely important in constructing approximate solutions to the problems and often treated as an exploratory tool to help qualitatively understand the properties of the solution. For numerically solving the nonlinear initial value problems, the present technique with the piecewise constant variable of controllable time intervals can be shown as a useful tool for obtaining a reasonably accurate solution.

Consider the dynamic system governed by equation (3.66). For such a system with cubic nonlinearity, closed form analytical solutions are usually not available and recourse must inevitably be made to numerical methods. Replacing the cubic term $Ax^3(t)$ in equation (3.66) by $Ax^3([\tau])$, and through a procedure similar to what has been employed for producing the solutions of equations (3.84) and (3.86), solution to equation (3.66) is given by

$$x_{[\tau]}(t) = e^{-\frac{\theta}{p}(t-[t])} \left\{ \begin{array}{l} d_{[\tau]}\left(1-\frac{Ad_{[\tau]}^2}{k}\right)\cos\left[\frac{\xi}{p}(t-[t])\right] \\ +\frac{1}{\xi}\left[v_{[\tau]}+\theta d_{[\tau]}\left(1-\frac{Ad_{[\tau]}^2}{k}\right)\right]\sin\left[\frac{\xi}{p}(t-[t])\right]+\frac{Ad_{[\tau]}^3}{k} \end{array} \right\}$$

$$(3.90)$$

The recurrence relations corresponding to the system are derived as

$$d_{[\tau]} = e^{-\frac{\theta}{p}} \left\{ \begin{array}{l} d_{[\tau]-1}\left(1-\frac{Ad_{[\tau]-1}^2}{k}\right)\cos\frac{\xi}{p} \\ +\frac{1}{\xi}\left[v_{[\tau]-1}+\theta d_{[\tau]-1}\left(1-\frac{Ad_{[\tau]-1}^2}{k}\right)\right]\sin\frac{\xi}{p} \end{array} \right\} + \frac{Ad_{[\tau]-1}^3}{k}. \qquad (3.91)$$

$$v_{[\tau]} = -e^{-\frac{\theta}{p}}\left(\sin\frac{\xi}{p}\right)\left[\left(\frac{\theta}{\xi}(v_{[\tau]-1}+\theta d_{[\tau]-1}-\frac{\theta Ad_{[\tau]-1}^3}{k}\right)\right.$$
$$\left.+\xi\left(d_{[\tau]-1}-\frac{Ad_{[\tau]-1}^3}{k}\right)\right] + e^{-\frac{\theta}{p}}v_{[\tau]-1}\cos\frac{\xi}{p} \qquad (3.92)$$

where ξ and θ are given by equation (3.80).

These relations are manageable and convenient for computer programming in evaluating numerical results. The numerical calculation starts at the initial state with the conditions, $x(0) = d_0$ and $\dot{x}(0) = v_0$. Motion of the system in the first interval of time is calculated by equation (3.90) with the initial displacement and velocity. Conditions at the end of this interval, d_1 and v_1, can also be obtained by equation (3.90). Once the values of d_1 and v_1 are obtained, the end conditions for the second interval, therefore, the starting conditions for the third interval can be produced by the recurrence relations shown in equations (3.91) and (3.92). Again, the calculation for the motion in the second interval may be accomplished by equation (3.90) with the local initial conditions d_1 and v_1. By continuing this step-by-step procedure, the continuous solution for the nonlinear system represented by equation (3.66) can be obtained over any desired length of time.

The comparison of the solutions corresponding to equations (3.87) and (3.90) with the linear and nonlinear vibrations under continuous forces expressed by equations (3.63) and (3.66) are shown in Figure 3.20 and Figure 3.21.

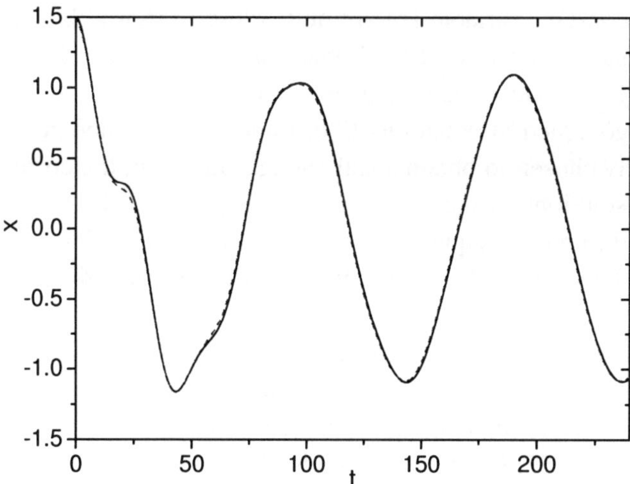

Figure 3.20.   Comparison of solutions of $\ddot{x} + 2.5\dot{x} + 196x = 196\cos(4[\tau])$ for $p = 60$ (solid line) and $\ddot{x} + 2.5\dot{x} + 196x = 196\cos(4\tau)$ (dashed line). Initial conditions: $d_0 = 1.5$, $v_0 = 0$.

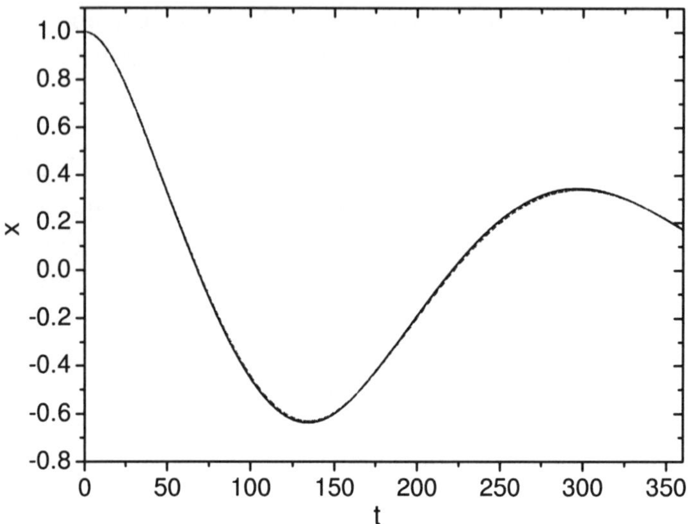

Figure 3.21. Comparison of solutions to nonlinear systems of $\ddot{x} + 0.5\dot{x} + x + 2x^3([\tau]) = 0$ for $p = 60$ (solid line) and $\ddot{x} + 0.5\dot{x} + x + 2x^3(\tau) = 0$ (dashed line). Initial conditions: $d_0 = 1.0$, $v_0 = 0$.

It can be observed from the graphs that the solutions obtained with a piecewise constant variable $[\tau]$ are quite close to those obtained by the classical analytical approaches (Lomen and Mark 1988, Rosear 1987) and numerical solution (Christopher 1973) for the linear and nonlinear vibrations governed by equations (3.63) and (3.66). When the parameter p is properly chosen to obtain small enough time unit, the corresponding numerical solutions of the dynamic systems governed by the linear and nonlinear differential equations, such as equations (3.84), (3.87) and (3.90), can be sufficiently close to those of the corresponding continuous systems.

As a further example, chaotic oscillations governed by Duffing's equations with numerical solutions which are sensitive to the initial conditions and the time integration step (Lichtenberg and Lieberman 1992, Tongue 1987, Mahfouz and Badrakhan 1990), are also undertaken below. Ueda (1980), who first explored chaotic motion of stable Duffing's systems, reported the numerical solutions of the similar results for the chaos governed by the equation of motion:

$$\ddot{x}(t) + bx(t) + x^3(t) = \beta \cos t \tag{3.93}$$

According to the technique discussed above, and by using the smaller time unit, the above equation is rewritten as:

$$\ddot{x}(\tau) + bx(\tau) + x^3(\tau) = \beta \cos([\tau]) \tag{3.94}$$

The corresponding solution can be obtained as

$$x(\tau) = \frac{1}{b}\left[b \quad 1 - e^{-\frac{b}{p}(\tau - [\tau])}\right] D_{3p}$$

$$+ \frac{1}{b^2}\left[d_{[\tau]}^3 - \beta \cos\left(\frac{[\tau]}{p}\right)\right]\left[1 - e^{-\frac{b}{p}(\tau - [\tau])} - b(\tau - [\tau])\right] \tag{3.95}$$

where the recurrence relation in the matrix form of $\mathbf{D}_{3p}$ is

$$D_{3p} = \sum_{j=1}^{[\tau]} \frac{d_{[\tau]}^3 - \beta \cos\left(\frac{[\tau] - j}{p}\right)}{b}\left[\begin{array}{cc} 1 & \frac{1}{b}(1 - e^{-\frac{b}{p}}) \\ 0 & e^{-\frac{b}{p}} \end{array}\right]^{j-1}\left[\begin{array}{c} \frac{1}{b}(1 - e^{-\frac{b}{p}}) - \frac{1}{p} \\ e^{-\frac{b}{p}} - 1 \end{array}\right]$$

$$+ \left[\begin{array}{cc} 1 & \frac{1}{b}(1 - e^{-\frac{b}{p}}) \\ 0 & e^{-\frac{b}{p}} \end{array}\right]^{[\tau]}\left[\begin{array}{c} d_0 \\ v_0 \end{array}\right] \tag{3.96}$$

Obviously, the solution described above is also continuous over the range $\tau > 0$. The solution given by equation (3.95) is graphically shown in Figure 3.22 which compares very well with the numerical solution reported by Ueda (1980a, b).

Based on the analysis and graphical comparisons above, the present technique provides continuous solutions for the dynamic systems under piecewise constant forces of small interval, and significantly, the technique can also be used as a numerical method in approximately solving the linear and nonlinear vibration problems.

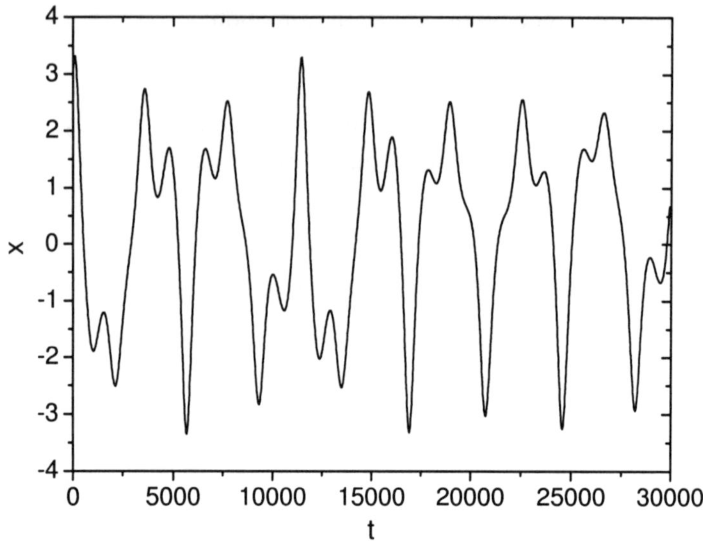

Figure 3.22.   Chaotic motion of a nonlinear system governed by $\ddot{x} + 0.05\dot{x} + 2x^3(\tau) = 7.5\cos[\tau]$ for $p = 600$. Initial conditions: $d_0 = 3.0, v_0 = 4.0$.

The essential feature of the above investigation is a study of the response of a vibration system when subjected to a piecewise constant force i.e. a force which is constant in an interval of time and changes only at the beginning and at the end of the interval. In analyzing such a system, it is important to keep in mind the following points.

1. The system is subjected to a piecewise constant force which is known a priori in its form.
2. The piecewise constant force is represented by a nonlinear unknown function of the given equation.
3. The magnitude of the piecewise constant force is unknown but is expressible in a linear form in terms of the unknown function.
4. When in the case of a linear or nonlinear system subjected to an unknown piecewise constant force, the interval of the piecewise constant force can be made small enough such that the solution can be a good approximation to the system subjected to continuous excitation.

References

Aftabizadeh, A. R. and Wiener, J., "Oscillatory and Periodic Solutions for Systems of Two First Order Linear Differential Equations with Piecewise Constant Argument," Applicable Analysis, Vol. 26, pp. 327-338, 1988.

Boyce, W. E. and DiPrima, R. C., Elementary Differential Equations, John Wiley & Sons, Inc., New York, 1969.

Burden, R. L. and Faires, J. D., Numerical Analysis, PWS-KENT Publishing Company, Boston, 1989.

Busenberg, S. and Cooke, K. L., Models of Vertically Transmitted Disease with Sequential Continuous Dynamics, in "Nonlinear Phenomena in Mathematical Sciences" (V. Lakshmikantham, Ed.), Academic Press, New York, pp. 179-189, 1982.

Christopher, P. A. T., "An Approximate Solution to a Strongly Non-Linear, Second-Order, Differential Equation," International Journal of Control, Vol. 17, No. 3, pp. 597-608, 1973.

Cooke, K. L. and Wiener, J., "Retarded Differential Equations with Piecewise Constant Delays," Journal of Mathematical Analysis and Applications, Vol. 99, No. 1, pp. 256-297, 1984.

Cooke, K. L. and Wiener, J., "An Equation Alternate of Retarded and Advanced Type," Proceedings of the American Mathematical Society, Vol. 99, pp. 726-732, 1987.

Dai, L. and Singh, M. C., "On Oscillatory Motion of Spring-Mass Systems Subjected to Piecewise Constant Forces," Journal of Sound and Vibration, Vol. 173, pp. 217-233, 1994.

Den Hartog, J. P., "Recent Developments in Dynamics and Vibration," Applied Mechanics Reviews, Vol. 4, No. 1, pp. 2-3, 1951.

Duffing, G., Erzwungene Schwingungen bei veränderlicher Eigenfrequenz und ihre technische Bedeutung, Braunsweig, F. Vieweg u. Sohn, 1918.

Fang, T. and Dowell, E. H., "Numerical Simulations of Periodic and Chaotic Responses in a Stable Duffing's System," International Journal of Nonlinear Mechanics, Vol. 22, No. 5, pp. 401-425, 1987.

Farmer, J. D., "Chaotic Attractors of an Infinite-Dimensional Dynamical System," Physica, Vol. 40, pp. 366-393, 1982.

Forsythe, G. E., Malcolm, M. A. and Moler, C. B., Computer Methods for Mathematical Computations, Prentice-Hall, Englewood Cliffs, New Jersey, 1977.

Gantmacher, F. R., the Theory of Matrices, Chelsea Publishing Company, New York, 1960.

Inglis, C. E., "Mechanical Vibrations: Their Cause and Prevention," Institution of Civil Engineers (London), Vol. 22, No. 8, pp. 312-357, 1944.

Jayasree, K. N. and Deo, S. G., "Variation of Parameters Formula for the Equation of Cooke and Wiener," Proceedings of the American Mathematical Society, Vol. 112, No. 1. pp. 75-80, 1991.

Klitter, K., "Non-Linear Vibration Problems Treated by the Averaging Method of W. Rotz," Proceedings of First United States National Congress of Applied Mechanics, pp. 125-131, 1951.

Kutta, W., "Beitrag zur näherungsweisen Integration totaler Differentialgleichungen," Zeitschrift fur Mathematik und Physik, Vol. 46, pp. 435-453, 1901.

Lichtenberg, A. J. and Lieberman, M. A., Regular and Chaotic Dynamics, 2nd Edition, Springer-Verlag, New Jersey, 1992.

Lomen, D. and Mark, J., Differential Equations, Prentice-Hall, Englewood Cliffs, New Jersey, 1988.

Mahfouz, I. A. and Badrakhan, F., "Chaotic Behavior of Some Piecewise-Linear Systems, Part I: Systems with Set-Up Spring or with Unsymmetric Elasticity," Journal of Sound and Vibration, Vol. 143, pp. 255-288, 1990.

Morino, L., Leech, J. W. and Witmer, E. A., "Optimal Predictor-Corrector Method for Systems of Second-order Differential Equations," AIAA Journal, Vol. 12, No. 10, pp. 1343-1347, 1974.

Newmark, N. M., "A Method of Computation for Structural Dynamics," ASCE Journal of the Engineering Mechanics Division, Vol. 85, pp. 67-94, 1959.

Runge, C., "Ueber die numerische Auflösung totaler Differential gleichungen," Mathematische Annalen, Vol. 46, pp. 167-178, 1895.

Rosear, M., Vibration in Mechanical Systems, Spring-Verlag, New York, 1987.

Sanchez, D. A., Allen, R. C., Jr. and Kyner, W. T., Differential Equations, Addison-Wesley Publishing Company, New York, 1988.

Thomson, W. T., Theory of Vibration with Applications, Prentice-Hall, Inc., New Jersey, 1981.

Thompson, J. M. T. and Stewart, H. B., Nonlinear Dynamics and Chaos, John Wiley and Sons Ltd., New York, 1986.

Tongue, B. H., "Characteristics of Numerical Simulations of Chaotic Systems," Journal of Applied Mechanics, Vol. 54, pp. 695-699, 1987.

Ueda, Y., "Steady Motions Exhibited by Duffing's Equation: A Picture Book of Regular and Chaotic Motions," in New Approaches to Nonlinear Problems in Dynamics, (editor P.J. Holmes), Philadelphia: SIAM., pp. 311-322, 1980.

Ueda, Y., Explosion of Strange Attractors Exhibited by Duffing's Equation, in Nonlinear Dynamics, Helleman, R. H. G. (ed.), New York Academy of Sciences, New York, 1980.

Wang, Y. and Yan, J., "Oscillation of a Differential Equation with Fractional Delay and Piecewise Constant Arguments," Computers and Mathematics with Applications, Vol. 52, pp. 1099-1106, 2006.

Weaver, W. J., Timoshenko, S. and Young, D. H., Vibration Problems in Engineering, John Wiley & Sons, Inc., New York, 1990.

Wiener, J. and Aftabizadeh, A. R., "Differential Equations Alternately of Retarded and Advanced Types," Journal of Mathematical Analysis and Applications, Vol. 129, pp. 243-255, 1988.

Zhang, B. G, and Parhi, N. "Oscillatory and Nonoscillatory Properties of First Order Differential Equations with Piecewise Constant Deviating Arguments," Journal of Mathematical Analysis and Applications, Vol. 139, pp. 23-25, 1989.

CHAPTER 4

Analytical and Semi-Analytical Solution Development with Piecewise Constant Arguments

4.1. Introduction

It has been described in Chapter 3 that the analytical solutions of the continuous dynamic systems can be obtained by utilizing some piecewise constant functions of the form $[t]$. The solutions with piecewise constant variables are firstly developed and then letting some of the systems parameters to approach to zero or vanish with a limiting case, as those discussed in Sections 3.4 and 3.5. The complete analytical solutions are then obtained for the linear systems. This is still an indirect method for developing the analytical solutions for linear continuous systems, and the relationship between piecewise constant systems and their corresponding continuous systems are unable to be revealed on the basis of the approaches discussed previously. In the previous chapter, moreover, an approximate technique is introduced for solving the linear and nonlinear dynamic problems. However, the technique is not valid for generating complete analytical solutions.

As described in the previous chapters, the piecewise constant systems are different in general from the continuous systems. The primary distinguishable characteristics of the two types of systems can be summarized as follows.

1. Piecewise constant variables or their derivatives are involved in the mathematical model and governing differential equations, whereas

117

the variables of the conventional continuous systems are all continuous.

2. The piecewise constant systems are categorized as discrete systems, as the variables of the systems vary piecewise constantly. A piecewise constant system may only be considered as a linear system in a specified interval.

3. The behavior of piecewise constant systems are different from that of continuous systems, as demonstrated in the previous chapter. Extraordinary and nonlinear-like behavior may occur to nonlinear even linear piecewise constant systems.

Obviously, it will be theoretically and practically significant if the relationship between the piecewise constant systems and the corresponding continuous systems can be established. In this chapter, a novel piecewise constant argument $[Nt]/N$ will first be introduced for establishing such a relationship. A methodology with implementing the piecewise constant argument will be presented for bridging the gap between the piecewise constant systems and their corresponding continuous systems. Moreover, a new approach of directly solving for linear and nonlinear dynamic problems for analytical, semi-analytical and approximate solutions is also to be demonstrated in this chapter.

4.2.   A New Piecewise Constant Argument $[Nt]/N$

The piecewise constant systems such as the retarded and advanced functional differential equations with piecewise constant argument in the form $[t]$ or $(t \pm n[t])$ have attracted a great attention from the research workers since early 1980s, as indicated in Chapter 1. It has been noticed that the analysis of the piecewise constant systems were initiated on the basis of constructing the sequential-continuous models for vertically transmitted diseases in which the continuous dynamics for intervals of the form $([t], [t] + 1)$ was considered (Busenberg and Cooke 1982). Piecewise constant systems or systems close to piecewise constant systems can also be found in many fields such as engineering and biology. However, most of the piecewise systems found in the literature are modeled with first-order differential equations studied from the mathematical point of view with proofs of existence and uniqueness of the solutions and oscillatory properties of the systems. Dai and Singh (1994) presented the solutions of several second-order differential equations representing motions of a spring-mass system disturbed by a piecewise constant force in the form of $f([t])$ or $f(x([t]))$. The oscillatory, nonoscillatory and periodic properties of motion of the spring-mass system were also investigated.

In previous chapter, a technique making use of the piecewise constant argument of small interval is employed as a numerical method to solve dynamical problems. According to the numerical results presented in Chapter 3 with the variable τ, one may reasonably assume that the numerical results may progressively approach the exact solutions representing the continuous systems as the value of p increases or the size of the interval in between the two points of time decreases. The smaller the interval between the neighboring two points, the closer the numerical result to the exact solution. However, the unit of τ is different from that of t and it is finite no matter how small the interval between two neighboring points of time is. Furthermore, as discussed in the previous chapter and according to the studies reported by the researchers in this area (Busenberg and Cooke 1982, Shah and Wiener 1983, Cooke and Wiener 1984, Aftabizadeh and Wiener 1985, Jayasree and Deo 1992, Cooke and Wiener 1987, Zhang and Parhi 1989, Aftabizadeh and Wiener 1988, Wiener and Aftabizadeh 1988, Wiener and Cooke 1989, Leung 1988, Dai and Singh 1991, Dai and Singh 1994), the behavior of a system with piecewise constant arguments is unique and demonstrate different behavior from that with continuous arguments. Nevertheless, there are also some similarities between a piecewise constant system and its corresponding continues system with respect to the governing equations, and also the behavior to certain extents. With these considerations, one may naturally ask whether there exists a relationship between a system with piecewise constant arguments and its corresponding system with continuous arguments; a relationship that may smoothly and seamlessly connect the two systems of piecewise constant and continuous. Consequently, how this relationship is to be established is another question to be asked, should the relationship exist. Undoubtedly, it will be very significant to link a continuous system to its corresponding piecewise constant system in analyzing the characteristics of piecewise constant system and in relating them with that of the continuous systems. Dai and Singh have been searching for such relationship for years and reported a technique relating the two systems (1993, 2003). In their research, a new piecewise constant argument $[Nt]/N$ was introduced. With the implementation of the piecewise constant argument, filling the gap between a continuous system and a corresponding piecewise constant system becomes readily available. In this chapter, the relationship between the two systems is to be further investigated and the establishment of the relationship with utilization of the piecewise constant argument is to be demonstrated in a systematical manner.

Taking the mathematical symbol $[\cdot]$ as the greatest-integer function, and assume n as an integer and y as a value in the range $n \leq y < n + 1$, one may then have $[y] = n$. This implies that $[y]$ keeps a constant integer value n as y varies in the range of $n \leq y < n + 1$. If $[y]$ is further employed as an argument, when y increases continuously, the corresponding value of $[y]$ increases in a piecewise constant fashion.

Let $[Nt]$ be a greatest integer function of N and t, where N designates as a parameter which may take any fixed positive or negative numerical value. However, for the sake of numerical calculations and clarity in theoretical analysis, it is convenient for N to be selected as a positive integer.

Introduce a new piecewise constant argument $[Nt]/N$ for using it as a function in differential equations governing the equations of motion. The governing equations with piecewise constant argument $[t]$, studied in the prior chapters, can then be considered as a special case of the differential equations of dynamics with piecewise constant argument $[Nt]/N$. It is easy to show that $[Nt]/N$ yields a value close to t as N takes a large value. The larger the value N is, the closer is the value $[Nt]/N$ to t. Obviously, N is a value controlling how close that $[Nt]/N$ is to t. When the parameter N approaches infinity, as will be shown theoretically later in this chapter, the argument $[Nt]/N$ tends to become t. The difference between a piecewise constant and a continuous system is consequently vanished in a limiting case. More significantly, the second-order differential equations which govern the dynamic or vibration problems, can be directly solved through a procedure of constructing a piecewise constant system corresponding to the continuous system represented by the governing equations. The differential equations with the argument $[Nt]/N$ are therefore more general and fall into a category covering both of the classical continuous systems and the systems with piecewise constant arguments.

It is also true that a function $f([Nt]/N)$ with the argument $[Nt]/N$ is a good approximation to the given continuous function $f(t)$ with argument t, if the parameter N is sufficiently large. This makes the procedure of solving linear and nonlinear dynamic problems an efficient semi-analytical and numerical method. An important distinction between the present method and existing numerical methods is that the solutions given by the existing numerical methods are discrete, whereas the solutions and their first derivatives produced by the approach of implementing the piecewise constant argument are continuous everywhere along the entire time range considered. Moreover, since the function $f([Nt]/N)$ is piecewise constant and $[Nt]$ is an integer, the recurrence relation necessary for performing a numerical calculation can be easily derived and the numerical calculation can be conveniently carried out on a computer.

4.3. Solving for Dynamic Systems with Implementation of Piecewise Constant Arguments

In order to solve a dynamic problem governed by a second-order differential equation with classical analytical methods such as Euler's method (Bear 1962, Euler 1743, Salvadori 1954), conventionally, a form of the sought solution is conjectured in advance with undetermined constants. Using the conjectured solution, a purely algebraic equation known as the characteristic equation has to be developed, and the characteristic equation together with the initial conditions then yields the complete solution in the conjectured form. However, in solving the second-order differential equations by the piecewise-constant procedure, continuous solutions without utilizing a presumed solution of any form can be expected. For developing the approach with implementing the piecewise constant argument on a theoretically sound basis, several mathematical concepts need to be established first with considerations of analytically solving the dynamical problems.

Many phenomena in dynamics associated with complex dynamic systems in engineering applications can be simplified to a linear system with a motion of one degree of freedom and described by an equation of motion of the form

$$m\ddot{x} + c\dot{x} + kx = f(t), \quad t > 0 \tag{4.1}$$

where m is the mass, $x(t)$ is a dependent variable of time t, c is the damping coefficient, and k is a parameter relating to material or structural characteristics of the system, and the time dependent function $f(t)$ expresses the continuous external excitation acting on the system.

In order to solve the dynamic system governed by above equation via implementation of the piecewise constant argument, and thus to establish the relationship between a dynamic system with piecewise constant argument $[Nt]/N$ and the continuous systems described by equation (4.1), the following theorem needs to be introduced.

Theorem 1. *Suppose an argument $[Nt]$ is the greatest-integer function of the product of time t and parameter N, where N is a real positive integer (for the sake of convenience, though N is not necessary to be an integer), then, ratio $[Nt]/N$ approaches t as N goes to infinity.*

Proof. For any given ε, there exists a N_0 such that $1/N_0 < \varepsilon$. Since $[Nt]/N - t = ([Nt] - Nt)/N$ and $-1 \leq [Nt] - Nt \leq 0$, hence, $|[Nt]/N - t| \leq 1/N$. Accordingly for any $\varepsilon > 0$, take $N_0 \geq [1/\varepsilon$, when $N \geq N_0$. It follows that

$$\left|\frac{[Nt]}{N} - t\right| \leq \frac{1}{N} \leq \frac{1}{N_0} < \varepsilon \tag{4.2}$$

By the arbitrariness of ε, a result of limiting case can be given as follows.

$$\lim_{N \to \infty} \frac{[Nt]}{N} = t \tag{4.3}$$

Theorem 1 and the conclusion given in equation (4.3) are extremely important for the investigation on the relationship between the piecewise constant systems and their corresponding continuous systems, and most of the derivations in the subsequent sections will depend on them.

Based on Theorem 1, practically, any variable of a governing equation in the form of $f(t)$ can be replaced by a piecewise constant function of $g([Nt]/N)$. With the replacement, the piecewise-constant procedure described previously can be applied. A continuous solution of the reversed governing equation with the replacement can then be generated on a small time segment $[Nt]/N \leq t \leq ([Nt] + 1)/N$ where one or more of the terms in the governing equation can be treated as constants. Therefore, the governing equation on this time segment is simplified in a way such that the corresponding analytical solution can be easily obtained (direct integration becomes available).

This implies that the argument $[Nt]/N$ tends to t when N approaches infinity in such a way that the governing differential equations with piecewise constant argument $[Nt]/N$ tend to become the differential equations with continuous argument t. With the introduction of the piecewise constant argument, the second-order differential equations which govern the dynamical problems, can be directly solved through a piecewise-constant procedure. Under the conditions of continuity, an approximate solution can be consequently constructed by combining all solutions corresponding to the small time segments over the range from zero to t. When the parameter N approaches infinity and the length of the time segment tends to zero, as will be proved in the following sections, the solution corresponding to a given equation of motion uniquely converges to the classical solution of the corresponding continuous system. This is to state that the dynamic systems governed by the differential equation in the form of (4.1) can solved through this piecewise-constant procedure for closed form solutions.

Applying the present method for solving the dynamic problems, in contrast with classical methods, there is no need to conjecture a solution of any form in advance and no need to construct a characteristic equation either. To demonstrate the procedure of solving the problems of dynamics by employing the piecewise constant argument $[Nt]/N$, a typical general governing equation in dynamics will be employed. The existence and uniqueness of the solutions of the dynamic system with piecewise constant arguments will also be demonstrated together with the necessary mathematical derivations and proofs.

Replacing the variable t of the continuous function $f(t)$ in equation (4.1) with the piecewise constant variable $[Nt]/N$, the following piecewise constant system corresponding to the continuous system indicated in equation (4.1) can be constructed:

$$m\ddot{X} + c\dot{X} + kX = g\left(\frac{[Nt]}{N}\right)$$

$$[Nt]/N \le t < ([Nt]+1)/N$$

(4.4)

For the purpose of distinction between the continuous system and the piecewise constant system indicated above, and for convenience of further discussion, x here is designated for the displacement of the system with piecewise constant argument $[Nt]/N$, and the function $g([Nt]/N)$ stands for the piecewise constant force acting on the system. The exerting forcing function $g([Nt]/N)$ is discontinuous on the entire range of $t \in [0,\infty)$, and only varies its value at the moment $t = i/N$, $i = 1, 2, 3, \ldots, [Nt]$. The value of $g([Nt]/N)$ is constant in the interval $i/N \le t < (i+1)/N$ and $g([Nt]/N)$ has the same values as $f(t)$ at $t = i/N, i = 1,2,3,\ldots,[Nt]$, i.e., $g(i/N) = f(i/N)$. This implies that the length of the interval is $1/N$. The length of the interval is less than a unit of time if the parameter N is taken a value greater than one. Theoretically, according to Theorem 1, the length of the interval approaches zero as N tends to infinity.

Based on the discussion above, equation (4.4) represents a simplified continuous dynamic system of linear vibration within a time segment of length $1/N$. When the parameter N is made large enough, the time interval can be made correspondingly as small as desired; the motion governed by equation (4.4) can then approximately represent the vibration governed by equation (4.1). It will be shown in the following sections that the solutions of equations (4.1) and (4.4) will be identical if N tends to infinity. The procedure of mutating a term in equation (4.1) into a constant over $i/N \le t < (i+1)/N$ is referred as "piecewise-constant procedure" hereafter.

The function $X(t)$ can be considered as the solution of equation (4.4) if and only if the following conditions are satisfied.

Definition 1. *Solution $X(t)$ of equation (4.4) on the range of $t \in [0,\infty)$ satisfies the following conditions.*

(i) *$X(t)$ and its derivative $\dot{X}(t)$ are continuous over $t \in [0,\infty)$;*

(ii) *$\ddot{X}(t)$ exists at each point of time with the possible exception at the points i/N where the left-sided derivatives exist;*

(iii) *On each time interval $i/N \le t < (i+1)/N$, $X(t)$ satisfies equation (4.4);*

(iv) *the general solution $X(t)$ is the combination of any particular solution of the inhomogeneous equation (4.4) and the solution of the corresponding homogeneous equation.*

With this Definition, existence and uniqueness of the solution of equation (4.4) can now be proved. To show the existence and uniqueness of the solution, the Lipschitz condition (Moulton 1930, Lipschitz 1877, Johnson and Riess 1982) stated below must be satisfied.

Definition 2. *A function $h(x_1, x_2, ..., x_n)$ is said to satisfy a Lipschitz condition in the neighborhood of $\left(x_1^0, x_2^0, ..., x_n^0\right)$ if there exists a constant L such that*

$$\left| h(x_1, x_2, ..., x_n) - h\left(x_1^0, x_2^0, ..., x_n^0\right)\right| \le L\left|x_i - x_i^0\right| \tag{4.5}$$

for any point $h(x_1, x_2, ..., x_n)$ in the neighborhood of $\left(x_1^0, x_2^0, ..., x_n^0\right)$, where $i = 1, 2, 3, ..., n$.

Lemma 1. If h is a continuous function and satisfies a Lipschitz condition for all variables except t in the neighborhood of $t_0, y_0, y_0', y_0^{(n-1)}$, then, there is a unique function $y(t)$ such that

$$y(n) = h\left(t, y, y', ..., y^{(n-1)}\right) \tag{4.6}$$

$$y(t_0) = y_0, \quad y'(t_0) = y_0', \quad ..., \quad y^{(n-1)}(t_0) = y_0^{(n-1)} \tag{4.7}$$

According to the classical existence and uniqueness theorem for continuous linear differential equations (Moulton 1930, Reid 1971), in equation (4.1), if $m>0$ (mass can not be zero or negative), c and k are bounded, there must exist a unique solution to equation (4.1) of continuous form with initial conditions

$$x(0)=d_0 \quad \text{and} \quad \dot{x}(0)=v_0 \tag{4.8}$$

With Theorem 1, Lemma 1 and the existence and uniqueness of the continuous linear differential equations, the following theorem for existence and uniqueness of the solution $X(t)$ can be proved.

Theorem 2. *If $m>0$, c and k are bounded, then on $[i/N,(i+1)/N)$ there exists a unique solution $X_i(t)$ to*

$$\ddot{X}_i + \frac{c}{m}\dot{X}_i + \frac{k}{m}X_i = \frac{1}{m}g\left(\frac{i}{N}\right) \tag{4.9}$$

with the conditions

$$X_i\left(\frac{i}{N}=d_i\right) \quad \text{and} \quad \dot{X}_i=\left(\frac{i}{N}=v_i\right) \tag{4.10}$$

where $i=0,1,2,3,...,[Nt]$. Consequently, there must exist a unique solution $X(t)$ satisfying equation (4.4).

Proof. Rewrite equation (4.9) as

$$\ddot{X}_i = -\frac{c}{m}\dot{X}_i - \frac{k}{m}X_i + \frac{1}{m}g\left(\frac{i}{N}\right) = \eta\left(t, X_i, \dot{X}_i\right) \tag{4.11}$$

Since c/m and k/m are bounded, the function $\eta\left(t, X_i, \dot{X}_i\right)$ satisfies a Lipschitz condition for X_i and $\dot{X}_i$. Then by Lemma 1, there must exist a unique solution for equation (4.9). Now let $X(t) = X_i(t)$ on $i/N \leq t < (i+1)/N$. According to Definition 1, $X(t)$ is obviously a solution of equation (4.4). As for uniqueness, since the restriction of $X(t)$ on $i/N \leq t < (i+1)/N$ is the same as $X_i(t)$, due to the uniqueness of $X_i(t)$, $X(t)$ must be unique.

4.4. Analytical Solutions of Free Vibration Systems via Piecewise Constantization

The intention of the approach in developing the true analytical solutions with implementation of piecewise constant arguments for dynamic systems is to search for the solutions with utilization of neither the existing theoretical methods for solving differential equations nor assumptions for solutions of any form, except direct integration. The approach is therefore innovative and independent of the conventional methods for developing for the solutions of the dynamic systems.

As discussed above, the complete analytical solution for $X(t)$ in equation (4.1) can be derived by employing the piecewise constant argument $[Nt]/N$, on the basis of existence and uniqueness of the solution $X(t)$ on $[0,\infty)$ and $X_i(t)$ for the interval $i/N \le t < (i+1)/N$. To demonstrate the approach for the analytical solutions with implementation of the piecewise constant arguments, the typical governing equations in dynamics are to be considered. For the sake of clarification, a free vibration without damping, i.e., $c = 0$ and $f(t) = 0$ in equation (4.1), will first be solved by the piecewise-constant procedure.

Consider an equation of motion of an undamped vibratory system:

$$\ddot{x}(t) + \omega^2 x(t) = 0 \tag{4.12}$$

where $\omega^2 = k/m$.

The initial conditions are given by equation (4.8). A piecewise constant system corresponding to that governed by equation (4.12) can be constructed by replacing the term $\omega^2 x(t)$ in equation (4.12) with a piecewise constant function over an arbitrary time interval $i/N \le t < (i+1)/N$. The corresponding equation of motion is expressible in the form

$$\ddot{X}_i(t) + \omega^2 X_i\left(\frac{i}{N}\right) = 0 \tag{4.13}$$

for any time interval $i/N \leq t < (i+1)/N$, $i = 0,1,2,3,...,[Nt]$. Equation (4.13) is now a simplified linear differential equation with equation (4.10) as an auxiliary condition.

In equations (4.13) and (4.10), the subscript i represents the ith time interval $i/N \leq t < (i+1)/N$ from the origin $t = 0$ Equation (4.13) is now a simplified linear differential equation with equation (4.12) as an auxiliary condition. The term $\omega^2 X_i(i/N)$ in the piecewise constant equation is constant over the time segment $i/N \leq t < (i+1)/N$. To a dynamic system such constructed, with equation (4.13), a solution to the piecewise constant differential equation can be generated through direct integration. The first derivative of the solution is integrated over the time interval as

$$\dot{X}_i(t) = \int -\omega^2 X_i\left(\frac{i}{N}\right) dt \tag{4.14}$$

Letting the local initial conditions be

$$d_i = X_i\left(\frac{i}{N}\right) \quad \text{and} \quad v_i = \dot{X}_i\left(\frac{i}{N}\right) \tag{4.15}$$

which are actually the displacement and velocity respectively at $t = i/N$ corresponding to the interval $i/N \leq t < (i+1)/N$. With the local initial conditions in the interval, the integration shown in equation (4.14) gives the first derivative of the solution in the following complete form.

$$\dot{X}_i(t) = -\omega^2\left(t - \frac{i}{N}\right) d_i + v_i \tag{4.16}$$

Further integration of the first derivative yields the complete solution at this time interval as

$$X_i(t) = \left[1 - \frac{\omega^2}{2}\left(t - \frac{i}{N}\right)^2\right] d_i + \left(t - \frac{i}{N}\right) v_i \tag{4.17}$$

Again, the local initial conditions are used.

Similarly, denoting X_{i-1} as the solution on the interval $(i-1)/N \leq t < i/N$, the direct integration gives

$$X_{i-1}(t) = \left[1 - \frac{\omega^2}{2}\left(t - \frac{i-1}{N}\right)^2\right]d_{i-1} + \left(t - \frac{i-1}{N}\right)v_{i-1} \qquad (4.18)$$

As a rational hypothesis for the actual responses of a piecewise constant system in practice, there should be no jump or discontinuity of the displacement $X(t)$ and velocity $\dot{X}(t)$ on $t \in [0,\infty)$. This implies the continuity of X and $\dot{X}$ with time t. The following conditions of continuity should therefore be satisfied for the solutions of the piecewise constant system over all of the time intervals $i/N \leq t < (i+1)/N$, $i = 0,1,2,3,...,[Nt]$.

$$X_i\left(\frac{i}{N}\right) = X_{i-1}\left(\frac{i}{N}\right), \qquad \text{and} \qquad \dot{X}_i\left(\frac{i}{N}\right) = \dot{X}_{i-1}\left(\frac{i}{N}\right) \qquad (4.19)$$

With the conditions of continuity, a recursive relation is obtained by combining equations (4.17) and (4.18), such that

$$\begin{bmatrix} d_i \\ v_i \end{bmatrix} = \begin{bmatrix} 1 - \omega^2/2N^2 & 1/N \\ -\omega^2/N^2 & 1 \end{bmatrix} \begin{bmatrix} d_{i-1} \\ v_{i-1} \end{bmatrix} \qquad (4.20)$$

As a consequence of an iterative procedure, d_i and v_i can be connected to the initial displacement d_0 and initial velocity v_0 in the following form.

$$\begin{bmatrix} d_i \\ v_i \end{bmatrix} = \begin{bmatrix} 1 - \omega^2/2N^2 & 1/N \\ -\omega^2/N^2 & 1 \end{bmatrix}^i \begin{bmatrix} d_0 \\ v_0 \end{bmatrix} \qquad (4.21)$$

It is clear that the displacement and velocity of the system at any given point of time i/N can be calculated by using equation (4.21).

Considering that the ith time interval is arbitrarily chosen, the complete solution $X(t)$ can be rewritten on the basis of equations (4.17) and (4.21), such that

$$X(t) = \left[1 - \frac{\omega^2}{2}\left(t - \frac{[Nt]}{N}\right)^2 \quad t - \frac{[Nt]}{N}\right]\begin{bmatrix} 1 - \omega^2/2N^2 & 1/N \\ -\omega^2/N^2 & 1 \end{bmatrix}^{[Nt]} \begin{bmatrix} d_0 \\ v_0 \end{bmatrix}$$

$$(4.22)$$

This solution is to the piecewise constant dynamic system governed by (4.13). Now a limit can be taken as N approaches infinity. Note that the solution is valid on the time range $0 \le t \le [Nt]$, and it is continuous on any interval $i/N \le t < (i+1)/N$ over the range $[0, \infty)$. For any real value of t there is a value of $X(t)$ corresponding to it.

Let $\mathbf{Q}$ represent the square matrix in equation (4.22)

$$\mathbf{Q} = \begin{bmatrix} 1 - \dfrac{\omega^2}{2N^2} & \dfrac{1}{N} \\ -\dfrac{\omega^2}{N} & 1 \end{bmatrix} \tag{4.23}$$

or, in the form in (Lancaster 1966)

$$\mathbf{Q} = \mathbf{EDE}^{-1} \tag{4.24}$$

where $\mathbf{E}$ is a set of linearly independent eigenvectors of $\mathbf{Q}$ and $\mathbf{D}$ a diagonal matrix of the eigenvalues of $\mathbf{Q}$. Designating ϕ as the eigenvalue of the matrix $\mathbf{Q}$ and noting that $\mathbf{EE}^{-1}$ is equivalent to an identity matrix, the exponential matrix of $\mathbf{Q}$ may now be written as

$$\mathbf{Q}^{[Nt]} = \mathbf{ED}^{[Nt]}\mathbf{E}^{-1} = \mathbf{E}\begin{bmatrix} \phi_1 & 0 \\ 0 & \phi_2 \end{bmatrix}^{[Nt]} \mathbf{E}^{-1} = \mathbf{E}\begin{bmatrix} \phi_1^{[Nt]} & 0 \\ 0 & \phi_2^{[Nt]} \end{bmatrix} \mathbf{E}^{-1} \tag{4.25}$$

where the eigenvalues ϕ_1 and ϕ_2 are determined from equation (4.23) as

$$\phi_1 = 1 - \frac{\omega^2}{4N^2} + \frac{\omega^2}{4N^2}\left(1 - \frac{16N^2}{\omega^2}\right)^{(1/2)} \tag{4.26}$$

$$\phi_2 = 1 - \frac{\omega^2}{4N^2} + \frac{\omega^2}{4N^2}\left(1 - \frac{16N^2}{\omega^2}\right)^{(1/2)} \tag{4.27}$$

Corresponding to these eigenvalues, $\mathbf{E}$ and the inverse of $\mathbf{E}$ can be calculated as follows

$$\mathbf{E} = \begin{bmatrix} \dfrac{1}{4N}\left[1 - \left(1 - \dfrac{16N^2}{\omega^2}\right)^{1/2}\right] & \dfrac{1}{4N}\left[1 - \left(1 - \dfrac{16N^2}{\omega^2}\right)^{1/2}\right] \\ 1 & 1 \end{bmatrix} \tag{4.28}$$

$$\mathbf{E}^{-1} = \begin{bmatrix} -\dfrac{2N}{\left(1-\dfrac{16N^2}{\omega^2}\right)^{1/2}} & \dfrac{1}{2}+\dfrac{1}{2\left(1-\dfrac{16N^2}{\omega^2}\right)^{1/2}} \\[4ex] \dfrac{2N}{\left(1-\dfrac{16N^2}{\omega^2}\right)^{1/2}} & \dfrac{1}{2}-\dfrac{1}{2\left(1-\dfrac{16N^2}{\omega^2}\right)^{1/2}} \end{bmatrix} \qquad (4.29)$$

Taking these facts into consideration, the general solution (4.22) can be given in the following form.

$$X(t) = \frac{1}{a_{11}-a_{12}}\begin{bmatrix} b_{11} & b_{12}\end{bmatrix}\mathbf{A}\begin{bmatrix} d_0 \\ v_0\end{bmatrix} \qquad (4.30)$$

where

$$\mathbf{A} = \begin{bmatrix} a_{11}\phi_1^{[Nt]} - a_{12}\phi_2^{[Nt]} & -a_{11}a_{12}\phi_1^{[Nt]} + a_{11}a_{12}\phi_2^{[Nt]} \\[2ex] \phi_1^{[Nt]} - \phi_2^{[Nt]} & -a_{12}\phi_1^{[Nt]} + a_{11}\phi_2^{[Nt]} \end{bmatrix} \qquad (4.31)$$

and

$$b_{11} = 1 - \frac{\omega^2}{2}\left(t-\frac{[Nt]}{N}\right)^2, \qquad b_{12} = t - \frac{[Nt]}{N} \qquad (4.32)$$

$$a_{11} = \frac{1}{4N} - \left(\frac{1}{16N^2}-\frac{1}{\omega^2}\right)^{1/2}, \qquad a_{12} = \frac{1}{4N} + \left(\frac{1}{16N^2}-\frac{1}{\omega^2}\right)^{1/2} \qquad (4.33)$$

In calculating $\lim\limits_{N\to\infty} X(t)$ with the help of l'Hopital's rule, the following useful results can be obtained from equation (4.30).

$$\lim_{N\to\infty} \phi_1^{[Nt]} = e^{it\omega} \qquad (4.34)$$

$$\lim_{N\to\infty} \phi_2^{[Nt]} = e^{-it\omega} \qquad (4.35)$$

$$\lim_{N\to\infty} \frac{a_{11}\phi_1^{[Nt]} - a_{12}\phi_2^{[Nt]}}{a_{11}-a_{12}} = \frac{e^{i\omega t} + e^{-i\omega t}}{2} = \cos(\omega t) \qquad (4.36)$$

and

$$\lim_{N \to \infty} \frac{-a_{11}a_{12}\phi_1^{[Nt]} + a_{11}a_{12}\phi_2^{[Nt]}}{a_{11} - a_{12}} = \frac{1}{\omega}\sin(\omega t) \tag{4.37}$$

Substituting all of these results into equation (4.30), the final result is obtained as follows.

$$\lim_{N \to \infty} X(t) = \begin{bmatrix} \cos(\omega t) & \dfrac{1}{\omega}\sin(\omega t) \\ 0 & 0 \end{bmatrix} \begin{bmatrix} d_0 \\ v_0 \end{bmatrix} \tag{4.38}$$

which is the analytical solution in exactly the same form as the complete classical analytical solution of closed form obtained via the conventional method (Weaver *et al.* 1990) for an undamped free vibration system governed by equation (4.12), i.e.,

$$\lim_{N \to \infty} X(t) = x(t) = d_0 \cos \omega t + \frac{v_0}{\omega}\sin \omega t \tag{4.39}$$

The difference between the solution of the continuous system governed by equation (4.12) and the piecewise constant system expressed in equation (4.13) vanishes in the limiting case as the parameter N in equation (4.13) approaches infinity. Equation (4.13) may therefore be considered as a more general equation of motion covering both the piecewise constant and continuous systems. It can be seen from equation (4.13) that as N takes on a finite value, the system is piecewise constant; but when N tends to infinity, the corresponding system become continuous representing a linear vibration. It may also be concluded from the above discussion that the linear vibration problem governed by equation (4.12) is analytically solved through an approach independent of the classical analytical methods for solving the linear dynamic systems. In comparing with the conventional approach for solving these dynamic systems (Weaver *et al.* 1990), following should be emphasized.

1. With conventional approach as indicated in Sections 1.3 and 2.2, a solution (usually in the form of Ce^{rt}) is assumed before an analytical solution is to be generated. In developing for the solution (4.39) with

    piecewise-constant procedure, however, no solution form is assumed in advance.

2. With the conventional approach, a characteristic equation needs to be constructed for the solution to be obtained. However, no need for a characteristic equation when the approach with piecewise-constant procedure in developing for the solution.

3. Both the solutions obtained by the conventional and piecewise-constant procedure are analytical solutions of closed form.

4. The analytical solutions created by the approach with implementation of piecewise constant arguments rely on the availability of the limits of the solution as N tends to infinity.

4.5. Analytical Solutions to Undamped Systems with Piecewise Constant Excitations

An undamped free vibration problem is solved in the previous section with implementation of the approach with piecewise constant argument, an approach independent from the existing analytical method for solving differential equations. Based on the results obtained in the previous section, a forced vibration problem may also be solved analytically via the piecewise-constant procedure. Consider a dynamic system with the equation of motion for an undamped linear spring-mass system, subjected to a sinusoidal excitation.

$$\ddot{x} + \omega^2 x = F \cos \Omega t \tag{4.40}$$

where F refers to the amplitude of the excitation force and Ω is the angular frequency.

    Since the solution of the homogeneous part of equation (4.40), $\ddot{x} + \omega^2 x = 0$, has been obtained previously, the equation of motion is expressed in the following piecewise constant form.

$$\ddot{X}_t + \omega^2 X_i = F \cos\left(\Omega \frac{i}{N}\right) \tag{4.41}$$

which is valid on an ith interval of $i/N \le t < (i+1)/N$.

By the same procedure as discussed in the previous section, with identical initial and continuity conditions of the identical forms, the general solution of equation (4.41) can be expressed on the basis of the solution (4.38) as

$$X(t) = \Psi_1 + \Psi_2 + \Psi_3 \tag{4.42}$$

where

$$\Psi_1 = \left[\cos\left[\omega\left(t - \frac{[Nt]}{N}\right)\right] \frac{1}{\omega}\sin\left[\omega\left(t - \frac{[Nt]}{N}\right)\right]\right] \mathbf{T}[Nt] \begin{bmatrix} d_0 \\ v_0 \end{bmatrix}; \tag{4.43}$$

$$\Psi_2 = \left[\cos\left[\omega\left(t - \frac{[Nt]}{N}\right)\right] \frac{1}{\omega}\sin\left[\omega\left(t - \frac{Nt}{N}\right)\right]\right]$$

$$\times \sum_{r=1}^{[Nt]} \left\{ \frac{F}{\omega^2}\mathbf{T}^{r-1} \begin{bmatrix} \cos\dfrac{\omega}{N} - 1 \\ -\omega\sin\dfrac{\omega}{N} \end{bmatrix} \cos\left[\Omega\left(\frac{[Nt]-r}{N}\right)\right] \right\}; \tag{4.44}$$

$$\Psi_3 = \frac{F}{\omega^2}\cos\left(\Omega\frac{[Nt]}{N}\right)\left[1 - \cos\left(\omega t - \omega\frac{[Nt]}{N}\right)\right]. \tag{4.45}$$

In the above equations, $\mathbf{T}$ is a square matrix of the form

$$\mathbf{T} = \begin{bmatrix} \cos\dfrac{\omega}{N} & \dfrac{1}{\omega}\sin\dfrac{\omega}{N} \\ -\omega\sin\dfrac{\omega}{N} & \cos\dfrac{\omega}{N} \end{bmatrix} \tag{4.46}$$

with the eigenvalues

$$\phi_1 = e^{i\frac{\omega}{N}} \quad \text{and} \quad \phi_2 = e^{-i\frac{\omega}{N}} \tag{4.47}$$

The exponential matrix of $\mathbf{T}$ with power $[Nt]$, similar in form to that in equation (4.25), may then be represented by the following expression with the corresponding eigenvectors of the square matrix $\mathbf{T}$.

$$T^{[Nt]} = \frac{1}{2}\begin{bmatrix} -\dfrac{i}{\omega} & \dfrac{i}{\omega} \\ 1 & 1 \end{bmatrix}\begin{bmatrix} e^{i\omega\frac{[Nt]}{N}} & 0 \\ 0 & e^{-i\omega\frac{[Nt]}{N}} \end{bmatrix}\begin{bmatrix} -\dfrac{\omega}{i} & 1 \\ \dfrac{\omega}{i} & 1 \end{bmatrix} \tag{4.48}$$

When the parameter N approaches infinity, from equation (4.43) it can be shown that

$$\lim_{N\to\infty}\Psi_1 = \begin{bmatrix} \cos\omega t & 0 \\ 0 & \dfrac{1}{\omega}\sin\omega t \end{bmatrix}\begin{bmatrix} d_0 \\ v_0 \end{bmatrix} \tag{4.49}$$

This is a free vibration in exactly the same form as that of the classical solution (Weaver *et al.* 1990) for the homogeneous part of the continuous governing equation (4.40).

Noting that the sum after the symbol $\sum_{r=1}^{[Nt]}$ in equation (4.44) is with respect to the argument r, the summation of all the terms with argument r can be carried out to give the following result.

$$M = \sum_{r=1}^{[Nt]}\frac{1}{2}\begin{bmatrix} -\dfrac{i}{\omega} & \dfrac{i}{\omega} \\ 1 & 1 \end{bmatrix}\begin{bmatrix} e^{i\omega\frac{[Nt]}{N}} & 0 \\ 0 & e^{-i\omega\frac{[Nt]}{N}} \end{bmatrix}\begin{bmatrix} -\dfrac{\omega}{i} & 1 \\ \dfrac{\omega}{i} & 1 \end{bmatrix}\cos\left[\Omega\left(\frac{[Nt]-r}{N}\right)\right]$$

$$= \frac{1}{2}\begin{bmatrix} -\dfrac{i}{\omega} & \dfrac{i}{\omega} \\ 1 & 1 \end{bmatrix}\begin{bmatrix} L_1 & 0 \\ 0 & L_2 \end{bmatrix}\begin{bmatrix} -\dfrac{\omega}{i} & 1 \\ \dfrac{\omega}{i} & 1 \end{bmatrix} \tag{4.50}$$

where

$$L_1 = \frac{1}{2}\left[\frac{e^{\frac{i}{N}([Nt]\Omega-\Omega)} - e^{\frac{i}{N}([Nt]\omega-\Omega)}}{1-e^{\frac{i}{N}(\omega-\Omega)}} + \frac{e^{-\frac{i}{N}([Nt]\Omega-\Omega)} - e^{\frac{i}{N}([Nt]\omega+\Omega)}}{1-e^{\frac{i}{N}(\omega+\Omega)}}\right] \tag{4.51}$$

$$L_2 = \frac{1}{2}\left[\frac{e^{\frac{i}{N}([Nt]\Omega-\Omega)} - e^{\frac{i}{N}([Nt]\omega+\Omega)}}{1-e^{-\frac{i}{N}(\omega+\Omega)}} + \frac{e^{-\frac{i}{N}([Nt]\Omega-\Omega)} - e^{\frac{i}{N}([Nt]\omega-\Omega)}}{1-e^{\frac{i}{N}(\omega-\Omega)}}\right] \tag{4.52}$$

Making use of the results given by equations (4.3) and (4.50) together with l'Hopital's rule, the limit of the second term in equation (4.42) may be obtained to have the following simple form.

$$\lim_{N \to \infty} \Psi_2 = \frac{F}{\omega^2 - \Omega^2} \left(\cos \Omega t - \cos \omega t \right) \qquad (4.53)$$

The limit of the last term,

$$\lim_{N \to \infty} \Psi_3 = \lim_{N \to \infty} \frac{F}{\omega^2} \cos \left(\Omega \frac{[Nt]}{N} \right) \left[1 - \cos \left(\omega t - \omega \frac{[Nt]}{N} \right) \right] = 0 \quad (4.54)$$

Substituting the results of equations (4.49), (4.53) and (4.54) into equation (4.42) and taking a limit as $N \to \infty$, the limit of the solution to equation (4.41) is now completed in the following form.

$$\lim_{N \to \infty} X(t) = \begin{bmatrix} \cos \omega t & 0 \\ 0 & \dfrac{1}{\omega} \sin \omega t \end{bmatrix} \begin{bmatrix} d_0 \\ v_0 \end{bmatrix} + \frac{F}{\omega^2 - \Omega^2} \left(\cos \Omega t - \cos \omega t \right) \quad (4.55)$$

As expected, this solution is identical to the analytical solution of equation (4.40) obtained by using the classical analytical methods (Weaver *et al.* 1990, Marley 1945).

4.6. Development of General Analytical Solutions for Linear Vibration Systems

The approach of implementing the piecewise-constant procedure can be employed to solve for more general dynamical problems.

From equation (4.1), the equation of motion for free vibration with viscous damping is expressible as

$$\ddot{x} + 2n\dot{x} + \omega^2 x = 0 \qquad (4.56)$$

where $2n = c/m$, and the corresponding initial conditions are as shown in equation (4.8). For utilizing the piecewise-constant procedure, the second term in equation (4.56) is selected and the governing equation is expressed in the time interval $i/N \le t < (i+1)/N$ as

$$\ddot{X}_t + \omega^2 X_i = -2n\dot{X}_i\left(\frac{i}{N}\right) \tag{4.57}$$

Considering that the term $-2n\dot{X}_i(i/N)$ is a constant over $i/N \le t < (i+1)/N$, the results obtained in the precious sections may be employed to yield the homogeneous part of the solution to equation (4.57). On the basis of solution (4.38), and taking into account the conditions given by equation (4.19), the general solution of equation (4.57) is expressible in the form

$$X_i(t) = d_i \cos\left[\omega\left(t - \frac{i}{N}\right)\right] + \frac{v_i}{\omega}\sin\left[\omega\left(t - \frac{i}{N}\right)\right]$$
$$- v_i \frac{2n}{\omega^2}\left\{1 - \cos\left[\omega\left(t - \frac{j}{N}\right)\right]\right\} \tag{4.58}$$

By the same procedure as that for obtaining equation (4.22), and with the continuity conditions in equation (4.19), the complete solution $X(t)$ of equation (4.57) may be derived as

$$X_i(t) = \begin{bmatrix} \cos\left[\omega\left(t - \frac{[Nt]}{N}\right)\right] \frac{1}{\omega}\sin\left[\omega\left(t - \frac{[Nt]}{N}\right)\right] \\ -\frac{2n}{\omega^2}\left\{1 - \cos\left[\omega\left(t - \frac{[Nt]}{N}\right)\right]\right\} \end{bmatrix} \begin{bmatrix} d_i \\ v_i \end{bmatrix} \tag{4.59}$$

where

$$\begin{bmatrix} d_i \\ v_i \end{bmatrix} = \begin{bmatrix} \cos\dfrac{\omega}{N} & \dfrac{1}{\omega}\sin\dfrac{\omega}{N} - \dfrac{2n}{\omega^2}\left(1 - \cos\dfrac{\omega}{N}\right) \\ -\omega\sin\dfrac{\omega}{N} & \cos\dfrac{\omega}{N} - \dfrac{2n}{\omega}\sin\dfrac{\omega}{N} \end{bmatrix}^{[Nt]} \begin{bmatrix} d_0 \\ v_0 \end{bmatrix} \tag{4.60}$$

The procedure in taking the limits is all the same, and the methods employed in obtaining the limits are well known. By the same procedure for finding the limits as applied earlier, the limit of the closed-form result indicated in equation (4.59) as N approaches infinity can be found. The calculated limit is obtained as:

$$\lim_{N\to\infty} X(t) = x(t) = e^{-nt}\left[\cos\xi t \quad \frac{1}{\xi}\sin\xi t\right]\begin{bmatrix} d_0 \\ v_0 \end{bmatrix} + e^{-nt}\frac{nd_0}{\xi}\sin\xi t \qquad (4.61)$$

where $\xi = \sqrt{\omega^2 - n^2}$ and $n < \omega$. This solution is of the same form as that produced by the classical analytical solution (Weaver, Timoshenko and Young 1990) to equation (4.56). The other two solutions for the cases $n > \omega$ and $n = \omega$ may also be derived in a similar manner as for deriving equation (4.61). Again, the solutions obtained are identical to the corresponding solutions given by the classical analytical methods.

If a harmonic excitation force is further considered, the corresponding vibratory problem may be described by

$$\ddot{x} + 2n\dot{x} + \omega^2 x = F\cos\Omega t \qquad (4.62)$$

with initial conditions shown in equation (4.8). Since the homogeneous equation corresponding to equation (4.62) has been solved previously, the governing equation may be expressed in a piecewise constant form over $i/N \le t < (i+1)/N$ as

$$\ddot{X}_i + 2n\dot{X}_i + \omega^2 X_i = F\cos\left(\Omega\frac{i}{N}\right) \qquad (4.63)$$

and, with the same procedure as employed for obtaining the analytical solution, the complete solution $X(t)$ can be found for the case $n < \omega$ as

$$\begin{aligned}
X(t) = {}& e^{-n\left(t - \frac{[Nt]}{N}\right)}\left[\cos\left(\xi t - \xi\frac{[Nt]}{N}\right)\right. \\
&+ \frac{n}{\xi}\sin\left(\xi t - \xi\frac{[Nt]}{N}\right)\frac{1}{\xi}\sin\left(\xi t - \xi\frac{[Nt]}{N}\right)\right]\mathbf{W} \\
&+ \frac{F}{\omega^2}e^{-n\left(t - \frac{[Nt]}{N}\right)}\left[e^{-n\left(t - \frac{[Nt]}{N}\right)} - \cos\left(\xi t - \xi\frac{[Nt]}{N}\right)\right. \\
&\left.- \frac{n}{\xi}\sin\left(\xi t - \xi\frac{[Nt]}{N}\right)\right]\cos\left(\Omega\frac{[Nt]}{N}\right) \qquad (4.64)
\end{aligned}$$

where the matrix $\mathbf{W}$ is

$$\mathbf{W} = e^{-n\frac{[Nt]}{N}}\,\mathbf{G}^{[Nt]}\begin{bmatrix} d_0 \\ v_0 \end{bmatrix}$$

$$+ \sum_{r=1}^{[Nt]} e^{-\frac{nr}{N}}\,\mathbf{G}^{r-1}\begin{bmatrix} e^{\frac{n}{N}} - \cos\dfrac{\xi}{N} - \dfrac{n}{\xi}\sin\dfrac{\xi}{N} \\[2mm] \left(\dfrac{n^2}{\xi} + \xi\right)\sin\dfrac{\xi}{N} \end{bmatrix}\dfrac{F}{\omega^2}\cos\left[\Omega\left([Nt]-r\right)\right] \qquad (4.65)$$

in which the square matrix $\mathbf{G}$ has the form

$$\mathbf{G} = \begin{bmatrix} \cos\dfrac{\xi}{N} + \dfrac{n}{\xi}\sin\dfrac{\xi}{N} & \dfrac{1}{\xi}\sin\dfrac{\xi}{N} \\[3mm] -\left(\dfrac{n^2}{\xi} + \xi\right)\sin\dfrac{\xi}{N} & \cos\dfrac{\xi}{N} - \dfrac{n}{\xi}\sin\dfrac{\xi}{N} \end{bmatrix} \qquad (4.66)$$

The limit of $X(t)$ in equation (4.64), as $N \to \infty$, can be found with the use of equations (4.65) and (4.66) as

$$\lim_{N\to\infty} X(t) = e - nt\left[\cos\xi t + \dfrac{n}{\xi}\sin\xi t \quad \dfrac{1}{\xi}\sin\xi t\right]\begin{bmatrix} d_0 \\ v_0 \end{bmatrix}$$

$$+ \dfrac{F}{\left(\omega^2 - \Omega^2\right)^2 + \left(2n\Omega\right)^2}[\left(\omega^2 - \Omega^2\right)\left(\cos\Omega t - e^{-nt}\cos\xi t\right)$$

$$+ 2n\Omega\sin\Omega t - e^{-nt}\left(\omega^2 - \Omega^2\right)\dfrac{n}{\xi}\sin\xi t] \qquad (4.67)$$

As it should be, the solution above is identical to the corresponding analytical solution $x(t)$ of equation (4.56) produced by the classical analytical methods (Weaver *et al.* 1990).

For vibration problems governed by equation (4.1), solutions corresponding to free and forced vibrations with or without damping have been obtained through the piecewise-constant procedure and all the solutions have been proved to converge to the corresponding classical solutions of equation (4.1) as $N \to \infty$. Thus, it may be stated that, under the conditions of

(i) $X(t)$ satisfies equations (4.4) and (4.19) in $\left[i/N, (i+1)/N\right]$

(ii) $d_i = \lim_{t \to (i/N)} X(t)$, and $v_i = \lim_{t \to (i/N)} \dot{X}(t)$

as $N \to \infty$, $X(t)$ in equation (4.4) must converge to $x(t)$ which satisfies equation (4.1) with the corresponding initial conditions.

From the discussions above, it can be concluded that the piecewise-constant procedure can be used to solve for the piecewise constant systems in nonlinear dynamics and the complete solutions can be obtained for the systems. Significantly, the procedure can also be implemented for solving the dynamic systems governed by the general equation (4.1) for closed form analytical solutions. This implies that the gap between continuous systems and the piecewise constant systems can be filled with the implementation of the piecewise constant argument $[Nt]/N$.

In solving for dynamic problems with implementing the piecewise-constant procedure, following points need to be kept in mind.

1. In contrast with the conventional approach for analytically solving for the dynamic problems, neither a solution needs to be assumed in advance nor is a characteristic equation required for developing for the solution.

2. In applying the piecewise-constant procedure for a given dynamic system, one or more terms within the equation governing the system can be piecewise constantized with the piecewise constant argument.

3. In addition to a new method for solving dynamic problems discussed in this chapter, with the introduction of the argument $[Nt]/N$, the gap between a continuous system and its corresponding piecewise constant system has been filled.

4. The solution generated by piecewise-constant procedure is continuous and complete in a closed form. However, for obtaining the complete solution, the governing equation with piecewise constant arguments should be solvable to yield a solution of closed

form in the time interval of $[Nt]/N \leq t < ([Nt]+1)/N$. With this requirement, the original governing equation should be simplified into a form that may lead to an analytical solution in the piecewise constant interval.

5. In order to gain analytical solutions of closed form for a dynamic problem, the final solution with piecewise constant arguments should be constructed in such a way that a limit case can be taken.

6. The governing equations discussed in this chapter are fundamental and of great practical importance in connection with dynamical problems.

4.7. Semi-Analytical and Approximate Solutions for Nonlinear Piecewise Constant Dynamic Systems

As can be seen from the discussion in the previous section, piecewise constant approach is effective for solving the dynamic systems to which the limit case can be taken on the solutions generated by the piecewise-constant procedure. The analytical solutions such obtained are for the linear systems discussed in the previous sections of this chapter. For most nonlinear dynamical problems, however, the analytical solutions are difficult to develop with application of the same approach. Nevertheless, it would be convenient to generate approximate solutions via the implementation of piecewise constant arguments. For generating such approximate solutions, a natural choice would be linearization of the nonlinear dynamical systems considered through the piecewise-constant procedure. One of the advantages of implementing piecewise constant approach is that the approximate solution such generated contains as much as possible the nature of all the unchanged terms in the original governing equation, as only the term or terms which bring difficulties for obtaining analytical solutions in the interval $[Nt]/N \leq t < ([Nt]+1)/N$ should be linearized for solutions. With the application of the piecewise-constant procedure such designed, high accuracy and reliability of the solution for the nonlinear dynamic system concerned can be expected. As further mathematical manipulations can be applied on the complete and continuous solutions created by the piecewise-constant procedure, in comparing with the numerical solutions obtained per the existing numerical methods, the approximate solutions such obtained are considered as semi-analytical.

To demonstrate the application of the piecewise-constant procedure for the approximate or semi-analytical solutions, a nonlinear oscillatory system governed by the following well studied Duffing's equation with a linear and cubic stiffness can be considered.

$$\ddot{x}(t) + 2\theta\dot{x}(t) + \omega^2 x(t) + \beta x^3(t) = 0 \tag{4.68}$$

This equation of motion may be used to model, for example, a spring-mass system having a hardening spring discussed by Lancaster (1966) or a buckled beam under harmonic excitation studied by Dowell and Pezeshki (1986). By the piecewise-constant procedure, the above equation of motion is converted into a piecewise constant differential equation in the following form.

$$\ddot{X}_i(t) + 2\dot{X}_i\theta(t) + \omega^2 X_i(t) + \beta X_i^3\left(\frac{[Nt]}{N}\right) = 0 \tag{4.69}$$

over the time range in the ith interval $[Nt]/N \leq t < ([Nt]+1)/N$. As such, the governing equation refers to a linear dynamic system on each of the constant time intervals. Displacement of the piecewise constant system on the time interval $[Nt]/N \leq t < ([Nt]+1)/N$ can be created through the same procedure as discussed in previous sections, which gives

$$X_i(t) = e^{-\theta(t-[Nt]/N)}\left[\left(d_i + \frac{\lambda_i}{\omega^2}\right)\cos\left(\xi t - \xi\frac{[Nt]}{N}\right)\right.$$
$$\left. + \frac{1}{\xi}\left(v_i + \theta d_i + \frac{\theta\lambda_i}{\omega^2}\right)\sin\left(\xi t - \xi\frac{[Nt]}{N}\right) - \frac{\lambda_i}{\omega^2}\right] \tag{4.70}$$

and the velocity of the piecewise constant system

$$\dot{X}_i(t) = -\theta e^{-\theta(t-[Nt]/N)} \left[\left(d_i + \frac{\lambda_i}{\omega^2} \right) \cos\left(\xi t - \xi \frac{[Nt]}{N} \right) \right.$$

$$+ \frac{1}{\xi} \left(v_i + \theta d_i + \frac{\theta \lambda_i}{\omega^2} \right) \sin\left(\xi t - \xi \frac{[Nt]}{N} \right) \right]$$

$$+ e^{-\theta(t-[Nt]/N)} \left[-\xi \left(d_i + \frac{\lambda_i}{\omega^2} \right) \sin\left(\xi t - \xi \frac{[Nt]}{N} \right) \right.$$

$$+ \left(v_i + \theta d_i + \frac{\theta \lambda_i}{\omega^2} \right) \cos\left(\xi t - \xi \frac{[Nt]}{N} \right) \right] \qquad (4.71)$$

where

$$\lambda_i = \beta d_i^3, \quad \text{and} \quad \xi = (\omega^2 - \theta^2)^{1/2}$$

under the condition $\omega^2 > \theta^2$.

From the displacement and velocity solutions and making use of the initial conditions shown in equation (4.8) and conditions of continuity described in equation (4.19), a general solution of the problem can be obtained in the following form on the entire range for $t \geq 0$:

$$X(t) = e^{-\theta(t-[Nt]/N)} \left[\cos\left(\xi t - \xi \frac{[Nt]}{N} \right) \right.$$

$$+ \frac{\theta}{\xi} \sin\left(\xi t - \xi \frac{[Nt]}{N} \right) + \frac{1}{\xi} \sin\left(\xi t - \xi \frac{[Nt]}{N} t \right) \right] M$$

$$+ \frac{\lambda_i}{\omega^2} e^{-\theta(t-[Nt]/N)} \left[\cos\left(\xi t - \xi \frac{[Nt]}{N} \right) \right.$$

$$+ \frac{\theta}{\xi} \sin\left(\xi t - \xi \frac{[Nt]}{N} \right) - e^{-\theta(t-[Nt]/N)} \right] \qquad (4.72)$$

where the matrix $\mathbf{M}$ is

$$\mathbf{M} = e^{-\theta[Nt]/N}\mathbf{T}^{[Nt]}\begin{bmatrix} d_0 \\ v_0 \end{bmatrix}$$

$$+ \sum_{j=1}^{[Nt]} e^{\theta j/N}\mathbf{T}^j \begin{bmatrix} \cos\dfrac{\xi}{N} + \dfrac{\theta}{\xi}\sin\dfrac{\xi}{N} - e^{\theta/N} \\[2ex] -\left(\dfrac{\theta^2}{\xi} + \xi\right)\sin\dfrac{\xi}{N} \end{bmatrix}\lambda_{[Nt]-j} \tag{4.73}$$

in which the square matrix

$$\mathbf{T} = \begin{bmatrix} \cos\dfrac{\xi}{N} + \dfrac{\theta}{\xi}\sin\dfrac{\xi}{N} & \dfrac{1}{\xi}\sin\dfrac{\xi}{N} \\[2ex] -\left(\dfrac{\theta^2}{\xi} + \xi\right)\sin\dfrac{\xi}{N} & \cos\dfrac{\xi}{N} - \dfrac{\theta}{\xi}\sin\dfrac{\xi}{N} \end{bmatrix} \tag{4.74}$$

Equation (4.72) is actually a complete solution to equation (4.69) and the solution is continuous everywhere for $t > 0$. However, the solution such obtained is approximate to the analytical solution of the continuous governing equation (4.68), as the linearization on the original non-linear system is performed over the controllable time interval $[Nt]/N \le t < ([Nt]+1)/N$.

The accuracy of the complete approximate solution (4.72) depends on the size of the time interval and the interval size is controlled by the value N, as described previously. In other words, a solution with desired accuracy can be yielded with the piecewise constant approach described above, should a proper N value is selected. When N approaches infinity, theoretically, the solution such developed may lead to the analytical solution of closed form for the Duffing's equation shown in equation (4.68).

Further development of the approaches for the numerical and improved semi-analytical solutions without utilizing the conventional theoretical methods for solving differential equations will be discussed in Chapter 5. An innovative numerical method implementing piecewise constant arguments is also presented in Chapter 5 for improving the accuracy and reliability of the solutions.

References

Aftabizadeh, A. R., Wiener, J., "Oscillatory Properties of First Order Linear Functional Differential Equations," Applied Analysis, Vol. 20, pp. 165-187, 1985.

Aftabizadeh, A. R. and Wiener, J., "Oscillatory and Periodic Solutions for Systems of Two First Order Linear Differential Equations with Piecewise Constant Argument," Applicable Analysis, Vol. 26, pp. 327-338, 1988.

Bear, H. S., Differential Equations, Addison-Wesley Publishing Company, Inc., Reading, Massachusetts, 1962.

Busenberg, S., Cooke, K. L., "Models of Vertically Transmitted Diseases, in Nonlinear Phenomena in Mathematical Sciences (ed. Lakshmikantham, V.)," Academic Press, New York, pp. 179-187, 1982.

Cooke, K. L., Wiener, J., "Retarded Differential Equations With Piecewise Constant Delays," Journal of Mathematics and Analytical Applications, Vol. 99, pp. 265-297, 1984.

Cooke, K. L., Wiener, J., "An Equation Alternately of Retarded and Advanced Type," Proceedings of the American Mathematical Society, Vol. 99, pp. 726-732, 1987.

Dai, L. and Singh, M. C., "On Oscillatory Motion of a Spring-Mass System Subjected to Piecewise Constant Forces," Proceedings of Thirteenth Canadian Congress of Applied Mechanics, Vol. 1, pp. 288-289, 1991.

Dai, L. and Singh, M. C., "A New Approach in Solving Linear and Nonlinear Vibration Problems," Proceedings of 14th Canadian Congress of Applied Mechanics, Vol. 1, pp. 169-170, 1993.

Dai, L., Singh, M. C., "On Oscillatory Motion of Spring-Mass Systems Subjected to Piecewise Constant Forces," Journal of Sound and Vibration Vol. 173, pp. 217-232, 1994.

Dai, L. and Singh, M. C., "A New Approach with Piecewise-Constant Arguments to Approximate and Numerical Solutions of Oscillatory Problems," Journal of Sound and Vibration, Vol. 263, No. 3, pp. 535-548, 2003.

Dowell, E. H., Pezeshki, C., "On the Understanding of Chaos in Duffing's Equation Including a Comparison with Experiment," ASME Journal of Applied Mechanics, Vol. 53, pp. 5-9, 1986.

Euler, L., "De Integratione Aequationum Differentialium Altiorum Graduum," Miscellanea Berdinensia, Vol. 7, pp. 193-242, 1743.

Jayasree, K. N. and Deo, S. G., "Variation of Parameters Formula for the Equation of Cooke and Wiener," Proceedings of the American Mathematical Society, Vol. 112, No. 1. pp. 75-80, 1991.

Johnson, L. W. and Riess, R. D., Numerical Analysis, Addison-Wesley Publishing Company, Inc., Reading, Massachusetts, 1982.

Lancaster, P., Lambda-Matrices and Vibrating Systems, Pergamon Press, New York, 1966.

Leung, A. Y. T., "Direct Method for the Steady State Response of Structures," Journal of Sound and Vibration, Vol. 124, No. 1, pp. 135-139, 1988.

Lipschitz, R. O. S., Lehrbuch der Analtsis, Vol. 2, Bonn, 1877.

Moulton, F. R., Differential Equations, New York, 1930.

Marley, R. G., Waveform Analysis, A Guide to the Interpretation of Periodic Waves, Including Vibration Records, John Wiley & Sons, Inc., New York, 1945.

Reid, W. T., Ordinary Differential Equations, John Wiley & Sons, Inc., New York, 1971.

Salvadori, M. G. and Schwarz, R. J., Differential Equations in Engineering Problems, Prentice-Hall, Inc., New Jersey, 1954.

Shah, S. M. and Wiener, J., "Advanced Differential Equations with Piecewise Constant Argument Deviations," International Journal of Mathematics and Mathematical Sciences, Vol. 6, pp. 671-703, 1983.

Weaver, W. Jr., Timoshenko, S. and Young, D. H., Vibration Problems in Engineering, John Wiley & Sons Inc., New York, 1990.

Wiener, J. and Aftabizadeh, A. R., "Differential Equations Alternately of Retarded and Advanced Types," Journal of Mathematical Analysis and Applications, Vol. 129, pp. 243-255, 1988.

Wiener, J. and Cooke, K. L., "Oscillations in Systems of Differential Equations with Piecewise Constant Argument," Journal of Mathematical Analysis and Applications, Vol. 137, pp. 221-239, 1989.

Zhang, B. G., and Parhi, N. "Oscillatory and Nonoscillatory Properties of First Order Differential Equations with Piecewise Constant Deviating Arguments," Journal of Mathematical Analysis and Applications, Vol. 139, pp. 23-25, 1989.

CHAPTER 5

Numerical and Improved Semi-Analytical Approaches Implementing Piecewise Constant Arguments

5.1. Introduction

In this chapter, a novel semi-analytical and numerical methodology for the solutions of linear and nonlinear dynamical problems is to be developed. According to the investigations presented in the previous chapter, the difference between the solution of a continuous system and that of a piecewise constant system will vanish as the parameter N of the piecewise constant argument $[Nt]/N$ approaches infinity. If the value of N is taken to be large enough, the interval $[Nt]/N \leq t < ([Nt]+1)/N$, in which a solution for the system of piecewise constant is sought, can be as small as desired. The solution in this small interval can then be considered as an approximation to the continuous linear or nonlinear system, and the accuracy of the solution is proportional to the value of N. With this awareness, the solution derived for the piecewise constant systems discussed in Chapter 4 can be employed for a numerical purpose. Since the solutions with the piecewise constant argument $[Nt]/N$ and the corresponding recurrence relations are convenient for use in a computer program, the piecewise constant procedure is good for developing the numerical solutions of linear and nonlinear dynamic systems.

For numerically solving the continuous nonlinear dynamic problems with nonlinear differential equations, high accuracy and efficiency are always the goals to pursue. In considering that most nonlinear analyses in dynamics depend on numerical simulations and calculations on

147

computers, numerical methods for highly accurate and reliable numerical solutions are crucial. Moreover, behaviors of nonlinear dynamical systems are extremely sensitive to the accuracy of numerical calculations. Reliability of the numerical results is therefore greatly relying on the accuracy and efficiency of numerically methods to be used.

The piecewise constant procedure discussed previously is mainly for analytically and numerically solving for the piecewise constant dynamic systems. It will be demonstrated in this chapter that the numerical and approximate calculation with the piecewise constant argument $[Nt]/N$ can be made more efficient when the Taylor series expansion is jointly used. In this chapter, a new approach namely the P-T method will be established with the combination of the advantages of the piecewise constant procedure and Taylor series expansion, for approximately (or semi-analytically) and numerically solving the linear and nonlinear oscillatory problems. In approximately or numerically solving the dynamic problems by the P-T method, the terms of the original continuous differential equations remains mostly unchanged. Therefore, the accuracy of the solutions derived can be significantly improved with good convergence.

Unlike the discrete solutions produced by existing numerical methods, the solutions given by the methods to be presented are actually continuous on the entire time range from zero to t for any given value of N. Together with the other properties, the solutions are therefore considered as semi-analytical. The semi-analytical solutions to some linear and nonlinear vibration problems will be derived through the piecewise constant procedure and the P-T method. The corresponding recurrence relations for numerical calculations will also be derived. The numerical results of the semi-analytical solutions of the P-T method will be presented and compared with the exact and analytical solutions obtained through the conventional theoretical approaches. The numerical results calculated by using the P-T method will also be presented and compared with the results generated by the 4th-order Runge-Kutta method which is probably the most popular numerical method in solving the linear and nonlinear differential equations (Nakamura 1991, Zingg and Chisholm 1999, Abukhaled and Allen 1998) commonly seen in science and engineering. Formulae for numerical computation in solving various dynamic problems are to be provided and discussed in the present chapter.

Consistency of the numerical solutions to that of the original dynamical systems is important for the reliability of the numerical calculations. The consistency will therefore be analyzed in the present chapter.

Since the numerical approaches presented in this chapter are based on a single-step method, and the step length for numerical calculations can be varied freely for each time interval, a step size control technique therefore becomes possible. The application of this technique will be described in the present chapter.

Nonlinear characteristics of nonlinear dynamic systems can be analyzed conveniently by the numerical approaches with employment of the piecewise constant arguments. This will also be demonstrated in this chapter.

5.2. Numerical Solutions for Linear Dynamic Systems via Piecewise Constant Procedure

As discussed in Chapter 4, with a sufficiently large N, the set of solutions X_i of a piecewise constant system may approximately represent the resulting motion described by the governing equation of the corresponding continuous system considered, on the piecewise constant time segments over the range from zero to t. The smaller the time segments are, the more close is the approximate solution to the exact solution of the governing equation of the continuous system. When the parameter N approaches infinity, theoretically, the interval of time segment tends to zero and the solution corresponding to the dynamic system of piecewise constant arguments become the exact solution of the continuous system. The piecewise constant procedure therefore can be used for numerical solutions of linear and nonlinear dynamic systems, provide that the time segments are properly selected and the recurrence relations are properly determined. In fact, any variable or variables in a given governing equation of a continuous system can be replaced by a function or functions of piecewise constant arguments in such a way that the new governing equation modified with the piecewise constant arguments can be solved analytically for the solutions of complete form. These characteristics of the piecewise constant systems with the argument $[Nt]/N$ ensures desired accuracy for the numerical solution in the time interval of $[Nt]/N \leq t < ([Nt]+1)/N$ and consequently the entire time range considered. Moreover, due to the properties of the integer argument $[Nt]$ on the time range considered, the piecewise-constant approach is convenient to be applied for numerical calculations on computers.

To demonstrate the numerical approach with implementation of the piecewise constant arguments, a numerical solution for the undamped vibration system governed by the following differential equation may first be considered. This is the system identical to that considered in Chapter 4 for the continuous analytical solution.

$$\ddot{x}(t) + \omega^2 x(t) = 0 \tag{5.1}$$

where $\omega^2 = k/m$.

Assume that the initial conditions are the same as that given by equation (4.8). In order to generate a piecewise constant system to which a direct integration can be easily applied, as shown in Section 4.4, replacing the term $\omega^2 x(t)$ in equation (5.1) by a piecewise constant function over an arbitrary time interval $i/N \leq t < (i+1)/N$ is an obvious choice. The corresponding equation of motion can therefore be expressed in the form

$$\ddot{X}_i(t) + \omega^2 X_i\left(\frac{i}{N}\right) = 0 \tag{5.2}$$

for any time interval $i/N \leq t < (i+1)/N$, $i = 0,1,2,3,...,[Nt]$. To a dynamic system such constructed, with the piecewise-constant equation (5.2), as shown in Section 4.4, a solution can be generated through a direct integration, such that

$$X_i(t) = \left[1 - \frac{\omega^2}{2}\left(t - \frac{i}{N}\right)^2\right]d_i + \left(t - \frac{i}{N}\right)v_i \tag{5.3}$$

Its first derivative can be given by

$$\dot{X}_i(t) = -\omega^2\left(t - \frac{i}{N}\right)d_i + v_i \tag{5.4}$$

Notice that the selection of the ith time interval is arbitrary, with the continuity conditions shown in equation (4.19), the relations between the local initial conditions at the ith and $(i-1)$th intervals can be obtained as

$$d_i = \left[1 - \frac{\omega^2}{2N^2}\right]d_{i-1} + \frac{1}{N}v_{i-1} \tag{5.5}$$

$$v_i = -\left(\frac{\omega^2}{2N^2}\right)d_{i-1} + v_{i-1} \tag{5.6}$$

The local initial conditions d_i and v_i are the displacement and velocity respectively at $t = i/N$, and equations (5.3) and (5.4) such obtained are actually the recurrence relation which are practically important for numerical calculations.

Once the initial conditions are given, the displacement X_i and velocity $\dot{X}_i$ at the end of the first interval can be readily obtained from the above equations. Through a step-by-step procedure, equations (5.3), (5.4), (5.5) and (5.6) lead to a numerical solution for equation (5.1). From equations (5.3) and (5.4), however, it is evident that a solution of the piecewise-constant dynamical system and its first derivative are continuous in the interval, $i/N \le t < (i+1)/N$.

In performing the numerical calculations with the piecewise constant procedure, following points need to be taken into consideration.

1. For numerical results of a dynamic system, one may only need the recurrence relations to determine the solution at the right end of the time interval, provide that the initial conditions to the system is known.
2. Unlike the conventional numerical methods, from which the solution within the time interval $i/N < t < (i+1)/N$ is either not available or has to be determined through a linear regression process, the piecewise constant procedure may provide a continuous solution in the time segment of $i/N \le t < (i+1)/N$ and for the entire time

range of $t \leq 0$. The solution of any time segment is closely related to the original governing equation of the system, and the mathematical manipulations can be further taken to the solutions. Therefore, with the discussions provided previously, the solutions are considered as semi-analytical as shown in Chapter 4. However, for most numerical simulation in practice, the data in between the two ends of a time interval are not necessary, since the time intervals used in practice are usually very small. In this case, the numerical solutions are discrete.

3. Accuracy of the numerical solutions depends on the size of the time intervals, which is controlled by the value N. Without considering the accumulative and computer errors, larger N provides solutions of higher accuracy with the cost of longer calculation duration. For high efficiency, a step-size control technique is recommended for numerical calculations in practice.

On the basis of the results for the free vibration problem indicated above, numerical solutions to the following equation can be determined through a similar procedure.

$$\ddot{x} + \omega^2 x = F \cos \Omega t \tag{5.7}$$

where F refers to the amplitude of an excitation force and Ω is the angular frequency. Notice that the complete solution can be determined with implementation of the piecewise constant procedure as that discussed in Chapter 4, should this system is a piecewise constant system as indicated in equation (4.41). However, for numerical results of this continuous system, recurrence relations are needed, and the size of time segments needs to be determined for the accuracy desired.

For the forced vibration without damping governed by equation (5.7), the corresponding numerical solution can be obtained by the following formulae in the interval $i/N \leq t < (i+1)/N$.

$$X_i(t) = \left[d_{i-1} - \frac{F}{\omega^2} \cos\left(\Omega \frac{i}{N} \right) \right] \cos\left[\omega\left(t - \frac{i}{N} \right) \right] + \frac{v_i}{\omega} \sin\left[\omega\left(t - \frac{i}{N} \right) \right]$$
$$+ \frac{F}{\omega^2} \cos\left(\Omega \frac{i}{N} \right) \tag{5.8}$$

$$\dot{X}_i(t) = -\left[d_{i-1} - \frac{F}{\omega^2}\cos\left(\Omega\frac{i}{N}\right)\right]\omega\sin\left[\omega\left(t-\frac{i}{N}\right)\right] + v_i\cos\left[\omega\left(t-\frac{i}{N}\right)\right]$$

$$(5.9)$$

with recurrent relations

$$d_i = \left[d_{i-1} - \frac{F}{\omega^2}\cos\left(\Omega\frac{[Nt]-1}{N}\right)\right]\cos\frac{\omega}{N} + \frac{v_{i-1}}{\omega}\sin\frac{\omega}{N}$$
$$+ \frac{F}{\omega^2}\cos\left(\Omega\frac{[Nt]-1}{N}\right) \tag{5.10}$$

$$v_i = \left[\frac{F}{\omega^2}\cos\left(\Omega\frac{[Nt]-1}{N}\right) - d_{i-1}\right]\omega\sin\frac{\omega}{N} + v_{i-1}\cos\frac{\omega}{N} \tag{5.11}$$

For the free vibration with damping governed by the equation in the following form:

$$\ddot{x} + 2n\dot{x} + \omega^2 x = 0 \tag{5.12}$$

where $2n = c/m$, it can be numerically solved by the following equations and recurrence relations derived via the piecewise-constant procedure.

$$X_i(t) = d_i\cos\left[\omega\left(t-\frac{i}{N}\right)\right] + \frac{v_i}{\omega}\sin\left[\omega\left(t-\frac{i}{N}\right)\right]$$
$$- v_i\frac{2n}{\omega^2}\left\{1 - \cos\left[\omega\left(t-\frac{j}{N}\right)\right]\right\} \tag{5.13}$$

$$\dot{X}_i(t) = -\omega d_i\sin\left[\omega\left(t-\frac{i}{N}\right)\right] + v_i\cos\left[\omega\left(t-\frac{i}{N}\right)\right]$$
$$- v_i\frac{2n}{\omega}\sin\left[\omega\left(t-\frac{i}{N}\right)\right] \tag{5.14}$$

$$d_i = d_{i-1}\cos\frac{\omega}{N} + v_{i-1}\left[\frac{1}{\omega}\sin\frac{\omega}{N} - \frac{2n}{\omega^2}\left(1 - \cos\frac{\omega}{N}\right)\right] \tag{5.15}$$

$$v_i = -\omega d_{i-1} \sin\frac{\omega}{N} + v_{i-1}\left(\cos\frac{\omega}{N} - \frac{2n}{\omega^2}\sin\frac{\omega}{N}\right) \tag{5.16}$$

It has been found in the process of the numerical computations that the solutions together with the recurrence relations given above are concise and manageable on computers. In addition, the solutions and the corresponding recurrence relations are very easy to construct as shown in Chapter 4.

To evaluate the numerical method directly using piecewise-constant procedure discussed above, numerical calculations are performed with the numerical solutions shown in (5.13), (5.14), (5.15) and (5.16). The numerical results obtained by employing the piecewise-constant procedure are compared with that of Runge-Kutta method and exact analytical solution for this linear system. These numerical data are tabulated in Table 5.1. In order to stress the errors in the numerical calculations for these methods, a rather small values of N ($N = 2.5$, which is equivalent to a step length of $t = 0.4$) is used.

Table 5.1. Comparison of the analytical solution of $\ddot{x} + 2c\dot{x} + \omega^2 x = 0$ with the correspongding numerical solutions produced by the piecewise-constant method and second-order Runge-Kutta method. $c = 0.1$, $\omega = 2$, $d_0 = 2$, $v_0 = 3$. In the table, $h = 1/N$ is the integral step length.

t	Piecewise constant procedure $N = 2.5$	2nd Order Runge-Kutta $h = 0.4$	Piecewise constant procedure $N = 40$	Exact solution $h = 0.4$
0.0	2.0	2.00000	2.000000	2.000000
0.4	2.48582	2.512000	2.44645	2.443228
0.8	1.51742	1.223399	1.42975	1.428073
1.2	−0.428244	−0.942135	−0.355673	−0.341543
1.6	−2.25523	−2.432346	−1.81209	−1.775998
2.0	−2.86106	−2.179052	−2.10865	−2.064831
2.4	−1.79646	−0.363813	−1.15007	−1.127744
2.8	0.419323	1.712207	0.417360	0.394726
3.2	2.54089	2.560917	1.63437	1.570034
3.6	3.29109	1.573961	1.81182	1.739713
4.0	2.12357	−0.541210	0.916024	0.882164
4.4	−0.398314	−2.268442	−0.455765	−0.423724
4.8	−2.86024	−2.369654	−1.46787	−1.382196

As can be seen from the table, the value N has a significant influence to the accuracy of the numerical results. The numerical results are not acceptable if the N value is too small or the step length used is too large. As N becomes 40, the error of the solution provided by the piecewise constant procedure becomes much smaller. In numerically simulating a dynamic system with the method employing the piecewise constant procedure, therefore, the value N must be properly large to meet with the accuracy requirement and the time interval considered must be small enough ($\ll 1$).

As another practical example, consider the following forced vibration system with damping governed by equation

$$\ddot{x} + 2n\dot{x} + \omega^2 x = F\cos\Omega t \tag{5.17}$$

with the corresponding initial conditions as used above. The numerical solution of this system can be obtained with employment of the following formulae.

$$\begin{aligned}
X_i = e^{-n\left(t-\frac{i}{N}\right)}\Bigg\{&(d_i - \gamma_i)\cos\left[\xi\left(t-\frac{i}{N}\right)\right] \\
&+\frac{1}{\xi}[v_i + n(d_i - \gamma_i)]\sin\left[\xi\left(t-\frac{i}{N}\right)\right]\Bigg\} + \gamma_i
\end{aligned} \tag{5.18}$$

$$\begin{aligned}
\dot{X}_i = -ne^{-n\left(t-\frac{i}{N}\right)}\Bigg\{&(d_i - \gamma_i)\cos\left[\xi\left(t-\frac{i}{N}\right)\right] \\
&+\frac{1}{\xi}[v_i + n(d_i - \gamma_i)]\sin\left[\xi\left(t-\frac{i}{N}\right)\right]\Bigg\} \\
+e^{-n\left(t-\frac{i}{N}\right)}\Bigg\{&-\xi(d_i - \gamma_i)\sin\left[\xi\left(t-\frac{i}{N}\right)\right] \\
&+[v_i + n(d_i - \gamma_i)]\cos\left[\xi\left(t-\frac{i}{N}\right)\right]\Bigg\}
\end{aligned} \tag{5.19}$$

$$d_i = e^{-\frac{n}{N}}\left[(d_{i-1} - \gamma_{i-1})\cos\frac{\xi}{N} + \frac{1}{\xi}\left[(v_{i-1} + nd_{i-1} - n\gamma_{i-1})\sin\frac{\xi}{N}\right]\right] + \gamma_{i-1}$$

$$\tag{5.20}$$

$$v_i = e^{-\frac{n}{N}}\left(v_{i-1}\cos\frac{\xi}{N} - \left[(v_{i-1}+nd_{i-1}-n\gamma_{i-1})\frac{n}{\xi} + \xi(d_{i-1}-\gamma_{i-1})\right]\sin\frac{\xi}{N}\right)$$

$$(5.21)$$

where

$$\gamma_{i-1} = \frac{F}{\omega^2}\cos\left(\frac{[Nt]-1}{N}\right)$$

$$(5.22)$$

Starting with the initial state, the above relations are used in a computer program to obtain solution in each interval. The numerical results obtained by the application of the above equations are illustrated in Figure 5.1. As shown in the figure, the error of the numerical results from the exact solution is already small as N is given a value of 10. The numerical results are getting closer and closer to the exact solution of the system as the value of N increases. The parameter N obviously acts as a factor controlling the accuracy of the numerical solution, since it is directly related to the time interval $i/N \leq t < (i+1)/N$. In order to have a numerical solution of high accuracy, one may simply choose a sufficiently large N.

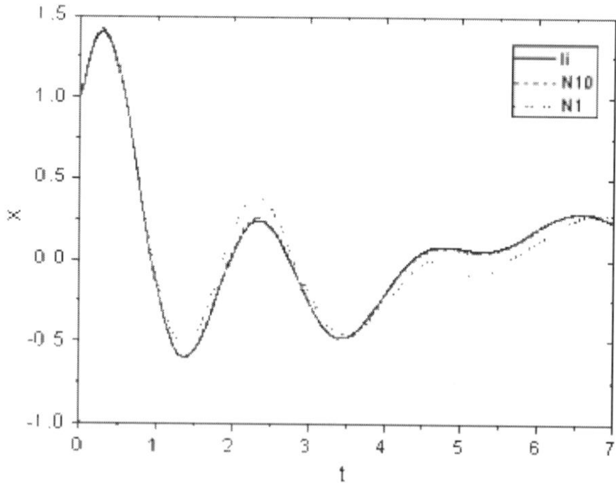

Figure 5.1.   Convergence of the solution of $\ddot{x}(t)+2n\dot{x}(t)+\omega^2 x([Nt]/N) = F\cos\Omega t$ to that of $\ddot{x}(t)+2n\dot{x}(t)+\omega^2 x(t) = F\cos\Omega t$ (solid line). $n=0.5$, $\omega=3$, $F=2$, and $\Omega=1$. In the figure, 'N10':N=10, 'N1':N=1.

5.3. Numerical Solutions of Nonlinear Systems

Nonlinear continuous systems can also be analyzed and numerically solved by the piecewise constant procedure in a similar manner as discussed above. In fact, the numerical simulations with implementation of piecewise constant procedure is especially suitable for solving nonlinear dynamic problems due to the continuity of the solutions.

Consider a nonlinear dynamic system subjected to a nonlinear piecewise constant force. If the piecewise constant force is cubic in the form of $x^3(t)$. For numerically solving for the dynamic problem utilizing the approach with piecewise constant procedure, this nonlinear term can be replaced by $x^3([Nt]/N)$, as this replacement will lead to the construction of following governing equation.

$$\ddot{x}(t) + 2\theta\dot{x}(t) + \omega^2 x(t) + \beta x^3\left(\frac{[Nt]}{N}\right) = 0 \tag{5.23}$$

Since there is only one piecewise constant term involved in the governing equation, this system is identical to the system shown in equation (4.68). The complete approximate solution for this nonlinear system has been developed in the previous chapter and shown in equation (4.72), and can be employed for determining the numerical solutions inside the time interval. The recurrence relations in this case can be developed on the basis of the equations (4.70) and (4.71) and are expressible as follows:

$$d_i = e^{-\theta/N}\left[\left(d_{i-1} + \frac{\lambda_{i-1}}{\omega^2}\right)\cos\left(\frac{\xi}{N}\right)\right.$$
$$\left. + \frac{1}{\xi}\left(v_{i-1} + \theta d_{i-1} + \frac{\theta\lambda_{i-1}}{\omega^2}\right)\sin\left(\frac{\xi}{N}\right)\right] - \frac{\lambda_{i-1}}{\omega^2} \tag{5.24}$$

$$v_i = e^{-\theta/N}\left\{v_{i-1}\cos\frac{\xi}{N} - \sin\frac{\xi}{N}\left[\frac{\theta}{\xi}\left(v_{i-1} + \theta d_{i-1} + \frac{\theta\lambda_{i-1}}{\omega^2}\right)\right]\right.$$
$$\left. + \xi\left(d_{i-1} + \frac{\lambda_{i-1}}{\omega^2}\right)\right\} \tag{5.25}$$

Although equation (4.72) is the complete solution to equation (5.23), equations (4.70) and (4.71) with the recurrence relations (5.24) and (5.25) are practical and manageable in using in a computer program to procure a numerical solution.

Notice that equation (5.23) is of a similar form as the well known Duffing's equation with a linear and cubic stiffness:

$$\ddot{x}(t) + 2\theta\dot{x}(t) + \omega^2 x(t) + \beta x^3(t) = 0 \qquad (5.26)$$

This equation of motion can be used to model, for example, a spring-mass system having a hardening spring (Weaver, Timoshenko and Young 1990) or a buckled beam under harmonic excitation (Dowell and Pezeshki 1986).

According to what has been discussed in section 3 of Chapter 4, the piecewise constant argument $[Nt]/N$ tends to be the continuous time t when N approaches infinity. Hence, as $N \to \infty$, equation (5.23) tends to be a continuous Duffing's equation, and theoretically, the solution in the form of equation (4.72) will be the accurate solution of the Duffing's equation (5.26), to which chaos may occur. The approach implementing the piecewise constant argument may therefore be employed to analyze and numerically solve the continuous nonlinear oscillation problems, in addition to solving the linear and nonlinear oscillatory systems subjected to piecewise constant forces. In fact, if the value of the integer N is properly chosen and large enough, the numerical results obtained from equations (5.24) and (5.25) will give the solution of the Duffing's equation (5.26) with sufficient accuracy.

In terms of numerical solutions, it is significant to note that the solutions given by the present technique are continuous in the time interval $[Nt]/N \leq t < ([Nt]+1)/N$ and the entire range of time $t \in [0,\infty)$. The time t in the solutions may be given any real value, and for any value of t there is a definite x value corresponding to it. In contrast to the present technique, the most existing numerical methods such as the average acceleration method, linear acceleration method (Klotter 1953, Newmark 1959, Morino, Leech and Witmer 1974), Euler's method and Runge-Kutta method (Phillips and Cornelius 1986, Gerald and Wheatley 1989, Runge 1895, Kutta 1901), provide only the solutions at the discrete points, t_n, $n = 1, 2, 3, ...$, such that the information in between t_n and t_{n+1} is not available, as indicated in Section 2.4. From the discussion above, it can also be seen that all the terms in the governing equations, except only one term which is piecewise-constant, remain unchanged during the derivation and calculation of the numerical solutions. For instance, the approximate solution of equation (5.26) is obtained on the basis of the governing equation (5.23) with a piecewise constant argument $\beta x^3 ([Nt]/N)$. Among the four terms in equation (5.23), the first three terms are identical with the first three terms in equation (5.26). The information carried by these three terms is retained in the piecewise constant solution (4.72) or an approximate solution to the system governed by (5.26). When the chosen value of N is sufficiently large, $\beta x^3 ([Nt]/N)$ can be as close as desired to $\beta x^3 (t)$, and equation (4.72) will be a good approximate solution with desired accuracy to equation (5.26). Similarly, because of the original physical information carried by the unchanged terms in the equations of motion, the continuous solutions shown in equations (5.13), (5.18), (4.70), and (4.72) can be considered　as good approximations to the exact analytical solutions of the corresponding continuous systems.

Employing the methods of average-acceleration and linear-acceleration, Weaver et $al.$ (1990) reported the numerical solutions for an equation of motion (5.26) without damping. Under the same conditions, the present technique is employed to solve the same equation of motion for a comparison. The numerical solution produced by using equations (4.70) and (4.71) with the recurrence relations (5.24) and (5.25) shows good convergence and matches well with the numerical results obtained by Weaver et $al.$ (1990). Figure 5.2 exhibits the convergence of the numerical solutions obtained by the present technique to the accurate solution as N increases.

In solving a nonlinear engineering problem in dynamics, according to the discussion and procedures showing above, governing equation in dynamic problems can be linearized on an arbitrary time interval $[Nt]/N \leq t < ([Nt]+1)/N$ by replacing the nonlinear terms in the form of $f(x(t), \dot{x}(t), t)$ (or the terms to which mathematical manipulations are difficult to perform) with the piecewise-constant terms in the form of $g(x([Nt]/N), \dot{x}([Nt]/N), [Nt]/N)$.

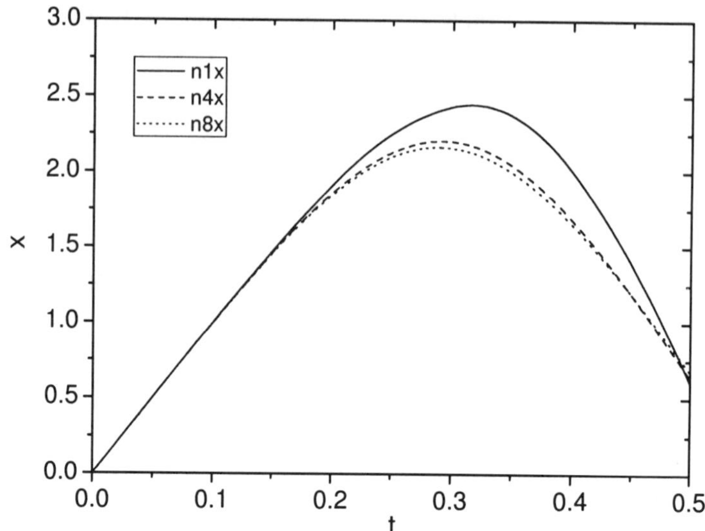

Figure 5.2. Convergence of the solution of $m\ddot{x} + k\{x + ax([Nt]/N)^3\} = 0$. $m = 100$, $k = 400$, $a = 2$, $d_0 = 0$, $v_0 = 10$. 'n4x':N = 40; 'n8x':N = 80.

In linearizing the governing equations, two or more terms or variables can be set in a piecewise-constant form so long as the considered governing equation can be solved in the chosen interval by the existing theories for solving linear or nonlinear equations in dynamics. As an example, consider the following equation of motion which represents the oscillations of a particle attached to a spring under the influence of quadratic and cubic damping.

$$\ddot{x}(t) + \omega^2 x(t) + \mu_1 \dot{x}(t) + \mu_2 \dot{x}(t)\,|\,\dot{x}(t)\,| + \mu_3 \dot{x}^3(t) = 0 \qquad (5.27)$$

where μ_1, μ_2 and μ_3 are constants. This equation may be expressed in piecewise-constant form in interval $[Nt]/N \leq t < ([Nt]+1)/N$ as

$$\ddot{x}_i(t) + \omega^2 x_i(t) + \mu_1 \dot{x}_i(t) + \mu_2 \dot{x}_i\left(\frac{[Nt]}{N}\right)\left|\dot{x}_i\left(\frac{[Nt]}{N}\right)\right| + \mu_3 \dot{x}_i^3\left(\frac{[Nt]}{N}\right) = 0$$

$$(5.28)$$

The approximate solution corresponding to the system (5.27) together with the recurrence relations can be produced through a procedure similar to that used for the system described in equation (5.23). The displacement, velocity and the recurrence relations for numerical calculations are derived as

$$
x_i = -e^{-\frac{\mu_1\left(t-\frac{[Nt]}{N}\right)}{2}}\left\{\left(d_i - \frac{F_i}{\omega^2}\right)\cos\left[\xi\left(t-\frac{[Nt]}{N}\right)\right]\right.
$$
$$
\left. + \frac{1}{\xi}\left[v_i + \frac{\mu_i}{2}\left(d_i - \frac{F_i}{\omega^2}\right)\right]\sin\left[\xi\left(t-\frac{[Nt]}{N}\right)\right]\right\} + \frac{F_i}{\omega^2} \qquad (5.29)
$$

$$
\dot{x}_i = -\frac{\mu_1}{2}e^{-\frac{\mu_1\left(t-\frac{[Nt]}{N}\right)}{2}}\left\{\left(d_i - \frac{F_i}{\omega^2}\right)\cos\left[\xi\left(t-\frac{[Nt]}{N}\right)\right]\right.
$$
$$
\left. + \frac{1}{\xi}\left[v_i + \frac{\mu_1}{2}\left(d_i - \frac{F_i}{\omega^2}\right)\right]\sin\left[\xi\left(t-\frac{[Nt]}{N}\right)\right]\right\}
$$
$$
+ e^{-\frac{\mu_1\left(t-\frac{[Nt]}{N}\right)}{2}}\left\{-\left(d_i - \frac{F_i}{\omega^2}\right)\xi\sin\left[\xi\left(t-\frac{[Nt]}{N}\right)\right]\right.
$$
$$
\left. + \left[v_i + \frac{\mu_1}{2}\left(d_i - \frac{F_i}{\omega^2}\right)\right]\cos\left[\xi\left(t-\frac{[Nt]}{N}\right)\right]\right\} \qquad (5.30)
$$

$$
d_i = -e^{-\frac{\mu_1}{2N}}\left\{\left(d_{i-1} - \frac{F_{i-1}}{\omega^2}\right)\cos\frac{\xi}{N}\right.
$$
$$
\left. + \frac{1}{\xi}\left[v_{i-1} + \frac{\mu_1}{2}\left(d_{i-1} - \frac{F_{i-1}}{\omega^2}\right)\right]\sin\frac{\xi}{N}\right\} + \frac{F_{i-1}}{\omega^2} \qquad (5.31)
$$

$$
v_i = -e^{-\frac{\mu_1}{2N}}\left\{v_{i-1}\cos\frac{\xi}{N} - \left[\frac{\mu_1}{2}v_{i-1} + \left(d_{i-1} - \frac{F_{i-1}}{\omega^2}\right)\left(\frac{\mu_1^2}{4\xi} + \xi\right)\right]\sin\frac{\xi}{N}\right\} \qquad (5.32)
$$

where

$$
\omega^2 > \mu_1^2/4, \quad F_i = -\mu_2 v_i |v_i| - \mu_3 v_i^3, \quad \xi^2 = \omega^2 - \frac{\mu_1^2}{4} \qquad (5.33)
$$

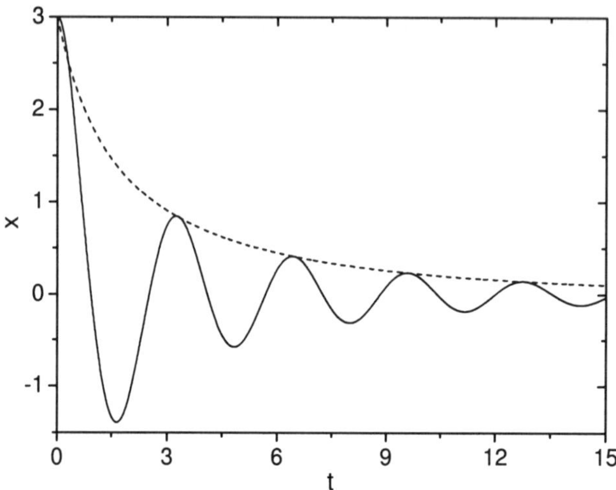

Figure 5.3.   Numerical solution of $\ddot{x} + A\dot{x} + Bx + D\dot{x}\,|\,\dot{x}\,| = 0$ (solid line) and the first approximation for amplitude (dashed line). $A = 0.25$, $B = 4$, $D = 0.2$, $d_0 = 3, v_0 = 0$.

The numerical solution for this system is illustrated in Figure 5.3 with a comparison between the numerical solution given by the piecewise constant procedure and the first approximation for the amplitude of motion given by Baum (1972).

5.4.   Chaotic Behavior of Numerical Solutions for Nonlinear Systems

In analyzing the nonlinear systems with chaotic behavior, the numerical solutions are known to be very sensitive to initial conditions and time integral steps used for numerical computations (Rasband 1990, Baker and Gollub 1990). In solving the chaotic problems, the method implementing the piecewise constant arguments can still be employed for the purpose of analysis and numerical computation. For revealing the nonlinear behavior of nonlinear systems and demonstrating the applications of the piecewise constant procedure in numerical simulations, some typical nonlinear dynamic systems are solved numerically with the implementation of the piecewise constant arguments.

(a) *Forced motions of a nonlinear pendulum*

Consider the following governing equation which may lead to chaotic motion:

$$\ddot{x} + b_1 \dot{x} + b_2 \sin x = b_3 \cos b_4 t \tag{5.34}$$

where b_1, b_2, b_3 and b_4 are parameters of the system. The system governed by the above equation represents a nonlinear, damped, sinusoidally driven pendulum. Express this equation in a piecewise constant form in interval $[Nt]/N \le t < ([Nt] + 1)/N$ as

$$\ddot{X} + b_1 \dot{X} + b_2 \sin X \left(\frac{[Nt]}{N} \right) = b_3 \cos b_4 t \tag{5.35}$$

The displacement and velocity for the system governed by equation (5.35) can be derived through a procedure similar to that used for obtaining equations (5.29) and (5.30) as

$$X_i = A_i + B_i e^{-b_i(t-[Nt]/N)} - \frac{b_3}{b_4^2 + b^2} \cos b_4 t$$
$$+ \frac{b_1 b_3}{b_4^3 + b_1^2 b_4} \sin b_4 t - \frac{b_2}{b_1} \left(t - \frac{[Nt]}{N} \right) \sin d_i \tag{5.36}$$

$$\dot{X}_i = -b_1 B_i e^{-b_i(t-[Nt]/N)} - \frac{b_3 b_4}{b_4^2 + b_1^2} \sin b_4 t$$
$$+ \frac{b_1 b_3}{b_4^3 + b_1^2} \cos b_4 t - \frac{b_2}{b_1} \sin d_i \tag{5.37}$$

where

$$B_i = \frac{1}{b_1} \left(\frac{b_3 b_4}{b_4^2 + b_1^2} \sin b_4 \frac{i}{N} + \frac{b_1 b_3}{b_4^3 + b_1^2} \cos b_4 \frac{i}{N} - \frac{b_2}{b_1} \sin d_i - v_i \right) \tag{5.38}$$

$$A_i = d_i - B_i + \frac{b_3}{b_4^2 + b_1^2} \cos b_4 \frac{i}{N} - \frac{b_1 b_3}{b_4^3 + b_1^2 b_4} \sin b_4 \frac{i}{N}. \tag{5.39}$$

The corresponding recurrence relations are

$$d_i = A_{i-1} + B_{i-1}e^{-b_i/N} - \frac{b_3}{b_4^2 + b_1^2}\cos b_4 \frac{[Nt]}{N}$$
$$+ \frac{b_1 b_3}{b_4^3 + b_1^2 b_4}\sin b_4 \frac{[Nt]}{N} - \frac{b_2}{b_1 N}\sin d_{i-1} \tag{5.40}$$

$$v_i = -b_1 B_{i-1}e^{-b_i/N} + \frac{b_3 b_4}{b_4^2 + b_1^2}\sin b_4 \frac{[Nt]}{N}$$
$$+ \frac{b_1 b_3}{b_4^3 + b_1^2}\cos b_4 \frac{[Nt]}{N} - \frac{b_2}{b_1}\sin d_{i-1} \tag{5.41}$$

Making use of the data generated by employing solutions (5.36) and (5.37) together with the recurrence relations (5.40) and (5.41), the numerical results are graphically presented in Figures 5.4 and 5.5.

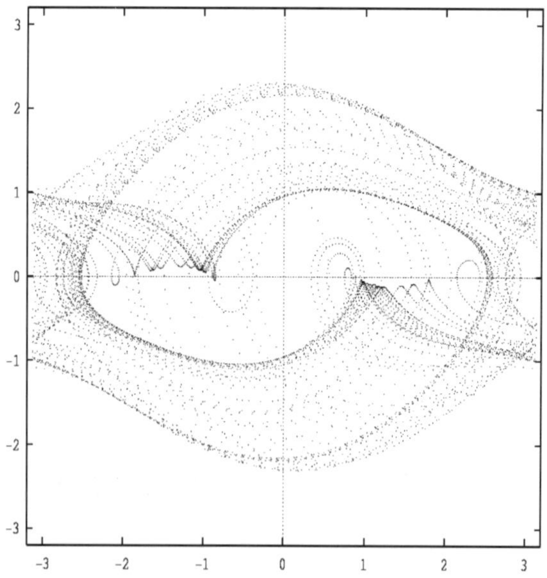

Figure 5.4.   Phase trajectory of a steady state solution for $\ddot{x} + b_1\dot{x} + b_2 \sin x = b_3 \cos b_4 t$. $b_1 = 0.5$, $b_2 = 1.0$, $b_3 = 1.15$, $b_4 = 2/3$, $d_0 = -2.5$, $v_0 = 0$.

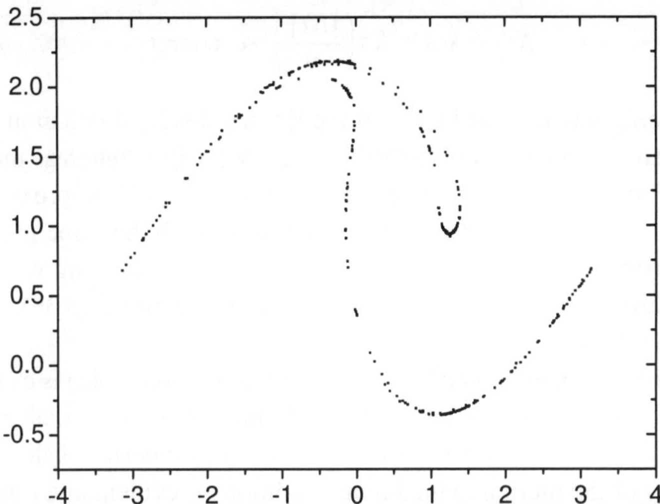

Figure 5.5.   Poincare map corresponding to the phase trajectory in Figure 5.4.

Figure 5.4 illustrates the phase trajectory and Figure 5.5 shows the corresponding Poincare map of a chaotic case for the motion governed by equation (5.34), which represents a nonlinear, damped, sinusoidally driven pendulum.. Under the same conditions, Figures 5.4 and 5.5 are almost identical to the phase trajectory and the Poincare map provided by Gwinn (1986), Blackbum *et al.* (1989) and Baker *et al.* (1990). In performing the numerical simulations for the nonlinear or chaotic systems, piecewise constant approach shows great reliability and efficiency with high accuracy.

(b) *Duffing's equation with periodic excitation*

A Duffing's equation in the following form was investigated by Ueda (1980)

$$\ddot{x} + c_1\dot{x} + c_2 x^3 = c_3 \cos c_4 t \tag{5.42}$$

where c_1, c_2, c_3 and c_4 are constants. To solve this nonlinear equation with employment of piecewise constant argument, the continuous governing equation is converted to the following piecewise constant differential equation.

$$\ddot{X} + c_1 \dot{X} + c_2 X^3 \left(\frac{[Nt]}{N} \right) = c_3 \cos c_4 t \tag{5.43}$$

Noticing that the third term on the left-hand-side of equation (5.43) is considered as a constant in $[Nt]/N \leq t < ([Nt] + 1)/N$, whereas the other terms are identical to the corresponding terms in (5.42), approximate or numerical solution of (5.42) can be derived through the same procedure as demonstrated previously. The solution and corresponding recurrence relations are in similar forms as those shown in equations (5.36), (5.37), (5.40) and (5.41).

A chaotic case governed by the equation of motion (5.42) is examined by using the solution obtained through the piecewise constant procedure and chaotic response was found in the numerical calculations. Sensitivity of the motion to the initial conditions is evident and illustrated in Figure 5.6.

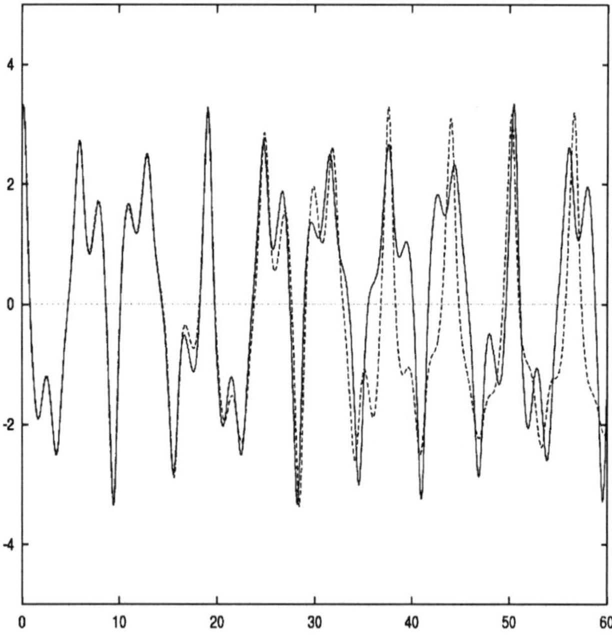

Figure 5.6. Demonstration of sensitivity of a nonlinear system governed by $\ddot{x} + c\dot{x} + x^3 = F \cos t$ to initial conditions.: $c = 0.05$, $F = 7.5$. Initial conditions: $d_0 = 3.0$, $v_0 = 4.0$ (solid line); $d_0 = 3.01$, $v_0 = 4.01$ (dashed line).

Under certain conditions the motion corresponding to equation (5.42) may be periodic or nonperiodic. Figure 5.7 exhibits the trajectory of a periodic case and Figure 5.8 gives the Poincare map of a quasiperiodic motion of the system.

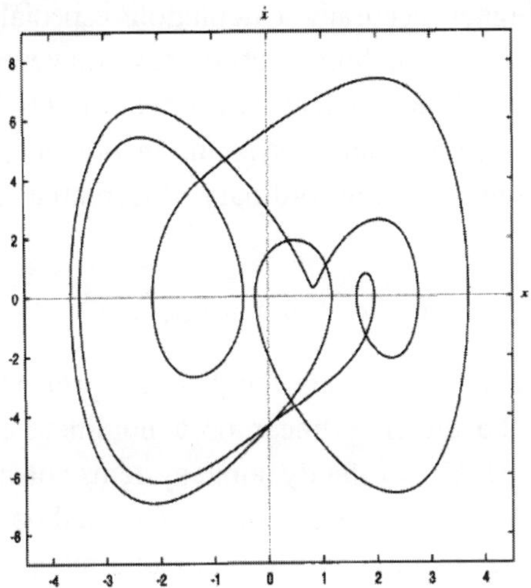

Figure 5.7.   Phase trajectory corresponding to the motion governed by $\ddot{x} + c_1\dot{x} + c_2x^3 = c_3\cos c_4 t$. $c_1 = 0.026$, $c_2 = 1.0$, $c_3 = 11.4$, $c_4 = 1.0$.

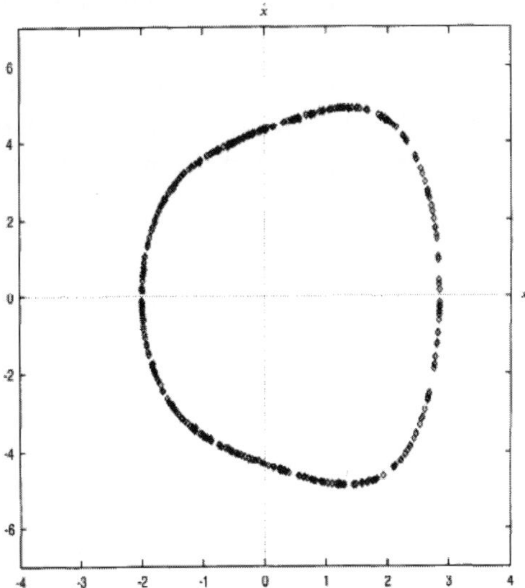

Figure 5.8.   Poincare map of nonperiodic motion governed by $\ddot{x} + c_1\dot{x} + c_2x^3 = c_3\cos c_4 t$. $c_1 = 0.00001$, $c_2 = 1.0$, $c_3 = 2.0$, $c_4 = 1.0$.

5.5. Development of P-T Method

The semi-analytical and numerical solutions discussed in the previous sections are the direct implementation of the piecewise constant arguments. For higher accuracy calculation especially for nonlinear numerical simulations with higher efficiency, however, more efficient numerical methods with higher accuracy will definitely be beneficial.

For frequently used dynamic systems in practice, it is common to express such systems by the ordinary differential equations in the following form,

$$\ddot{x} + 2c\dot{x} + \omega^2 x = f(t, x, \dot{x}) \tag{5.44}$$

where the coefficients c and ω are the physical properties of the system and $f(t, x, \dot{x})$ can be either a linear or a nonlinear function of time, displacement and velocity of the dynamic systems considered. In solving the above equation by the existing numerical methods such as Euler's method, Taylor-series method and Runge-Kutta method, conventionally, the second-order differential equation shown in equation (5.44) is transformed into a system of first-order differential equations, as discussed in Chapter 2. The differential equation governing the dynamic systems is usually simplified or linearized during the transformation for the consequent numerical calculations as described in Section 2.4. On the basis of the first-order differential equations, a recurrence relation for numerical calculation is then developed in the following form.

$$x_i = F(x_{i+1}, \dot{x}_{i+1}) \tag{5.45}$$

Obviously the numerical solution given by the above equation can only be discrete, and the solutions at the ith point, x_i, rely upon the initial conditions and the calculated solutions corresponding to the discrete points $i = 1, 2, 3,\ldots, i\text{-}1$, prior to the ith point considered. In addition, due to the mathematical operations such as the simplification and linearization, or Taylor expansion with first few terms required by the numerical methods used, the physical meaning embedded in the original equation of motion is usually lost in the mathematical manipulations. Therefore, the numerical results would reflect more closely to the reality, should the nature of the original governing equation can be maintained to the possible utmost level. With this consideration, implementation of piecewise constant procedure is a natural choice, as the piecewise constant arguments can be used to simplify or linearize the desired term or terms of the governing equation, instead of simplify the entire governing equation of the system. Remaining the advantage of the piecewise constant procedure in developing the new numerical method is certainly valuable and constructive.

It has demonstrated in the previous chapter that the following condition is satisfied as N approaches infinity.

$$\lim_{N \to \infty} \frac{[Nt]}{N} = t \tag{5.46}$$

Consequently, the approximate solutions produced by a piecewise-constant procedure employing the argument $[Nt]/N$ have been shown to become the corresponding accurate solutions when N tended to infinity. In numerically solving a dynamical problem by the piecewise-constant technique, the accuracy of the numerical results and the time step used for the numerical calculations can be conveniently adjusted by a single numerical value N. The piecewise-constant procedure produces reasonably accurate results and has shown effectiveness.

The semi-analytical and numerical approaches with piecewise constant arguments would be practically more valuable if their accuracy and efficiency can be further increased. It has been noticed that a variable of a governing equation replaced by a piecewise constant function is considered as a constant in the time interval $[Nt]/N \leq t < ([Nt]+1)/N$. The accuracy of the numerical calculation implementing the semi-analytical and numerical approaches would be significantly increased if the original variable can be linearized or replaced by a piecewise constant function of higher order. Taylor series expansion is essentially an expression which can generally be used to approximate any function to any desired degree of accuracy. In considering this, a more accurate approximate or numerical solution may be anticipated if a linear or a nonlinear governing equation can be replaced by a second order ordinary differential equation with piecewise constant arguments together with a power series of finite terms. Taylor series expansion is therefore a natural selection to join with the piecewise constant approach, for combining the advantages of the approach with implementation of piecewise constant argument and Taylor expansion. We here therefore develop a new numerical and semi-analytical method combining the piecewise constant approach and Taylor expansion. Due to the implementation of the piecewise constant approach and Taylor expansion, this new method is called the P-T method hereafter. The detailed development of the method is to be presented in the following.

To describe the development of the P-T method systematically, let us consider the general dynamic systems shown in equation (5.44) and start with the methodology of approximation and numerical approach employing Taylor series expansion. Making use of the piecewise constant argument $[Nt]/N$ and Taylor series expansion of order n, equation (5.44) can be approximately expressed as the follows

$$\ddot{x}_i = g_i + g_i' t + \frac{1}{2!} g_i'' t^2 + \cdots \frac{1}{n!} g_i^{(n)} t^n \tag{5.47}$$

where $g = f(t, x, \dot{x}) - 2c\dot{x} - \omega^2 x$, and the subscript $i = 1,2,3, \ldots$ represents an arbitrary time interval of $[Nt]/N \leq t < ([Nt]+1)/N$, and further

$$g_i = g\left(\frac{[Nt]}{N},\ x_i\left(\frac{[Nt]}{N}\right),\ \dot{x}_i\left(\frac{[Nt]}{N}\right)\right) \tag{5.48}$$

$$g_i' = \left[\frac{d}{dt} g(t, x_i, \dot{x}_i)\right]_{t=\frac{[Nt]}{N}} \tag{5.49}$$

$$g_i'' = \left[\frac{d^2}{dt^2} g(t, x_i, \dot{x}_i)\right]_{t=\frac{[Nt]}{N}} \tag{5.50}$$

$$\cdots\cdots\cdots\cdots\cdots$$

$$g_i^{(n)} = \left[\frac{d^n}{dt^n} g(t, x_i, \dot{x}_i)\right]_{t=\frac{[Nt]}{N}} \tag{5.51}$$

Notice that the right-hand side of the equal sign of equation (5.47) has been transferred to an implicit function of time t. The solution to equation (5.47) is hence available and can be expressed in the following form.

$$x_i(t) = d_i + v_i t + \frac{1}{2!} g_i t^2 + \frac{1}{3!} g_i' t^3 + \cdots + \frac{1}{(n+2)!} g_i^{(n)} t^{n+2} \qquad (5.52)$$

This solution may be used for numerical calculation purpose if it is expressed in the following form.

$$x_{i+1}(t) = d_i + v_i \frac{1}{N} + \frac{1}{2!} g_i \frac{1}{N^2} + \frac{1}{3} g_i' \frac{1}{N^3} + \cdots + \frac{1}{(n+2)!} g_i^{(n)} \frac{1}{N^{n+2}} \qquad (5.53)$$

where the displacement and velocity of the system at $t = [Nt]/N$ are given as

$$d_i = x_i\left(\frac{[Nt]}{N}\right) \quad \text{and} \quad v_i = \dot{x}_i\left(\frac{[Nt]}{N}\right). \qquad (5.54)$$

The numerical solution calculated by equation (5.53) is similar in principle to that given by the numerical method of direct application of Taylor-series of order n upon which the popular Runge-Kutta method is developed (Dahlquist and Bjorck 1974, Isaacson and Keller 1966, Nakamura 1991, Morries 1983, Dai and Singh 1994, 1997, 1998). The accuracy of the solution depends merely on the order n or the number of terms $n+1$ in equation (5.47). Nevertheless, making use of the piecewise-constant approach, the accuracy of the numerical calculation is controlled by the parameter N.

A better numerical solution may be expected if the governing equation (5.44) can be expressed in a form such that a portion of it can be made analytically solvable, as such the majority of the terms in the governing equation can be kept unchanged. With this consideration, regard the following governing equation of a general dynamic system.

$$\ddot{x} + 2c\dot{x} + \omega^2 x = \varphi(t) + \psi(t, x, \dot{x}) \qquad (5.55)$$

In the above equation, the left-hand-side of the equation together with the first term $\varphi(t)$ on the right-hand-side is a nonhomogeneous ordinary second-order differential equation for which an analytical solution is readily available. $\varphi(t)$ is a monotonic and explicit function of t. In order to numerically or approximately solve this equation for the entire time range with utilizing the piecewise constant arguments and the Taylor series expansion, the governing equation needs to be modified into a linear one for an analytical solution in the small time interval considered. For this purpose, $\psi(t, x, \dot{x})$ is expressed as a function of t by Taylor series expansion, such that

$$\ddot{x}_i + 2c\dot{x}_i + \omega^2 x_i = \varphi_i(t) + \psi_i + \dot{\psi}_i t + \frac{1}{2!}\ddot{\psi}_i t^2 + \cdots + \frac{1}{n!}\psi_i^{(n)} t^n \quad (5.56)$$

which is valid on an arbitrarily small ith time interval, $[Nt]/N \le t < ([Nt]+1)/N$. In comparison with equation (5.47), the above equation contains much more of the original physical information embedded in the governing equation (5.44), it is therefore likely to have a solution close to the accurate solution of equation (5.44), should all the solutions of equation (5.56) can be combined with the conditions of continuity as described previously.

For a numerical method, it is important to ensure that the convergence of the numerical solution with the original dynamical system is satisfied. To analyze the convergence of the numerical solutions derived through the present technique and evaluate the truncation error caused by the P-T method, the governing equation of the dynamic problem in a general form showing in equation (5.55) can be considered. With the P-T method, the solution of this system within the interval $[Nt]/N \le t < ([Nt]+1)/N$, can be obtained analytically by the following governing equation with piecewise constant arguments:

$$\ddot{x} + 2c\dot{x} + \omega^2 x = \varphi(t) + \psi\left(\frac{[Nt]}{N}, x_i\left(\frac{[Nt]}{N}\right), \dot{x}_i\left(\frac{[Nt]}{N}\right)\right) \quad (5.57)$$

Thus, within this interval, the difference between the continuous system with solution $x(t)$ to equation (5.55), and the system governed by this equation of piecewise-constant system is expressible as

$$\ddot{x} + 2c\dot{x} + \omega^2 x - \varphi(t) - \psi\left(\frac{[Nt]}{N}, x_i\left(\frac{[Nt]}{N}\right), \dot{x}_i\left(\frac{[Nt]}{N}\right)\right) = R_N \qquad (5.58)$$

Usually, the truncation error R_N is not zero. Employing Taylor expansion, the derivatives shown in the above equation can be given as

$$\ddot{x}(t) = \ddot{x}\left(\frac{[Nt]}{N}\right) + \dddot{x}\left(\frac{[Nt]}{N}\right)\left(t - \frac{[Nt]}{N}\right) + 0\left(\left(t - \frac{[Nt]}{N}\right)^2\right) \qquad (5.59)$$

$$\dot{x}(t) = \dot{x}\left(\frac{[Nt]}{N}\right) + \ddot{x}\left(\frac{[Nt]}{N}\right)\left(t - \frac{[Nt]}{N}\right) + 0\left(\left(t - \frac{[Nt]}{N}\right)^2\right) \qquad (5.60)$$

and

$$x(t) = x\left(\frac{[Nt]}{N}\right) + \dot{x}\left(\frac{[Nt]}{N}\right)\left(t - \frac{[Nt]}{N}\right) + 0\left(\left(t - \frac{[Nt]}{N}\right)^2\right) \qquad (5.61)$$

From equation (5.55), at a specified pint of $[Nt]/N$, the following expression must be satisfied.

$$\ddot{x}\left(\frac{[Nt]}{N}\right) + 2c\dot{x}\left(\frac{[Nt]}{N}\right) + \omega^2 x\left(\frac{[Nt]}{N}\right) - \phi\left(\frac{[Nt]}{N}\right)$$
$$- \Psi\left(\frac{[Nt]}{N}, x\left(\frac{[Nt]}{N}\right), \dot{x}\left(\frac{[Nt]}{N}\right)\right) = 0 \qquad (5.62)$$

Hence,

$$R_N = \left[\dddot{x}\left(\frac{[Nt]}{N}\right) + 2c\ddot{x}\left(\frac{[Nt]}{N}\right) + \omega^2\dot{x}\left(\frac{[Nt]}{N}\right)\right]\left(t - \frac{[Nt]}{N}\right) + 0\left(\left(\frac{1}{N}\right)^2\right) \qquad (5.63)$$

whereas $\left|t - [Nt]/N\right| \le 1/N$.

Utilizing the conclusion indicated equation (5.46) and the proof described in the previous chapter, the truncation in the above equation is zero as $N \to \infty$ and the time integral step tends to zero, such that the difference between the exact solution and the numerical solution is vanished correspondingly.

In the proof above, c and ω are bounded, as $N \to \infty$, $\ddot{x}([Nt]/N) \to \ddot{x}(t)$, $\dot{x}([Nt]/N) \to \dot{x}(t)$, $x([Nt]/N) \to x(t)$, $\varphi([Nt]/N) \to \varphi(t)$, and $\psi([Nt]/N, x_i([Nt]/N), \dot{x}_i([Nt]/N)) \to \psi(t, x_i, \dot{x}_i)$, thus, the piecewise-constant system becomes the original continues system.

To expose the above point more clearly, consider the following equation of motion representing a damped linear oscillatory system.

$$\ddot{x}(t) + 2c\dot{x}(t) + a^2 x(t) = bx(t) \tag{5.64}$$

for which an analytical solution is available. Based on the discussion above, this equation is replaced by the following equation with Taylor expansion and the piecewise constant argument for the term $bx(t)$.

$$\ddot{x}_i(t) + 2c\dot{x}_i(t) + a^2 x_i(t) = bd_i + bv_i\left(t - \frac{[Nt]}{N}\right) \tag{5.65}$$

which is valid on the ith interval $[Nt]/N \le t < ([Nt]+1)/N$. In this equation, only the first two terms of the Taylor series expansion are considered and the rest of the higher order terms are neglected.

Solution of the linear equation (5.65) can be derived and expressed as

$$x_i = e^{-c\left(t - \frac{[Nt]}{N}\right)}\left\{B_1 \cos\left[\xi\left(t - \frac{[Nt]}{N}\right)\right] + B_2 \sin\left[\xi\left(t - \frac{[Nt]}{N}\right)\right]\right\}$$
$$+ A_1 + A_2\left(t - \frac{[Nt]}{N}\right) \tag{5.66}$$

in which

$$\xi^2 = a^2 - c^2 \tag{5.67}$$

$$A_2 = \frac{1}{a^2}, \qquad A_1 = \frac{1}{a^2}(bd_i - 2cA_2) \tag{5.68}$$

$$B_1 = d_1 - A_1, \qquad B_2 = \frac{1}{\xi}(v_i + cB_1 - A_2) \tag{5.69}$$

It may be noted in equation (5.66) that the solution $x(t)$ and its first derivative $\dot{x}(t)$ are continuous in the time interval considered. Because of the continuity of x and $\dot{x}$ on $t \in [0, \infty)$, the following conditions must be satisfied.

$$x_i\left(\frac{[Nt]}{N}\right) = x_{i-1}\left(\frac{[Nt]}{N}\right) \quad \text{and} \quad \dot{x}_i\left(\frac{[Nt]}{N}\right) = \dot{x}_{i-1}\left(\frac{[Nt]}{N}\right) \quad (5.70)$$

The above conditions of continuity lead to the recurrence relations for numerical calculations:

$$d_{i+1} = e^{-\frac{c}{N}}\left(B_1 \cos\frac{\xi}{N} + B_2 \sin\frac{\xi}{N}\right) + A_1 + \frac{A_2}{N} \quad (5.71)$$

$$v_{i+1} = -ce^{-\frac{c}{N}}\left(B_1 \cos\frac{\xi}{N} + B_2 \sin\frac{\xi}{N}\right)$$

$$+ e^{-\frac{c}{N}}\left(-\xi B_1 \sin\frac{\xi}{N} + \xi B_2 \cos\frac{\xi}{N}\right) + A_2 \quad (5.72)$$

With equation (5.66) and the recurrence relations (5.71) and (5.72), a numerical solution for the governing equation can be obtained through a step-by-step procedure. Should the solution within the two neighboring points of an interval is desired for any cases; solution (5.66) can be utilized.

As can be seen from the above discussion, the approximate or numerical solution of the governing equation is so developed that a linear dynamic system is established between any neighboring points of $[Nt]/N$ and $([Nt]+1)/N$, with implementation of the piecewise constant argument $[Nt]/N$ together with Taylor series expansion. With this arrangement, unlike the discrete solutions produced by the existing numerical methods, the approximate and numerical solutions produced by the P-T method are continuous everywhere on the entire time range from zero to t for any given value of N.

Though the methodology developed above is called the P-T method and it utilizes the piecewise constant arguments, we should note that the governing equation, displacement and its first and second derivatives are piecewise continuous over the entire time range considered for a given nonlinear dynamic system. This implies that there exists a linear dynamic system for each of the time intervals that are smoothly connected.

In numerically solving the dynamic problems by the P-T method, the major portion of the corresponding original differential equations remains unchanged, as can be seen from the processes above. This ensure that the solutions derived by the P-T method to be more accurate than that of the existing numerical methods such as Runge-Kutta method. Most significantly, the P-T method reveals the actual physical behavior of the oscillatory systems to the maximum possible level in comparison with Runge-Kutta method and the other numerical methods. In the following sections, the P-T method will be quantitatively compared with the analytical and Runge-Kutta methods.

5.6.   Analytical and Numerical Approaches and the Approaches Implementing P-T Method

After the development of the P-T method, it may be a good time to summarize and analyze the techniques and the characteristics of the techniques used for determining the analytical solutions obtained though the implementation of piecewise constant arguments.

1. Piecewise constant arguments can be used as a steppingstone for obtaining analytical solutions of dynamic systems. With this approach, the solutions can be expressed by piecewise constant functions with one of the terms in the governing equation to be constantized as the "steppingstone". Letting the coefficient of the steppingstone approach zero (through taking the limit for certain cases), the analytical solution of closed form can be obtained. Some cases following in this category can be found in Sections 3.4 and 3.5. This approach is useful to reveal the relationship between piecewise constant systems and the corresponding continuous systems. This approach can also be used for analyzing some weakly nonlinear systems. It is significant to notice that certain expressions with

piecewise constant functions are interchangeable with some continuous expressions, as they can also be expressed by the continuous expressions of different form. An example for this case can be found from Section 3.5.

2. Some analytical solutions can be generated by employing the piecewise constant argument $[Nt]/N$ and taking limits when N approaches infinity, as that indicated in Sections 4.4, 4.5 and 4.6. One or more terms in the governing equations are first replaced by the piecewise constant arguments in terms of $[Nt]/N$. The solutions such developed yield the exact analytical solutions then N tends to infinity, as the argument $[Nt]/N$ approaches t and the solution itself tends to analytical solution of closed form. The approach following in this category is considered as a new approach for analytically determining for the solutions, independent of the existing theoretical approaches available in the literature of this field. This approach is especially useful for determining the solutions for linear dynamic systems; though it can be used in wider areas, theoretically.

3. Since the solutions generated with the implementation of piecewise constant procedures are closely related to the original governing equation of the dynamic system considered, and the size of the time segment $[Nt]/N \leq t < ([Nt]+1)/N$ is adjustable with N as the control parameter, the accuracy of the solutions generated can be very high. Also, the solutions are complete and continuous over the time range considered. Therefore, the solutions such developed are good approximations to the original system, and they can be considered as semi-analytical solutions. The methods following in this category are the direct integration of piecewise constant arguments and the P-T method.

The main motivation of developing the P-T method is to increase the accuracy of the numerical or approximate results while maintain the reliability and efficiency in numerical calculations, in comparing with the other existing numerical or approximate methods. A quantitative comparison between the exact accurate solution and that of the P-T method is needed to evaluate the accuracy of the solutions of the P-T method. The system governed by equation (5.64) can be used to serve this purpose. In order to perform the comparison, the solution given by equation (5.66) corresponding to the system is expanded by Taylor series into the power series form as shown below.

$$
x_i = d_i + v_i \left(t - \frac{[Nt]}{N} \right) - \left[(a^2 - b)\frac{d_i}{2} + cv_i \right]\left(t - \frac{[Nt]}{N} \right)^2
$$

$$
+ \left[\frac{cd_i}{3}(a^2 - b) - \frac{v_i}{6}(a^2 - b - 4c^2) \right]\left(t - \frac{[Nt]}{N} \right)^3
$$

$$
+ \left[\frac{d_i}{24}(a^2 - b)(a^2 - 4c^2) + \frac{cv_i}{6}\left(a^2 - \frac{b}{2} - 2c^2 \right) \right]\left(t - \frac{[Nt]}{N} \right)^4 + \cdots
$$

$$(5.73)$$

The exact solution for equation (5.64) has a closed form (Weaver *et al.* 1990) and can be shown in the following equation

$$
x(t) = e^{-nt}\left[\cos t\sqrt{\omega^2 - d - n^2} \quad \frac{1}{\xi}\sin t\sqrt{\omega^2 - d - n^2} \right]\left[\begin{array}{c} d_0 \\ v_0 \end{array} \right]
$$

$$
+ e^{-nt}\frac{nd_0}{\xi}\sin t\sqrt{\omega^2 - d - n^2} \qquad (5.74)
$$

This analytical solution can also be expressed into a power series form on the ith time interval, with the employment of the piecewise constant argument $[Nt]/N$ and Taylor's series expansion. The solution in power series is given as the following, for the purpose of comparison.

It can be seen from the governing equation (5.47) and its solution (5.52) that the first two terms in the right hand side of equation (5.47) are necessary for obtaining the third order solution in equation (5.53), which is equivalent to that developed by the method of direct Taylor series expansion (Nakamura 1991). All the other terms in equation (5.53) that is higher than the third orders should be neglected, if only the two terms of the Taylor expansion are kept.

$$x_i = d_i + v_i\left(t - \frac{[Nt]}{N}\right) - \left[(a^2 - b)\frac{d_i}{2} + cv_i\right]\left(t - \frac{[Nt]}{N}\right)^2$$

$$+ \left[\frac{cd_i}{3}(a^2 - b) - \frac{v_i}{6}(a^2 - b - 4c^2)\right]\left(t - \frac{[Nt]}{N}\right)^3$$

$$+ \left[\frac{d_i}{24}(a^2 - b)(a^2 - b - 4c^2) + \frac{cv_i}{6}(a^2 - b - 2c^2)\right]\left(t - \frac{[Nt]}{N}\right)^4$$

$$+ \frac{1}{120}\{-4c(a^2 - b)(a^2 - b - 2c^2)d_i$$

$$+ [(a^2 - b)^2 - 4c^2(3a^2 - 3b - 4c^2)]v_i\}\left(t - \frac{[Nt]}{N}\right)^5 + \cdots \qquad (5.75)$$

The solution given by equation (5.66) is a close-form solution to equation (5.65) that also retains two terms of Taylor series expansion. However, in contrast with the solution shown in equation (5.53), all the higher order terms are still available in the solution given by equation (5.73) which is based on equation (5.66). As can be observed, the first four terms in equation (5.73) are identical to those of the exact solution presented in equation (5.75). Hence, the accuracy of the solution (5.73) equivalent to that of the Taylor-series method of order three is ensured. Significantly, the other higher order terms appearing in equation (5.73) are still very close to the corresponding terms in equation (5.75) of the exact solution. Therefore, the solution generated by equation (5.73) is a very good approximation to the exact solution. It can also be found from the two solutions, the solution of P-T method is better than that of the direct Taylor series expansion, under the condition of retaining the same number of the terms of the expansion as that of the P-T method for calculations. In comparing with the exact solution in equation (5.75), the difference between the solution provided by direct Taylor series expansion in equation (5.53) and the solution provided by equation (5.73) is significantly large, should the same order of accuracy is considered. It should be noted that the difference between the two solutions of P-T method and direct Taylor series expansion is independent of the parameter N.

It is clear, if the higher order terms of the Taylor expansion for the term bx are further considered in equation (5.65), a more accurate numerical solution can be obtained. Through the same procedure as discussed above, the equation of motion containing one more term of higher order can be considered. The numerical solution to this governing equation can be derived through the similar procedure.

$$\ddot{x}(t) + 2c\dot{x}(t) + a^2 x(t) = bd_i + bv_i\left(t - \frac{[Nt]}{N}\right)$$

$$+ \frac{b}{2!}(bd_i - 2cv_i - a^2 d_i)\left(t - \frac{[Nt]}{N}\right)^2 \quad (5.76)$$

This equation has a theoretical solution in the closed form similar to equation (5.66). Also, the solution for equation (5.76) is continuous therefore can be expressed in the following Taylor series expansion form for the convenience of comparison with the exact solution as shown in the following form. In comparing with the exact solution shown in (5.75) corresponding to the governing original equation, this solution has an accuracy equivalent to that of the Taylor-series method of order four in terms of displacement x, or of order three in terms of velocity $\dot{x}$.

$$x_i = d_i + v_i\left(t - \frac{[Nt]}{N}\right) - \left[(a^2 - b)\frac{d_i}{2} + cv_i\right]\left(t - \frac{[Nt]}{N}\right)^2$$

$$+ \left[\frac{cd_i}{3}(a^2 - b) - \frac{v_i}{6}(a^2 - b - 4c^2)\right]\left(t - \frac{[Nt]}{N}\right)^3$$

$$+ \left[\frac{d_i}{24}(a^2 - b)(a^2 - b - 4c^2) + \frac{cv_i}{6}(a^2 - b - 2c^2)\right]\left(t - \frac{[Nt]}{N}\right)^4$$

$$+ \frac{1}{120}\left\{-4c(a^2 - b)\left(a^2 - \frac{b}{2} - 2c^2\right)d_i\right.$$

$$+ [a^2(a^2 - b) - 4c^2(3a^2 - 2b - 4c^2)]v_i\bigg\}\left(t - \frac{[Nt]}{N}\right)^5 + \cdots \quad (5.77)$$

Again, as can be seen from equations (5.75) and (5.77), the coefficients in the fifth order term in equation (5.77) are very close to the corresponding terms in the exact solution in equation (5.75). Equations in the form of (5.77) are therefore the good continuous approximate solutions to the governing equation (5.64). It should be noticed that the complete semi-analytical solution for the governing equation (5.64) can be obtained though combining all the solutions in equation (5.77) with the conditions of continuity.

For numerical calculation, the recurence relations corrsponding to equation (5.76) can be determined by its solution together with the conditions of continuity demonstrated in equation (5.70). The procudures of determining for the recurrence relations are the same as those of the numerical approaches with implementing the piecewise constant arguments.

Also, it can be shown that the solution with an accuracy of order four in velocity can be developed in a similar manner as described above if the third order term

$$\frac{b}{3!}[2c(a^2-b)d_i-(a^2-b-4c^2)v_i]\left(t-\frac{[Nt]}{N}\right)^3 \qquad (5.78)$$

is added to the right hand side of the equal sign in equation (5.76), should this level of accuracy is desired.

As can be seen from the above discussion, the main advantage of the P-T method in solving oscillatory problems is to maintain as much as possible the original physical information of the governing equation into the solutions. Therefore, the solutions such obtained represent much closely the actual response of the physical dynamic system concerned, in comparing with the solutions generated by the other approximate or numerical methods which usually simplify and/or linearized the entire governing equation for the solutions. The piecewise constant procedure applied to as less as possible the terms in the governing equation allows a greater portion of the original information to be kept intact in the numerical calculation. In return, a higher accuracy solution is developed. The piecewise constant argument $[Nt]/N$, which has obvious advantage in bridging the continuous and piecewise constant systems, is also a contribution to simplifying the procedure of deriving the numerical solutions discussed above. One may note that the P-T method discussed above is indeed a technique combining the idea of piecewise-constant approach and Taylor-series method. It is also significant to state that the solutions generated by the P-T method for the nonlinear dynamic systems are actually the combinations of many solutions of piecewise constant linear systems on the small time intervals. Again, the solutions of P-T method are continuous and semi-analytical, and can be used as good approximates of the exact solutions of the original dynamic systems. Theoretically, the solutions of P-T method will be the exact analytical solution when N approaches to infinity. In other words, with the methodology developed in this chapter, the solution of a nonlinear system can be considered as the combination of infinite number of solutions corresponding to infinite number of linear governing equations at the intervals of $i/N \leq t < (i+1)/N$, $i = 1,2,3, \ldots$

5.7.   Numerical Solution Comparison between P-T and Runge-Kutta Methods

The P-T method provides approximate solutions of high accuracy for nonlinear dynamic systems. Making use of the approximate solutions provided by the P-T method, numerical solutions can be easily generated by the recurrent relations developed via the manipulations of the P-T method and the conditions of continuity. To quantitatively evaluate the accuracy and effectiveness of the P-T method in numerical simulation, it is significant to compare the numerical solutions of P-T method and that of the existing numerical methods, such as Runge-Kutta method. In the analyses of nonlinear dynamics and engineering practices, Runge-Kutta method of fourth order is probably the most popular numerical method in solving the dynamic problems, and it is a method better than the other numerical methods such as Taylor's expansion and Euler's method in solving for the dynamic problems. In order to compare the results produced by the P-T method with those generated by the existing numerical methods in analyzing for the problems in dynamics, both the P-T method and Runge-Kutta method with the identical order of accuracy are used for solving an oscillatory system governed by the following equation.

$$\ddot{x} + 0.2\dot{x} + 10x = x \qquad (5.79)$$

Equation (5.79) is linear and its exact analytical solution is available for comparison. One may probably note that Runge-Kutta method provides numerical solutions at discrete points, therefore, a continuous solution in Taylor series expansion form as those shown in equations (5.73) and (5.75) is not available.

The numerical solutions for equation (5.79) are determined by both the P-T method of fourth order accuracy and Runge-Kutta method of fourth order respectively. The numerical results are calculated with the solutions and plotted in Figure 5.9.

The vertical axis of the figure denotes the difference between a numerical solution and the exact solution of equation (5.79), and the horizontal axis represents the duration of the calculations, and pt4 and rk4 in the figure are the solutions of the P-T method of fourth order and Runge-Kutta method of fourth order respectively. All the other parameter values and conditions are identical to both the numerical calculations of the P-T method and Runge-Kutta method. It can be visualized from Figure 5.9, errors of both the methods are small. However, the fourth order P-T method supplies a more accurate numerical solution in comparison with the Runge-Kutta method of order four, as the errors of the P-T method calculations are much smaller than that of Runge-Kutta method. This is even true when the duration of the calculations is getting larger.

The numerical calculation for the system governed by equation (5.79) is also provided for another case with utilizing the fourth order P-T and Runge-Kutta methods. The results of the calculation are provided in Table 5.2, to compare with the numerical results of the exact analytical solution. (The results of the fourth order P-T method and that of the exact solution can hardly be distinguished if they are plotted in a diagram).

For the sake of identifying the differences between the results generated for the linear dynamic system considered with the identical parameter values and conditions, a small N value corresponding to a large step length is used for the calculations.

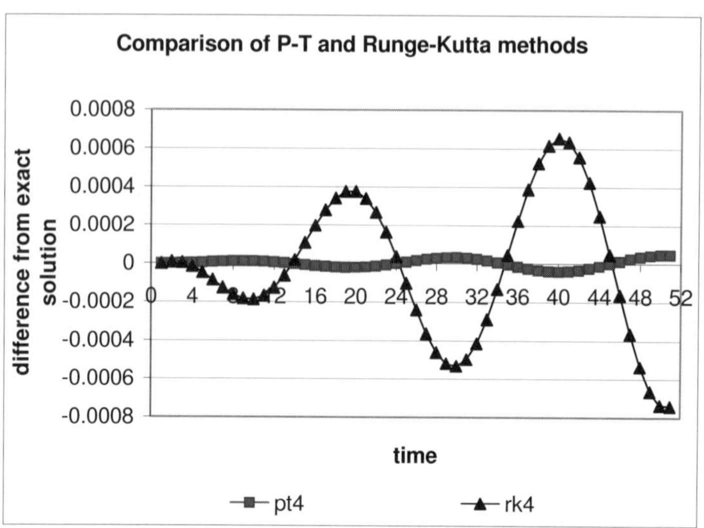

Figure 5.9. Comparison of numerical results generated by the P-T method of fourth order and Runge-Kutta method of fourth order for the governing equation $\ddot{x} + 0.2\dot{x} + 10x = x$ with $x(0) = 1$ and $\dot{x}(0) = 2$.

Table 5.2. Numerical solutions of $\ddot{x} + 0.2\dot{x} + 4x = 0$ given by the 4th order Runge-Kutta method and P-T method in comparison with the exact solution. $d_0 = 2$, $v_0 = 3$.

t	4th Order Runge-Kutta $h = 0.4$	4th Order P-T method $N = 2.5$	Exact solution $h = 0.4$
0.0	2.00000	2.00000	2.000000
0.400000	2.439040	2.44193	2.443228
0.800000	1.430458	1.42329	1.428073
1.20000	−0.321837	−0.338812	−0.341543
1.60000	−1.742200	−1.75454	−1.775998
2.00000	−2.036486	−2.02803	−2.064831
2.40000	−1.128086	−1.09641	−1.127744
2.80000	0.357421	0.395500	0.394726
3.20000	1.512597	1.53083	1.570034
3.60000	1.696587	1.67801	1.739713
4.00000	0.883713	0.834753	0.882164
4.40000	−0.372830	−0.423763	−0.423724
4.80000	−1.309170	−1.32895	−1.382196

As can be found from the table, the results given by the P-T method are closer to that of the exact solution as expected, in comparing with that of the fourth order Runge-Kutta method. Should the N value become larger, the results of much higher accuracy can be anticipated.

As demonstrated in the previous sections, the piecewise-constant procedure and the P-T method can be conveniently used to obtain numerical solutions of the dynamical systems for the linear and nonlinear continuous responses of the systems. In numerically solving the linear, nonlinear and chaotic dynamic problems, the techniques presented in the previous sections produce reasonably accurate results with good convergence in comparing with the other methods. However, the accuracy of the numerical solution calculated with employment of the piecewise-constant technique as $N = 1$ is not adequately high, especially in solving for nonlinear dynamic problems. When numerically solving the nonlinear problem of damped and forced pendulum, for example, the parameter N must be large enough ($N > 600$ for instance) or the time interval $[Nt]/N \leq t < ([Nt]/N + 1)/N$ must be sufficiently small to obtain a numerical solution with acceptable accuracy. It is also evident that the number of steps for solution over the entire time range is directly related to the value N, the greater the value of N, the larger is the number of steps needed for a reliable numerical solution, as indicated in the previous sections.

The speed of numerical calculation of the P-T method is slightly faster than that of Runge-Kutta method for this case. To compare the speed of the two methods, the CPU time of the actual calculations for solving equation (5.79) are tabulated in Table 5.3. It should be noted however, the CPU times spend for the two methods are almost the same for most of the numerical problems that have been performed. The

Table 5.3. Comparison of the CPU times for solving $\ddot{x} + 0.2\dot{x} + 10x = x$. For both the methods of P-T and Runge-Kutta of fourth order, the initial conditions are $x(0) = 1$ and $\dot{x}(0) = 2$; time range from 0.0 to 20,000; and step length 0.3.

Method	CPU time (seconds)
P-T	8.31
Runge-Kutta	8.33

contribution of the P-T method is mainly on accuracy of the numerical calculations, though the CPU times can also be compared for a specified identical accuracy.

Evidently, the P-T method can also be applied to nonlinear dynamical systems in providing numerical or approximate solutions. Due to the higher accuracy of the P-T method in numerical calculations, it is significant to use it for solving the nonlinear dynamic problems, which are very sensitive to the accuracy of the numerical methods used, the parameter values and the initial conditions.

A typical nonlinear or linear dynamic system can be expressed by the following differential equation.

$$\ddot{x}(t) + 2c\dot{x}(t) + \omega^2 x(t) = f(x, \dot{x}, t) \tag{5.80}$$

By the P-T method, expand function $f(x, \dot{x}, t)$ with Taylor series to the desired order of accuracy on an ith time interval $[Nt]/N \le t < ([Nt]+1)/N$, such that

$$\ddot{x}_i(t) + 2c\dot{x}_i(t) + \omega^2 x_i(t) = f_{[Nt]/N} + f'_{[Nt]/N}\left(t - \frac{[Nt]}{N}\right)$$

$$+ \frac{1}{2!} f''_{[Nt]/N}\left(t - \frac{[Nt]}{N}\right)^2 + \frac{1}{3!} f'''_{[Nt]/N}\left(t - \frac{[Nt]}{N}\right)^3 + \cdots \tag{5.81}$$

This is a general process with piecewise constant arguments per the P-T method. In performing the process of piecewise constantization with the P-T method, the following points need to be kept in mind.

1. For the function $f(x, \dot{x}, t)$, there is no need to apply the piecewise constant procedure to the variables which are explicit functions of t. Remaining the explicit functions of t may help to increase the accuracy and reliability of the solution to the system governed by equation (5.80).

2. Any individual term which is a linear function of x or $\dot{x}$ in $f(x, \dot{x}, t)$ should be combined with the term $2c\dot{x}$ or $\omega^2 x$ correspondingly in equation (5.80), before the piecewise constant procedure is applied. This is important as it may significantly increase the accuracy and reliability of the numerical or approximate solution of the system.

3. Nonlinear terms in $f(x,\dot{x},t)$ are actually converted to the explicit functions of t with the performance of piecewise constant procedure and Taylor series expansion. Such that a linear dynamic system is established in each of the time intervals.

4. The order of accuracy of the solutions of P-T method depends on the number of retained terms of Taylor series expansion.

A complete solution for equation (5.81) on the time interval $[Nt]/N \leq t < ([Nt]+1)/N$ can then be easily obtained with the desired order of accuracy, as the equation is converted to a linear ordinary second order differential equation to which an analytical solution is available. The recurrence relation for numerical calculation can be consequently developed through the procedure for numerical calculations, as demonstrated previously for the linear system.

In order to have an approximate solution with the same accuracy as that of Runge-Kutta method of fourth order, for the sake of comparison, employ the first four terms in equation (5.81) for the function $f(x,\dot{x},t)$ and truncate the higher order terms. The approximate solution in $[Nt]/N \leq t < ([Nt]/N +1)/N$ can then be developed for equation (5.81) as

$$x_i = e^{-c\left(t-\frac{[Nt]}{N}\right)}\left\{B_1\cos\left[\xi\left(t-\frac{[Nt]}{N}\right)\right]+B_2\sin\left[\xi\left(t-\frac{[Nt]}{N}\right)\right]\right\}+A_1$$

$$+ A_2\left(t-\frac{[Nt]}{N}\right)+A_3\left(t-\frac{[Nt]}{N}\right)^2+A_4\left(t-\frac{[Nt]}{N}\right)^3 \tag{5.82}$$

where

$$\xi^2 = \omega^2 + c^2 \tag{5.83}$$

$$A_4 = \frac{1}{6\omega^2}f'''_{[Nt]/N} \tag{5.84}$$

$$A_3 = \frac{1}{\omega^2}\left(\frac{1}{2}f''_{[Nt]/N}-6cA_4\right) \tag{5.85}$$

$$A_2 = \frac{1}{\omega^2}(f'_{[Nt]/N}-4cA_3-6A_4) \tag{5.86}$$

$$A_1 = \frac{1}{\omega^2}(f_{[Nt]/N} - 2cA_2 - 2A_3)$$

$$(5.87)$$

$$B_1 = d_i - A_1, \qquad\qquad B_2 = \frac{1}{\xi}(v_i + cB_1 - A_2).$$

$$(5.88)$$

It should be noted that the above formulae developed via the P-T method are suitable for solving any linear or nonlinear dynamic problems in a general form as exhibited in equation (5.80). Although the solution is derived for the accuracy of fourth order, the solution of higher order arrcuracy can be obtained by keeping more terms of Tayolr series expansion in equation (5.81). A semi-analytical solution for the systems (5.80) can also be obtained through the piecewise constant procedure as described before with the conditions of continuity and initial conditions specified, on the basis of equation (5.82) which is continuous with respect to time.

To compare the numerical results of the P-T method and Runge-Kutta method with accuracy of fourth order, consider the following nonlinear governing equation of a vibration system with a hardening spring.

$$\ddot{x}(t) + 2c\dot{x}(t) + \omega^2 x(t) = bx^3(t)$$

$$(5.89)$$

Piecewise constantized or linearize this equation by the P-T method with the implementation of the piecewise constant argument, in a time interval $[Nt]/N \le t < ([Nt]/N + 1)/N$, following governing equation can be considered.

$$\ddot{x}_i + 2c\dot{x}_i + \omega^2 x_i = bd_i^3 + 3d_i^2 v_i \left(t - \frac{[Nt]}{N} \right)$$

$$+ \frac{1}{2}[6d_i v_i^2 + 3d_i^2(bd_i^3 - 2cv_i - \omega^2 d_i)]\left(t - \frac{[Nt]}{N} \right)^2$$

$$+ \frac{1}{6}\{6bd_i v_i^3 + 18bd_i v(bd_i^3 - 2cv_i - \omega^2 d_i)$$

$$+ 3bd_i^2 v_i[3d_i^2 v_i - 2c(bd_i^3 - 2cv_i - \omega^2 d_i)$$

$$- \omega^2 v_i)]\}\left(t - \frac{[Nt]}{N} \right)^3$$

$$(5.90)$$

Actually, the linearized equation shown above can be generated by simply substituting $f(x) = bx^3$ into equations (5.80) and (5.81) with the following differentiations of $f(x)$ in the time interval $[Nt]/N \le t < ([Nt]/N + 1)/N$.

$$f_{[Nt]/N} = bd_i^3 \tag{5.91}$$

$$f'_{[Nt]/N} = 3bd_i^2 v_i \tag{5.92}$$

$$f''_{[Nt]/N} = 6bd_i v_i^2 + 3bd_i^2 G_1, \tag{5.93}$$

where

$$G_1 = bd_i^3 - 2cv_i - \omega^2 d_i \tag{5.94}$$

$$f'''_{[Nt]/N} = 6bd_i v_i^3 + 18bd_i v G_1 + 3bd_i^2 v G_2 \tag{5.95}$$

where

$$G_2 = f'_{[Nt]/N} - 2cG_1 - \omega^2 v_i \tag{5.96}$$

As such, the nonlinear governing equation (5.81) can be used as a general equation in $[Nt]/N \le t < ([Nt]/N + 1)/N$ for developing the governing equation of any given nonlinear system described in equation (5.80), so long as the function $f(x, \dot{x}, t)$ is given. It should be noticed, however, it is the solution of a dynamic system that is used in actual numerical calculations, not the description of equation (5.90).

Similarly, by substituting the equations from (5.91) to (5.96) into the solution of general form (5.82), the solution of (5.89) with the accuracy of fourth order can be generated. Consequently, the recurrence relation for the numerical calculation can be obtained with the solution on the time intervals, the continuity conditions and the equations from (5.91) to (5.96). In fact, the equations from (5.91) to (5.96) are to be repeatedly used in the numerical calculations, in the form listed above.

As demonstrated for the solution (5.66) of equation (5.65), it can be shown that the solution of equation (5.89) such developed indeed has an accuracy equivalent to that provided by Taylor-series method of order four. The solution developed is thus ready for numerical calculation to compare with that of the Runge-Kutta method.

Figure 5.10 exhibits the comparison of the numerical solutions for this system with the identical parameter values and conditions, calculated by both the P-T method and Runge-Kutta method. Numerical solution by Runge-Kutta method for the governing equation (5.80) is obtained by conventional Runge-Kutta method with the fourth order accuracy, to compare with the solution by the P-T method with the fourth order accuracy. In Figure 5.10, pt4_.3 and rk4_.3 are the curves representing the solutions by the P-T method and Runge-Kutta method respectively. The step length for these numerical calculations of both the solutions of the P-T and Runge-Kutta methods is kept to be 0.3. This step length is selected to distinguish the numerical values for the two methods, though it is much larger than what is required for a reliable numerical solution.

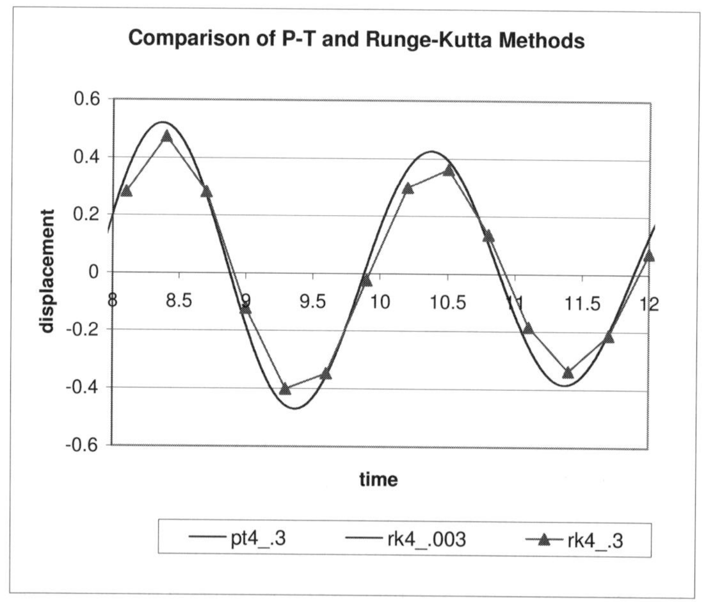

Figure 5.10.   Comparison of numerical results generated by P-T method of fourth order and Runge-Kutta method of fourth order for the governing equation $\ddot{x} + 0.2\dot{x} + 10x = x^3$, with $x(0) = 1$ and $\dot{x}(0) = 2$. In the figure, "pt4_.3" represents the P-T method with step length 0.3; "rk4_.3" Runge-Kutta method with step length 0.3 and "rk4_.003" Runge-Kutta method with step length 0.003.

As can be seen from the figure, the solution of Runge-Kutta method is discrete whereas the solution of the P-T method is continuous and smooth everywhere. Combining the continuous solutions within the time intervals and the continuity conditions shown in equation (5.70), the curve pt4_.3 is actually continuous and smooth in the entire time range considered. In fact, for each time interval of a chosen step length, there is a continuous solution of dynamics corresponding to it, as shown in the solution (5.82). The dynamic behavior of the system is governed by equation (5.82) in which the original physical information is maintained and protected to the utmost level.

It is well known that a smaller step length may yield a more accurate numerical solution (Nakamura 1991, Zingg 1999). Employing Runge-Kutta method of fourth order, with a much smaller step length of 0.003 unit, a more accurate numerical solution is obtained and represented by the curve rk4_.003 in Figure 5.10. This solution can be considered as the one with the accuracy of very high level in comparison with the solution corresponding to a step length a hundred times larger. As can be observed from the figure, the solution of the P-T method with step length 0.3 matches very well with that of the Runge-Kutta method with step length 0.003, the two curves are overlapped almost every where, though the curves are started to be plotted after 8 units of the time from $t = 0$. On the other hand, Runge-Kutta method with step length 0.3 provides a discrete solution with significantly lower accuracy in comparison with that of the P-T method of the same step length. Clearly, the P-T method provides a solution with higher accuracy in comparison with Runge-Kutta method. In addition, the solution calculated by the P-T method is continuous in the entire time range considered.

Although the curve rk4_.003 seems smooth, the curve is in fact formed by short straight lines connecting the end points at t = $[Nt]/N$ and $t = ([Nt]+1)/N$, no matter how small the step length is taken.   It is also significant to note that the curve segment of pt4_.3 generated by the P-T method, corresponding to a step length of 0.3 unit, yielded by the solution of equation (5.82) that reflects the physical characteristics of the oscillatory system, has the same shape as the corresponding curve segment of rk4_.003 with higher accuracy. Evidently, the solution by the P-T method is a good approximation to the motion of the governing equation. Solution of the P-T method not only generates a fairly accurate continuous curve, as against a straight line to connect the two points of a time interval $[Nt]/N \leq t < ([Nt]/N + 1)/N$, but also provides a curve that represents much closely the actual physical behavior of the original nonlinear dynamic system.

Though the difference between the curves pt4_.3 and rk4_.003 is hardly distinguishable by naked eyes, the numerical values corresponding to the two curves are actually different. To demonstrate the difference between the two curves, the detailed numerical data calculated by the P-T and Runge-Kutta methods corresponding to the curves are listed in the following table. From Figure 5.10 and the table, one may again conclude that the solution given by the P-T method is more accurate in comparison with Runge-Kutta method as expected. The high accuracy of the P-T method evidently owes to the advantage that the maximum possible original physical information is kept from variation in developing the equations for semi-analytical and numerical analysis via the P-T method.

Table 5.4. Accuracy comparisons for fourth order P-T method and Runge-Kutta method of the same order in solving for $\ddot{x}(t) + 2c\dot{x}(t) + \omega^2 x(t) = bx^3$. The initial conditions are $x(0) = 1$ and $\dot{x}(0) = 2$.

Time	P-T Step 0.3	Runge-Kutta Step 0.3	Runge-Kutta Step 0.003
8.1	0.347724	0.281461	0.343268
8.4	0.517027	0.473953	0.518208
8.7	0.267069	0.28183	0.272421
9.0	−0.18254	−0.1204	−0.17695
9.3	−0.45882	−0.40096	−0.45817
9.6	−0.35455	−0.34717	−0.35839
9.9	0.027678	−0.02265	0.021947
10.2	0.364802	0.298642	0.362458
10.5	0.390918	0.362595	0.393205
10.8	0.1028	0.135726	0.107953
11.1	−0.25117	−0.18411	−0.24755
11.4	−0.3827	−0.33687	−0.38347
11.7	−0.20041	−0.2129	−0.20458
12	0.132724	0.072054	0.128444

Table 5.5. Comparison of the CPU times for solving $\ddot{x} + 0.2\dot{x} + 10x = x^3$. For both the methods of P-T and Runge-Kutta of fourth order, the initial conditions are $x(0) = 1$ and $\dot{x}(0) = 2$; time range from 0.0 to 20,000; and step length 0.3.

Method	CPU time (seconds)
P-T	8.36
Runge-Kutta	8.37

The CPU times spent for the calculations employing both the P-T method and Runge-Kutta method are shown in Table 5.5.

Note that the CPU time of 8.37 seconds shown in Table 5.5 for Runge-Kutta method is obtained with step length of 0.3. The CPU time will be enormously longer than 8.37 seconds for Runge-Kutta method if the step length of 0.003 is applied.

As demonstrated previously, in numerically solving the dynamic problems by using P-T method, it is beneficial to retain as much as possible the original information in the given equation of motion. Since the solution generated by P-T method is higher in accuracy with respect to that of Runge-Kutta, it is worth to evaluate the accuracy of a T-P method solution of third order with that of Runge-Kutta of fourth order. For a forced nonlinear pendulum described by the equation of motion

$$\ddot{x}(t) + \theta\dot{x}(t) + \omega^2 \sin x(t) = A\cos\Omega t \tag{5.97}$$

the P-T method of third order accuracy can be applied with the consideration that the nonlinear term need to be made piecewise constant for generating a linear system in the time interval. With the P-T method, the above equation of motion is written in the following form on a time interval of $[Nt]/N \le t < ([Nt]/N + 1)/N$.

$$\ddot{x}_i + \theta\dot{x}_i = A\cos\Omega t - \omega^2 \sin d_i - \omega^2 v_i \left(t - \frac{[Nt]}{N} \right) \cos d_i$$

$$- \frac{\omega^2}{2}[(A\cos\Omega t - \theta v_i - \omega^2 \sin d_i) - v_i^2 \sin d_i]\left(t - \frac{[Nt]}{N} \right)^2$$

$$\tag{5.98}$$

Table 5.6. Numerical solutions calculated by using the fourth order Runge-Kutta method and third order P-T method in solving a nonlinear dynamical system with equation of motion $\ddot{x} + 2c\dot{x} + \omega^2 \sin x = B \cos \Omega t$, where $c = 0.25$, $\omega = 1$, $B = 0.8$, $\Omega = 2/3$ and the initial conditions are given by $d_0 = 0$, $v_0 = 1$.

t	4th Order Runge-Kutta $h = 0.1$	3rd Order P-T method $N = 10$	4th Order Runge-Kutta $h = 0.0001$
0.000000	0.000000	0.000000	0.000000
0.100000	0.101310	0.101308	0.101308
0.200000	0.204481	0.204466	0.204469
0.300000	0.308398	0.308340	0.308341
0.400000	0.411975	0.411829	0.411833
0.500000	0.514182	0.513893	0.513899
0.600000	0.614060	0.613555	0.613565
0.700000	0.710730	0.709928	0.709941
0.800000	0.803403	0.802211	0.802229
0.900000	0.891384	0.889703	0.889724
1.000000	0.974070	0.971791	0.971737

The corresponding solution and recurrence relations for numerical calculation can be obtained by the same procedure as used for the solution of equation of (5.65). The numerical solution derived from the above equation has the accuracy of third order whereas Runge-Kutta method has fourth order accuracy. The numerical results based on equation (5.98) are tabulated in Table 5.6 and compared with Runge-Kutta method of order four. As shown in Table 5.6, the numerical results given by P-T method with a fairly large step length are still more accurate than that of the Runge-Kutta method of the fourth order accuracy, as the results of P-T method are closer to that produced by using Runge-Kutta method with a very small step length.

5.8. Consistency Analysis of Numerical Solutions with Implementation of Piecewise Constant Arguments

For a numerical technique, it is important to ensure that the consistency of the numerical solution with the original dynamical system is satisfied

(Smith 1985, Dahlquist and Bjorck 1974). To analyze the consistency of the numerical solutions derived through the implementation of the piecewise constant arguments or the P-T method and the truncation error caused by the piecewise-constant approaches, the following governing equation of dynamic problem in a general form may be considered.

$$m\ddot{x} + d\dot{x} + kx = f(t, x, \dot{x}) \tag{5.99}$$

where d is the damping coefficient and $f(t, x, \dot{x})$ can be either a linear or a nonlinear function of time, displacement and velocity of the dynamic system. As demonstrated previously, the numerical solution of this system may be obtained with the piecewise-constant approach by expressing the governing equation in the following form.

$$m\ddot{X} + d\dot{X} + kX = f\left(\frac{[Nt]}{N}, x\left(\frac{[Nt]}{N}\right), \dot{x}\left(\frac{[Nt]}{N}\right)\right) \tag{5.100}$$

Solution of this linear piecewise-constant equation of motion is readily available. Within $([Nt]/N, ([Nt]/N+1)/N)$, the difference between the continuous system with solution $x(t)$ to equation (5.99) and the corresponding piecewise-constant system with solution $X(t)$ of equation (5.100) is expressible as

$$m\ddot{x} + d\dot{x} + kx - f\left(\frac{[Nt]}{N}, x\left(\frac{[Nt]}{N}\right), \dot{x}\left(\frac{[Nt]}{N}\right)\right) = R_N \tag{5.101}$$

The truncation error $R_N(t)$ is obviously not zero for the finite number of intervals. Using a Taylor series expansion, with the condition that the solution is continuously derivable for sufficiently many times, the derivatives shown in the above equation can then be given as

$$\ddot{x}(t) = \ddot{x}\left(\frac{[Nt]}{N}\right) + x'''\left(\frac{[Nt]}{N}\right)\left(t - \frac{[Nt]}{N}\right) + 0\left(\left(t - \frac{[Nt]}{N}\right)^2\right) \tag{5.102}$$

$$\dot{x}(t) = \dot{x}\left(\frac{[Nt]}{N}\right) + \ddot{x}\left(\frac{[Nt]}{N}\right)\left(t - \frac{[Nt]}{N}\right) + 0\left(\left(t - \frac{[Nt]}{N}\right)^2\right) \tag{5.103}$$

and

$$x(t) = x\left(\frac{[Nt]}{N}\right) + \dot{x}\left(\frac{[Nt]}{N}\right)\left(t - \frac{[Nt]}{N}\right) + 0\left(\left(t - \frac{[Nt]}{N}\right)^2\right) \quad (5.104)$$

From equation (5.99)

$$m\ddot{x}\left(\frac{[Nt]}{N}\right) + d\dot{x}\left(\frac{[Nt]}{N}\right) + kx\left(\frac{[Nt]}{N}\right) - f\left(\frac{[Nt]}{N}, x\left(\frac{[Nt]}{N}\right), \dot{x}\left(\frac{[Nt]}{N}\right)\right) = 0$$

$$(5.105)$$

Thus

$$R_K = \left[mx''' + d\ddot{x}\left(\frac{[Nt]}{N}\right) + k\dot{x}\left(\frac{[Nt]}{N}\right)\right] - \left(t - \frac{[Nt]}{N}\right) + 0\left(\left(\frac{1}{N}\right)^2\right) \quad (5.106)$$

whereas $|t - [Nt]/N| \le 1/N$.

Definition. *If the limiting value of the truncation error R_N is zero as $N \to \infty$ and the time integral step tends to zero, then equation (5.100) is said to be consistent with original differential equation (5.99) (Smith 1985).*

Theorem. *Under the conditions that $m > 0$, and c and k in equation (5.99) are bounded, considering that the piecewise constant systems will become continuous systems as N approaches infinite, the piecewise-constant equation (5.100) is consistent with equation (5.99).*

Proof. Since $(t - [Nt]/N)$ in equation (5.106) tends to zero as $N \to \infty$, hence, $\ddot{x}([Nt]/N) \to \ddot{x}(t)$, $\dot{x}([Nt]/N) \to \dot{x}(t)$, $x([Nt]/N) \to x(t)$, $f'([Nt]/N, x([Nt]/N), \dot{x}([Nt]/N) \to f'(t, x, \dot{x})$, and $f([Nt]/N, x([Nt]/N), \dot{x}([Nt]/N))) \to f(t, x, \dot{x})$. Considering the conditions that $m > 0$, and $c, k, x'''(t), \ddot{x}(t), x(t), f'(t, x, \dot{x})$ and $f(t, x, \dot{x})$ are all bounded, therefore, R_N tends to zero as $N \to \infty$.

For the numerical solutions derived above, there are a few points need to be emphasized.

(1) The numerical solutions are derived directly from the corresponding second-order differential equations;

(2) The solutions are valid on the time interval $i/N \leq t < (i+1)/N$;

(3) The solutions are continuous everywhere on $i/N \leq t < (i+1)/N$ (i.e. in between X_i and X_{i+1}) and independent of the length of the interval which is controlled by the parameter N.

(4) Corresponding to the numerical solutions and the conditions of continuity, the complete solutions generated by the piecewise constant procedures and their first derivatives are continuous everywhere on the range from zero to t, independent of the value of the parameter N.

5.9.   Step Size Control

Comparatively speaking, to most dynamic problems, the values of N used in Sections 5.3 and 5.4 are rather small for obtaining a numerical solution with a sufficient accuracy. In order to improve the accuracy of a numerical solution, one may choose a larger value for the parameter N. As discussed in the previous chapter, the corresponding approximate solutions will tend to the exact solutions for the continuous dynamic systems as N approaches infinity. However, the numerical calculation time is proportional to the N value taken.

In fact, as mentioned in Chapter 3, there is no need for the parameter N to be kept fixed over the entire time range during the numerical calculation. In addition, the techniques of piecewise-constant method and P-T method are actually the so called single-step methods because they use the information from the last step computed to perform the next step (Isaacson and Keller 1966, Morries 1983). Once the local initial values d_i and v_i are available, the numerical solutions can be calculated through a step-by-step procedure according to the formulas derived.

Based on the discussion in the precious sections, it is important to keep the parameter N large enough or the step length small enough to maintain acceptable accuracy of the numerical solution. However, if N is greater than necessary, the numerical computation may be excessive, and the accumulated round-off error may increase as the numerical calculation time goes on. Furthermore, the requisite value of N may vary with the rate at which the solution itself varies. Also, different accuracy requirement may need different minimum N value. These requirements suggest that the numerical value of N should be made suitable for each step during the calculation. One way to do this is to tie the value N to the computed difference $\Delta x = x_{i+1}^{(N)} - x_{i+1}^{(2N)}$ corresponding to an identical time, where $x_{i+1}^{(N)}$ is a solution calculated by using a value $1/N$ as the step length and $x_{i+1}^{(2N)}$ is calculated by a step length of $1/2N$, half of the prior one. If the Δx is greater than a certain predetermined value ε for the accuracy desired, then the numerical calculation will back up and use a greater value for N to evaluate x_{i+1} with a higher accuracy. There are many techniques in the current literature in determining the proper step length for numerical calculations. In performing the numerical calculations for the examples shown in the previous sections, as presented in the programs presented in Appendix C, a technique provided by Press *et al.* (1986) is employed. When calculated Δx is less than ε while N takes a specific value N_k, the following formula is used in the computer program to decrease the value of N from N_k to N_{k+1}.

$$N_{k+1} = \frac{N_k}{(0.95)\left(\dfrac{\varepsilon}{\Delta x}\right)^{\frac{1}{4}}} \tag{5.107}$$

If $\Delta x > \varepsilon$, then the formula below is used to shorten the step length.

$$N_{k+1} = \frac{N_k}{(0.95)\left(\dfrac{\varepsilon}{\Delta x}\right)^{\frac{1}{5}}} \tag{5.108}$$

The numerical results are examined for each of the calculation iterations with the accuracy desired. Making use of the P-T method and this step-size control technique with variable step lengths, numerical calculation can be more efficient. In performing the numerical calculation, ε is a factor to control the accuracy of the solution and the parameter N governs the step length to ensure the accuracy required.

5.10. Characteristics of the P-T method

It can be seen from the above discussion that the P-T method is an efficient numerical method which may provide sufficiently accurate results with a good convergence for solving the linear and nonlinear dynamic systems. In comparison with the Runge-Kutta method and the other existing numerical methods, following characteristics of the P-T method are significant for being emphasized.

(1) The P-T method is a single-step as well as explicit integration method with high efficiency and accuracy in comparing with that of Runge-Kutta method. With this single-step method, the advantages of the step-size control techniques can be easily taken for the efficiency of the numerical calculations.

(2) Accuracy of the solutions generated through the P-T method depends on the selections of the following two factors: a) the value of N; b) the number of terms retained as results of Taylor series expansion in the right side of equation (5.81) for the numerical or approximate calculations in practice. Theoretically, the larger the value N and the more terms retained, the more accurate the solutions can be expected.

(3) Though the accuracy of the solutions derived per the P-T method is counted by the order of retained terms generated per Taylor series expansion, the accuracy of the solutions of P-T method can be higher than that of direct Taylor expansion method or Runge-Kutta method, due to the meticulous retaining of the original information in the governing equations.

(4) Most existing numerical methods provide the solutions only at the discrete points, as indicated in Section 2.4. The solution in between the discrete points is usually not available. In contrast to these numerical methods, the numerical solution derived by the P-T method, such as the solutions shown in equations (5.66) and (5.82), are continuous in the time interval considered. Under the conditions of continuity at each starting point of a time interval, the solution given by P-T method is continuous everywhere along the entire time range.

(5) For any given nonlinear dynamic system, with the P-T method, the piecewise constant procedure makes significantly large number of linear dynamic systems corresponding to it. Theoretically, the number of linear systems can be infinite should an exact solution is expected. This implies that the nonlinear dynamic system considered is considered as a combination of infinite number of linear dynamic systems.

(6) In numerically solving the dynamic problems with conventional approaches, as indicated in Section 2.4, the second-order differential equation of a dynamic system is usually transformed into a system of first-order differential equations in the following form.

$$y' = f(x, y) \qquad (5.109)$$

Numerical solutions corresponding to the first-order differential equations are then developed by employing the mathematical operations such as linearization (Smith 1985) or Taylor expansion (Dahlquist and Bjorck 1974, Isaacson and Keller 1966). The solutions such developed can only be discrete. In performing the numerical calculations with Runge-Kutta method, for example, the displacement and its first and second derivatives rely on the approximate formulas shown in equations in Section 2.4. In doing so, the physical meaning involving in the original equation of motion is quite often lost in the manipulations of the pure mathematical expressions. In comparison with these numerical methods, the P-T method strives to keep more physical information in the original governing equation unchanged. In each time interval considered, there is a continuous linear dynamic system corresponding to it, thanks to the piecewise constant procedure. This not only gives the solutions at the two ends of a time interval but also the continuous approximate solution over the time interval with the desired accuracy. Due to the maintenance of the original physical information, the continuous solution given by the P-T method is a good approximation to the exact or accurate solution in the time interval and along the entire time range. Solutions obtained by the P-T method are therefore continuous approximate solutions to the original dynamical systems when N is chosen to be suitably large.

Theoretically, the approximate solutions produced by the P-T method become the exact solutions to the dynamical systems as N tends to infinity.

(7) Iteration is a major operation for the numerical calculations of many numerical methods. When the local initial conditions are given, the iteration must be repeatedly carried out to obtain the numerical solution at the end of the time interval. However, there is no iteration involved in the numerical calculations by using the P-T method. Once the local initial conditions are available, the continuous solution for the time interval including the two ends of the time interval can be directly calculated by the formulae derived in the P-T method.

References

Abukhaled, M. I. and Allen, E. J., "A Class of Second-Order Runge-Kutta Methods for Numerical Solution of Stochastic Differential Equations," Stochastic Analysis and Applications, Vol. 16, pp. 977-992, 1998.

Baum, H. R., "On the Weakly Damped Harmonic Oscillator," Quarterly of Applied Mathematics, Vol. 30, pp. 573-576, 1972.

Baker, G. L. and Gollub, J. P., Chaotic Dynamics an Introduction, Cambridge University Press, Cambridge, 1990.

Blackburn, J. A., Vik, S., Wu, B. and Smith, H. J. T., "Driven Pendulum for Studying Chaos," Review of Scientific Instruments, Col. 60, pp. 422-426, 1989.

Dai, L. and Singh, M. C., "On Oscillatory Motion of Spring-Mass Systems Subjected to Piecewise Constant Forces," Journal of Sound and Vibration, Vol. 173, pp. 217-233, 1994.

Dai, L. and Singh, M. C., "An Analytical and Numerical Method for Solving Linear and Nonlinear Vibration Problems," International Journal of Solids and Structures, Vol. 34, pp. 2709-2731, 1997.

Dai, L. and Singh, M. C., "Periodic, Quasiperiodic and Chaotic Behavior of a Driven Froude Pendulum," International Journal of Nonlinear Mechanics, Vol. 33, pp. 947-965, 1998.

Dahlquist, G. and Bjorck, A., Numerical Methods, Prentice-Hall, Inc., New Jersey, 1974.

Dowell, E. H. and Pezeshki, C., "On the Understanding of Chaos in Duffing's Equation Including a Comparison with Experiment," ASME Journal of Applied Mechanics, Vol. 53, pp. 5-9, 1986.

Gerald, C. F. and Wheatley, P. O., Applied Numerical Analysis, Addison-Wesley, Reading, 1989.

Gwinn, E. G. and Westervelt, R. M., "Intermittent Chaos and Low-Frequency Noise in the Driven Damped Pendulum," Physical Review, Letters, Vol. 55, No. 15, pp. 1613-1616, 1985.

Isaacson, E. and Keller, H. B., Analysis of Numerical Methods, John Wiley & Sons, Inc., New York, 1966.

Klotter, K., "Steady State Vibrations in Systems having Arbitrary Restoring and Damping Forces," in Symposium on Nonlinear Circuit Analysis, Polytechnic Institute of Brooklyn, pp. 234-257, 1953.

Kutta, W., "Beitrag zur näherungsweisen Integration totaler Differentialgleichungen," Zeitschrift fur Mathematik und Physik, Vol. 46, pp. 435-453, 1901.

Morino, L., Leech, J. W. and Witmer, E. A., "Optimal Predictor-Corrector Method for Systems of Second-order Differential Equations," AIAA Journal, Vol. 12, No. 10, pp. 1343-1347, 1974.

Morries, J. L., Computational Methods in Elementary Numerical Analysis, John Wiley & Sons, Inc., New York, 1983.

Nakamura, S., Applied Numerical Methods with Software. New Jersey. Prentice Hall, 1991.

Newmark, N. M., "A Method of Computation for Structural Dynamics," ASCE Journal of the Engineering Mechanics Division, Vol. 85, pp. 67-94, 1959.

Phillips, C. and Cornelius, B., Computational Numerical Methods, Ellis Horwood Limited, New York, 1986.

Press, W. H, Flannery, B. P., Teukolsky, S. A. and Vetterling, W. T., Numerical Recipes: the Art of Scientific Computing, Cambridge University Press, Cambridge, 1986.

Rasband, S. N., Chaotic Dynamics of Nonlinear Systems, John Wiley & Sons, New York, 1990.

Runge, C., "Ueber die numerische Auflösung totaler Differential gleichungen," Mathematische Annalen, Vol. 46, pp. 167-178, 1895.

Smith, G. D., Numerical Solution of Partial Differential Equations: Finite Difference Methods, Clarendon Press, Oxford, 1985.

Ueda, Y., "Steady Motions Exhibited by Duffing's Equation: A Picture Book of Regular and Chaotic Motions," in New Approaches to Nonlinear Problems in Dynamics, (editor P.J. Holmes), Philadelphia: SIAM, pp. 311-322, 1980.

Weaver, W. J., Timoshenko, S. and Young, D. H. Vibration Problems in Engineering. New York: John Wiley & Sons, Inc 1990.

Zingg, D. W. and Chisholm, T. T., "Runge-Kutta Methods for Linear Ordinary Differential Equations," Applied Numerical Mathematics, Vol. 31, pp. 227-238, 1999.

CHAPTER 6

Application of P-T Method on Multi-Degree-of-Freedom Nonlinear Dynamic Systems

6.1. Introduction

Multi-Degree-of-Freedom (MDOF) dynamic systems are commonly seeing in engineering practices and they can be classified into two categories, linear and nonlinear MDOF systems. A single rigid body moving in space can form an MDOF system of two or more dimensions. An example for the motion of this type of system is the vibration of a simplified auto-body. The motions of the components of a gearbox can be described by a dynamic system consisting multi-bodies moving in the box correlatively, provided that the bodies are considered as rigid bodies. Actual continuous systems in engineering such as mechanical structures commonly seeing in civil and mechanical engineering are difficult to solve as they theoretically have infinite number of degrees of freedom. Furthermore, the analyses of continuous systems require the solutions of partial differential equations that are more complex in comparing with ordinary differential equations. For the sake of solubility or simplicity of dynamic analyses, continuous systems are often simplified as MDOF systems or combination of several MDOF dynamic systems. The analysis of a linear MDOF system conventionally requires the solution of a set of ordinary differential equations, which is relatively simple. As a nonlinear system can be considered as linear dynamic systems over the tiny intervals with piecewise constant arguments as described in the previous chapters, it can be expected that the solutions of the nonlinear MDOF systems can be generated with the implementation of the piecewise constant arguments and P-T method combined with the conventional approaches for solving linear MDOF systems.

The dynamic systems that have been discussed in the previous chapters are linear and nonlinear systems of single degree of freedom. For describing the dynamic systems of multidimensional or the systems with multi-degree-of-freedom, such as dynamic motion of multi-bodies or vibrations of systems with multi-degree-of-freedom, a system of equations of dynamics are needed. These dynamic systems are usually nonlinear and can be expressed by second-order or higher order differential equations. Moreover, the systems are usually more complex in comparing with that of the one dimensional systems, and analytical solutions for most of the systems are hard to develop on the basis of the conventional theories available in literature. Numerical or approximate solutions for the dynamic systems are therefore inevitable and important. For solving these equations of motion, the existing numerical methods such as Euler's method, Taylor-series method and Runge-Kutta method (Friedman 1994, Iserles, Zingg and Chisholm 1999, Luo and Han 1997) are commonly used. In solving for the dynamic systems by these methods, the second-order differential equations are usually transformed into a system of first-order differential equations. Moreover, some mathematical manipulations such as linearization are usually needed in order to derive for the solutions. Based on the discussions and numerical analyses provided in the previous chapters, there are a few characteristics of the solutions developed by the numerical methods need to be emphasized before the development of the solutions for nonlinear MDOF systems with piecewise constant arguments.

1. The numerical solutions generated by the numerical methods can only be discrete.
2. Following mathematical manipulations may damage the physical information or structure of the original equations of motion. The damages may provide negative influences to the accuracy and reliability of the results created by the approximate or numerical methods used.

   - Linearization: for solving the equations of motion, linearization is a commonly used mathematical operation.

- Simplification: some times the original differential equations may have to be simplified for the applicability of the approximate or numerical methods to be used.
- Utilization of Taylor expansion: this operation may be applied to most of the terms in the governing equations. Theoretically, the physical meaning embedded in the original equation of motion is represented by the terms expanded with Taylor expansion series. Practically, however, only the first few terms of the Taylor expansion are used for approximate or numerical calculations.

The researchers in this field have been striving to develop more reliable methods for solving the multidimensional or MDOF systems dynamic problems with higher accuracy. As described in the previous chapter, the new approach namely the piecewise constant procedure and the P-T method for approximately and numerically solving the linear and nonlinear dynamic problems have shown advantages in comparing with the other methods, such as Runge-Kutta method. The strategy of the approaches with implementation of the piecewise constant arguments is to endeavor for maintaining as much as possible the physical properties embedded in the original equations of dynamics undamaged during the process of applying the piecewise constantization. With this strategy, the approximate or numerical results such developed with higher accuracy and reliability can be expected. With the introduction of a piecewise constant argument $[Nt]/N$, together with Taylor series expansion, the numerical solution of differential equations of a dynamic system can be developed such that a linear dynamic system can be established between the two points $[Nt]/N$ and $([Nt]+1)/N,$ which can be as small as one may desire. Unlike the discrete solutions produced by the existing numerical methods, the approximate and numerical solutions produced by implementing the piecewise constant arguments are continuous everywhere in the time interval and on the entire time range from zero to t for any given value of N.

In approximate or numerically solving the dynamic problems by the piecewise constant approach, since the major portion of the corresponding original differential equation remains unchanged, as shown in the previous examples for both linear and nonlinear problems, the solutions derived by the P-T method are more accurate than the existing numerical methods such as Runge-Kutta method. Most significantly, the P-T method better reveals the actual physical behavior of a dynamic system in comparison with Runge-Kutta method and the other numerical methods with discrete solutions. The P-T method has also been proven to be an efficient numerical technique that provides sufficiently accurate results with a good convergence in solving for the dynamic problems. However, concentration of the P-T method has been found on the areas of vibration and oscillation problems of single degree of freedom. For solving the MDOF problems with the P-T method, especially for solving the problems of nonlinear MDOF involving linear couplings, difficulties may be found and proper procedures are needed to be taken (Dai and Singh 2003, Dai *et al.* 2006). A theoretically and practically sound approach therefore needs to be developed with implementation of the piecewise constant arguments for solving the MDOF dynamic systems. Such a development for solving these problems, with the advantages of the P-T method to be taken, will certainly be beneficial to efficient analyses of nonlinear dynamic problems.

In this chapter, we will present a methodology for solving nonlinear MDOF problems with implementation of piecewise constant arguments and the P-T method. Three types of common nonlinear problems: problems without linear coupling terms; problems with linear coupling terms and proportional damping terms, and problems with linear coupling terms and general damping terms, are to be solved by implementing the methodology with piecewise constant arguments. General procedures and formulae for semi-analytical and numerical computations in solving these problems are presented. Practical nonlinear dynamic systems are to be solved for demonstrating the application and effectiveness of the methodology. As the conventional approaches for solving linear MDOF systems are to be jointly utilized in solving for nonlinear MDOF dynamic systems with piecewise constant arguments, for developing such methodology, it is necessary to describe the main concepts and characteristics of the existing approaches for solving linear MDOF and nonlinear dynamic systems.

6.2.   Existing Approaches for Solving Multi-Degree-of-Freedom Linear and Nonlinear Dynamic Systems

The systems of multi-degree-of-freedom are more involved and complex in format and solution development, as the systems consist of more differential equations with more variables. Moreover, the variables and system parameters may be coupled even for linear MDOF systems. Many unique mathematical manipulations such as decoupling usually need to be performed before the MDOF systems can be solved with conventional methods. For solving the multi-dimensional dynamic systems of piecewise constant or continuous variables, we may need to have a fundamental comprehension on the concepts and characteristics of the existing approaches for solving the linear MDOF dynamic systems.

6.2.1.   *Governing Equations and Solution Development of Linear MDOF Systems*

Linear MDOF systems are commonly seen in physics and engineering fields, though most of the linear systems generated from the real-world problems are based on certain simplifications or linearizations. Once a linear MDOF system is established with differential equations, some matured techniques can be used for analytically solving it. A general linear MDOF dynamic system of n degrees of freedom is usually described by the following matrix expression consisting of n linear differential equations.

$$M\ddot{X} + C\dot{X} + KX = G \tag{6.1}$$

X in the equation is a matrix of the variables or coordinates of $x_1, x_2, \ldots, x_n$, M is known as mass matrix consisting n masses or equivalent masses of the system, C designates a damping matrix and K is called a stiffness matrix, and the matrix G is generally considered as the external excitations acting on the system. The differential equations describing the system are usually consisting more than one variables in each of the differential equations, and are considered as coupled.

Also, the mass matrix and/or the coefficient matrices may have one or more off-diagonal terms. Therefore, the equations cannot be solved individually; they have to be solved simultaneously.

The methods for solving linear MDOF systems are usually different corresponding to the absence of any of the matrices of C, K and G in equation (6.1). The simplest case for a linear MDOF system is probably free vibration of an undamped system, i.e., C and G are zero in equation (6.1) and the elements in M and K are all constants. The most popular analytical solution for such system is based on the assumption that the particular solution for the system is in the following form.

$$x_j = A_j \sin(\omega t + \theta), \quad j = 1, 2, \ldots, n \tag{6.2}$$

where A_j, ω and θ are the constants. With the assumption and the governing equation, the natural frequencies and the general solutions for the system can be determined via the techniques of variable separation and a procedure of solving eigenvalue problems (Kelly 1993, Weaver *et al.* 1990).

Linear MDOF systems usually have the identical number of natural frequencies as the number of degrees of freedoms, for a system having n degrees of freedom. Each of the natural frequencies corresponds to a type of motion or vibration, known as modal. The vibration with the natural frequency is known as the principal vibration. The general solution is actually a linear combination of all the possible solutions for this case, such that

$$x = \sum_{j}^{n} \vec{X}_j F_j \cos(\omega_j t + \phi_j) \tag{6.3}$$

where $\vec{X}_j$ is the jth modal vector, ω_j is the corresponding natural frequency and F_j and ϕ_j are constants.

In the cases that damping and excitations are involved, solving procedures for the system are more complex and decoupling of the differentia equations is needed. This is usually performed by introducing n new variables which satisfy the following relationship.

$$X = PY \qquad (6.4)$$

where Y is a matrix consisting of the new variables of $y_1, y_2, ..., y_n$, which are time-dependent generalized coordinates known as principal coordinates; and P is known as the principal or modal matrix consisting of the normal modes determined through solving the eigenvalue problems corresponding to the system.

If the damping matrix C in equation (6.1) can be expressed as a linear combination of the mass and stuffiness matrices, such as

$$C = \alpha M + \beta K \qquad (6.5)$$

which is known as proportional damping. Substituting equations (6.4) and (6.5) into (6.1) and considering the orthogonality with $X^T M X = I$, the following n differential equations can be obtained.

$$\ddot{y}_j + (\alpha + \beta \omega_j^2)\dot{y}_j + \omega_j^2 y_j = (P^T G)_j, \qquad j = 1, 2, ..., n \qquad (6.6)$$

With the proportional damping defined and the orthogonality, the damping matrix can be diagonalized simultaneously with the mass and stiffness matrices. The differential equations in the above equation are obviously uncoupled and have the identical form as that of the single degree of freedom systems, and thus can be solved with the techniques used for solving the single degree of freedom problems.

In the case that general damping is considered, the matrix cannot be diagonalized simultaneously with the mass and stiffness matrices, therefore, the uncoupled governing equations can not be created as that in the proportional damping case. In order to solve for the linear MDOF systems with general damping, usually, the original governing equations consisting n second-order coupled differential equations are transferred into $2n$ uncoupled first-order differential equations. This is usually done by using $\dot{x}$ as an auxiliary variable.

In solving the linear MDOF dynamic systems with the conventional methods, the following can be seen from the discussions above.

1. The governing equations of the dynamic systems discussed must all be linear. The methods utilized above are not valid for solving nonlinear MDOF problems.

2. The external excitations acting on the linear systems are usually functions of time only. The methods can hardly be applied or invalid if the excitations are not functions of time or involve other variables.

3. The procedures for solving MDOF problems are much more complex in comparing with that for solving the dynamic problems of single degree of freedom.

6.2.2. *Solving for Nonlinear MDOF Systems*

The analytical solutions discussed above are for the linear systems which are generated with the assumption that the nonlinearity of the systems is negligible. For nonlinear MDOF dynamic systems, however, analytical solutions are hard to obtain if not impossible. Most solutions for nonlinear systems are generated through approximate or numerical approaches.

Many analytical methods are available for generating approximate solutions for nonlinear MDOF systems. The most commonly used methods in dynamics are Lindstedt's perturbation method, averaging method, harmonic balance method, multiscale method, and asymptotic method. However, these methods are suitable for solving weak nonlinear problems. For high nonlinear systems, usually certain solutions in analytical form are first determined which may lead to differential equations or algebraic equations with perturbation approach (Strogatz 2000, Nayfeh 1981, Kapitaniak 1996). However, these equations are still hard to solve and numerical approaches are usually needed. As indicated in the introduction part of this chapter, simplification and linearization are necessary for obtaining the analytical solutions. Lindstedt's perturbation method, for instance, assumes that the angular frequency varies as a function of constant amplitude and the solution to be periodic (Ku 1958).

Numerical methods such as Runge-Kutta method, central difference, Wilson, and Newmark methods can all be used to solve nonlinear MDOF problems (Strogatz 2000, Nayfeh and Mook 1979, Kapitaniak 1996). The characteristics of the numerical solutions such as discrete solutions

(Runge-Kutta etc.), linearizations (Wilson, Newmark etc.) have been summarized in the introduction part of this chapter.

6.3. Derivation of General Nonlinear MDOF Dynamic Systems with Piecewise Constant Arguments

Based on the discussion in Section 6.2, it is seen that the existing techniques for solving the nonlinear dynamic problems of multi-degree-of-freedom provide approximate or discrete solutions. Additionally, the process of obtaining the solutions is based on the simplifications and linearizations of the original differential equations. With this consideration, implementation of the P-T method is a nature selection for developing for more accurate and reliable solutions of nonlinear dynamic systems. By the approaches implementing piecewise constant arguments, nonlinear dynamic systems are converted into linear dynamic systems in the time intervals that can be theoretically infinitesimal. Making these linear systems available, the solutions for the nonlinear MDOF systems can be generated with P-T method together with the conventional approaches for linear MDOF systems.

Implementing the P-T method, same as that used in solving nonlinear systems of single degree of freedom, the accuracy of the numerical or approximate solutions such generated can be controlled by a) the number of terms of Taylor series expansion used and; b) the values of N, which is directly related to the time step used for numerical calculations. Detailed descriptions of the P-T method and its application on determining the approximate and numerical solutions can be found from Chapter 5.

6.3.1. *Solving Nonlinear Systems Directly Implementing P-T Method*

In solving for linear or nonlinear dynamic MDOF systems, unlike one dimensional systems, the variables of the system of governing equations may be related to each other. The simplest scenario in nonlinear dynamic analysis is that the dynamic system consisting of differential equations that are independent to each other, or the nonlinear systems without linear coupling. For nonlinear dynamic systems without linear coupling terms, the P-T method can be directly implemented as that used for solving the one dimensional problems of dynamics. More significantly, the P-T method can be applied directly to solve for the nonlinear dynamic MDOF systems in which the homogeneous equations corresponding to the differential equations of the system are all in uncoupled form but the expressions of the external excitations involve more than one variable.   Such a nonlinear dynamic MDOF system can be described by the governing equations in the following form.

$$\ddot{x}_j + 2\mu_j \dot{x}_j + \omega_j^2 x = g_j\left(x_1, x_2, \ldots, x_n, \dot{x}_1, \dot{x}_2, \ldots, \dot{x}_n, t\right)$$
$$j = 1, 2, \ldots, n \tag{6.7}$$

It should be noticed that the external excitations represented by g_j can be the functions of the variables, the first derivatives of the variables and time t, in comparing with the linear systems discussed in Section 6.2, in which the excitations are merely functions of time. As there is no linear coupling terms among the terms on the left side of the equations of the dynamic system, the procedure described in Chapter 5 can be utilized to solve the above system consisting of n equations. When the jth equation is being solved, for instance, the displacements and velocities of all other n-1 equations are assumed to be constants or functions of time in the jth equation. It should be noted that all the terms on the right sides of the above equations are the functions of time, displacements and velocities of all coordinates and external forces. For assuring the accuracy of the solutions, the function on the right side of the equation is expanded with Taylor series to the desired order of accuracy on an ith time interval $[Nt]/N \leq t \leq ([Nt]+1)/N$, such that

$$\ddot{x}_{ji} + 2\mu_j \dot{x}_{ji} + \omega_j^2 x_{ji} = g_{ji,[Nt]/N}$$

$$+ g'_{ji,[Nt]/N}\left(t - \frac{[Nt]}{N}\right) + \frac{1}{2!} g''_{ji,[Nt]/N}\left(t - \frac{[Nt]}{N}\right)^2$$

$$+ \frac{1}{3!} g'''_{ji,[Nt]/N}\left(t - \frac{[Nt]}{N}\right)^3 + \cdots, \qquad j = 1, 2, \ldots, n \tag{6.8}$$

With this manipulation, the equation above has an analytical solution in the ith time interval $[Nt]/N \le t \le ([Nt]+1)/N$. This analytical solution is continuous on the ith time interval. Utilizing the local initial conditions and the conditions of continuity, the solution for the system of differential equations can be developed with the procedures described in the previous chapter. Practically, only the first few terms on the right hand side of the equal sign are considered for developing the approximate or numerical solutions, though more terms for higher accuracy of solutions can be considered if so desire. Following is the solution in general form of the system, with the 4th order accuracy.

$$x_{ji} = e^{-\mu_j\left(t-\frac{[Nt]}{N}\right)}\left\{B_{j1}\cos\left[\xi_j\left(t-\frac{[Nt]}{N}\right)\right]+B_{j2}\sin\left[\xi_j\left(t-\frac{[Nt]}{N}\right)\right]\right\}$$

$$+A_{j1}+A_{j2}\left(t-\frac{[Nt]}{N}\right)+A_{j3}\left(t-\frac{[Nt]}{N}\right)^2+A_{j4}\left(t-\frac{[Nt]}{N}\right)^3 \tag{6.9}$$

where

$$\xi_j^2 = \omega_j^2 - \mu_j^2$$

$$A_{j4} = \frac{1}{6\omega_j^2}g'''_{ji,[Nt]/N}$$

$$A_{j3} = \frac{1}{\omega_j^2}\left(\frac{1}{2}g''_{ji,[Nt]/N} - 6\mu_j A_{j4}\right)$$

$$A_{j2} = \frac{1}{\omega_j^2}\left(g'_{ji,[Nt]/N} - 4\mu_j A_{j3} - 6A_{j4}\right)$$

$$A_{j1} = \frac{1}{\omega_j^2}\left(g_{ji,[Nt]/N} - 2\mu_j A_{j2} - 2A_{j3}\right)$$

$$B_{j1} = d_{ji} - A_{j1}$$

$$B_{j2} = \frac{1}{\xi_j}\left(v_{ji} + \mu_j B_{j1} - A_{j2}\right)$$

The solution shown in equation (6.9) is in a general form for all the solutions to the n governing equations.

The complete solution of the governing equation for the system (6.7) depends on the availability of the following:

1. application of the continuity conditions in the following form:

$$x_{ji}\left(\frac{[Nt]}{N}\right) = x_{ji-1}\left(\frac{[Nt]}{N}\right) \quad \text{and} \quad \dot{x}_{ji}\left(\frac{[Nt]}{N}\right) = \dot{x}_{ji-1}\left(\frac{[Nt]}{N}\right)$$

2. determination of the analytical solutions of each of the n differential equations;
3. derivatives of the function g obtained directly from equation (6.7).

6.3.2. *Nonlinear Systems with Linear Coupling and Proportional Damping*

As indicated previously, the dynamic system consisting of two or more differential equations that have coupling terms is much more complex and difficult to solve in comparing with that of the system formed by the differential equations that are independent to each other. For the dynamic systems consisting of differential equations with linear coupling terms, however, P-T method can not be directly applied. In order to solve for such dynamic systems with implementation of the P-T method, proper mathematical manipulations need to be performed in advance.

For solving a nonlinear system with linear coupling and proportional damping, in general, the general governing equation consisting of a system of n differential equations may also be expressed as follows for the nonlinear system.

$$M\ddot{X} + C\dot{X} + KX = G \tag{6.10}$$

where X is the matrix of the variables of $x_1, x_2, \ldots, x_n$, M is the mass matrix, C the damping matrix and K is the stiffness matrix. The governing equation is such managed that all linear terms are on the left side of the equal sign and all nonlinear terms and external excitations are on the right side. All the elements of vector G are expressible as functions of time, displacements and velocities corresponding to the coordinates of the system and the external forces acting on the system. Thus, the elements of the vector G can be expressed in the following general form.

$$g_j = g_j\left(x_1, x_2, \ldots, x_n, \dot{x}_1, \dot{x}_2, \ldots, \dot{x}_n, t\right) \qquad j = 1, 2, \ldots, n \qquad (6.11)$$

Introducing the principal coordinates of the linear parts of the governing equations and taking the following transformation,

$$X = PY \qquad (6.12)$$

where Y is the principal coordinates of $y_1, y_2, \ldots, y_n$ and P is the principal or modal matrix. The governing equation of the system can then be expressed with the principal coordinates.

$$\ddot{y}_j + 2\mu_j \dot{y}_j + \omega_j^2 y_j = (P^T G)_j$$
$$= f_j\left(y_1, y_2, \ldots, y_n, \dot{y}_1, \dot{y}_2, \ldots, \dot{y}_n, t\right) \qquad (6.13)$$
$$(j = 1, 2, \ldots, n)$$

where y_j is the jth principal coordinate and P is the principal matrix. The left sides of these equations are thus uncoupled and the coupling terms on the right sides of these equations are still the functions of time, displacements and velocities of all the principal coordinates and external forces.

With this mathematical manipulation consisting of decoupling, the procedure described in Section 6.3.1 can be employed to solve these n equations approximately or numerically with respect to the principal coordinates and consequently the response of the physical coordinates, by taking the transformation $X = PY$. It should be noticed; all the terms on the right side of these n equations become linear combinations of the nonlinear terms and external excitations. For solving the system with the P-T method described in Section 6.3.1, the right sides of the governing equation in principal coordinates are expanded by Taylor series such that the right sides are merely functions of time. With these mathematical manipulations, the P-T method can be applied.

6.3.3. *Nonlinear Systems with Linear Coupling and General Damping*

For many practical engineering problems, the nonlinear governing equations may involve linear coupling terms and general damping terms, and it is very difficult and some times impossible to perform the mathematical manipulations for obtaining the expressions as that shown in equations (6.13), through the transformation discussed in the previous section.

In order to apply the principle of the P-T method to obtain the solution for such a nonlinear dynamic system of multidimension with linear coupling terms and general damping terms, the system is rewritten as a system of 2n first-order equations.

$$\tilde{M}\dot{Z} + \tilde{K}Z = \tilde{G} \tag{6.14}$$

where the new matrices are defined as

$$\tilde{M} = \begin{bmatrix} 0 & M \\ M & C \end{bmatrix}, \quad \tilde{K} = \begin{bmatrix} -M & 0 \\ 0 & K \end{bmatrix}, \quad Z = \begin{bmatrix} \dot{X} \\ X \end{bmatrix}, \quad \tilde{G} = \begin{bmatrix} 0 \\ G \end{bmatrix} \tag{6.15}$$

The solution of equation (6.14) implements eigenvalues and eigenvectors of $\tilde{M}^{-1}\tilde{K}$. Let the modal matrix P be the matrix whose columns are the normalized eigenvectors of $\tilde{M}^{-1}\tilde{K}$. The principal coordinates may then be given by

$$Z = PY \tag{6.16}$$

Substituting equation (6.16) into equation (6.14) leads to

$$\dot{Y} - \Lambda Y = F \tag{6.17}$$

The diagonal matrix with the eigenvalues of $\tilde{M}^{-1}\tilde{K}$ makes the system of equations uncoupled on the left side. Thus, the equations of the system are expressible in the following form.

$$\dot{y}_j - a_j y_j = f_j\left(y_1, y_2, \ldots, y_n, y_{n+1}, y_{n+2}, \ldots, y_{2n}, t\right) \\ j = 1, 2, \ldots, n, n+1, \ldots, 2n \tag{6.18}$$

Expanding the function on the right side of the equation with Taylor series to the desired order of accuracy on an ith time interval $[Nt]/N \le t \le ([Nt]+1)/N$ to obtain

$$\dot{y}_{ji} - a_j y_{ji} = f_{ji,[Nt]/N} + f'_{ji,[Nt]/N}\left(t - \frac{[Nt]}{N}\right) + \frac{1}{2!}f''_{j,[Nt]/N}\left(t - \frac{[Nt]}{N}\right)^2$$

$$+ \frac{1}{3!}f'''_{j,[Nt]/N}\left(t - \frac{[Nt]}{N}\right)^3 + \cdots, \qquad j = 1, 2, \ldots, 2n \qquad (6.19)$$

The approximate solution of the system can then be generated by truncating the higher order terms. An analytical solution can be developed for the above equation with finite terms in the above equation, as the expression on the right side of equal sign of the above equation is a function of time. If maintain the first four terms of the Taylor series for the fourth order accuracy, the solution can be given in the following form.

$$y_{ji} = c_{ji}e^{a_j\left(t - \frac{[Nt]}{N}\right)} - f_{ji[Nt]/N}\frac{1}{a_j} - f'_{ji[Nt]/N}\frac{1}{a_j}\left(t - \frac{[Nt]}{N} + \frac{1}{a_j}\right)$$

$$-\frac{1}{2}f''_{ji[Nt]/N}\frac{1}{a_j}\left[\left(t - \frac{[Nt]}{N}\right)^2 + 2\frac{1}{a_j}\left(t - \frac{[Nt]}{N} + \frac{1}{a_j}\right)\right]$$

$$-\frac{1}{6}f'''_{ji[Nt]/N}\frac{1}{a_j}\left\{\left(t - \frac{[Nt]}{N}\right)^3 + 3\frac{1}{a_j}\left[\left(t - \frac{[Nt]}{N}\right)^2 \right.\right.$$
$$\left.\left. + 2\frac{1}{a_j}\left(t - \frac{[Nt]}{N} + \frac{1}{a_j}\right)\right]\right\} \qquad (6.20)$$

and

$$\dot{y}_{ji} = a_j c_{ji}e^{a_j\left(t - \frac{[Nt]}{N}\right)} - f'_{ji[Nt]/N}\frac{1}{a_j} - \frac{1}{2}f''_{ji[Nt]/N}\frac{2}{a_j}\left[\left(t - \frac{[Nt]}{N}\right) + \frac{1}{a_j}\right]$$

$$-\frac{1}{6}f'''_{ji[Nt]/N}\frac{3}{a_j}\left\{\left(t - \frac{[Nt]}{N}\right)^2 + \frac{2}{a_j}\left[\left(t - \frac{[Nt]}{N}\right) + \frac{1}{a_j}\right]\right\} \qquad (6.21)$$

where

$$c_{ji} = d_{ji} + \frac{f_{ji[Nt]/N}}{a_j} + \frac{f'_{ji[Nt]/N}}{a_j^2} + \frac{f''_{ji[Nt]/N}}{a_j^3} + \frac{f'''_{ji[Nt]/N}}{a_j^4} \qquad (6.22)$$

With this solution, the actual response of the system governed by the original equation (6.10) can be obtained by simply transform the solution in terms of the principal coordinate back to the variable of the dynamic system via the expression of the principal coordinates used.

One may have noticed from the development above, for all the three types of dynamic systems discussed, the methodology with implementation of the P-T method for MDOF nonlinear problems attempts to maintain the form of the governing equations and the physical information involved in the original governing equations to the utmost level in pursuing the semi-analytical and numerical solutions of the systems. As demonstrated in the previous chapters, the maintenance of the original physical information and the implementation of the piecewise argument lead to the solution with higher accuracy. Moreover, the solution generated per the implementation of the P-T method is continuous in the time interval and over the entire time range considered. Therefore, the solution such developed can also be used as an approximate solution to the exact or accurate solution on the time intervals and over the entire time range desired. Significantly, the accuracy of the solution can be controlled by the number of terms expanded by Taylor series and the value of the parameter N. In numerical calculations, once the local initial conditions in the time interval become available, the solution at the end of the time interval and at any point within the time interval can be directly calculated by the formulae derived in this chapter with employment of the P-T method without the needs of iteration.

6.4. Numerical Solutions via Piecewise Constantization

The following examples may help readers to better understand the methodology described in the previous section for solving the multi-dimensional dynamic systems with implementation of the P-T method.

Example 1. *Nonlinear systems without linear coupling terms in homogeneous equations and those with linear coupling and proportional damping*

As mentioned previously, the procedures of solving the nonlinear oscillation problems without linear coupling terms in homogeneous equations and the nonlinear problems with linear coupling terms and proportional damping terms fall in the same category, however, two transformations are needed for the latter. Following 2-DOF nonlinear oscillation system can be used to demonstrate the procedure of solving the problems of this category with the methodology developed. This nonlinear oscillation system had been used to describe the nonlinear coupling of the pitch and roll modes of the ship motions in regular seas (Nayfeh 1973, 1979).

$$\ddot{x}_1 + 2\mu_1\dot{x}_1 + \omega_1^2 x_1 = \alpha_1 x_1 x_2 + F_1\cos(\omega t + \theta_1) \tag{6.23}$$

$$\ddot{x}_2 + 2\mu_2\dot{x}_2 + \omega_2^2 x_2 = \alpha_2 x_1^2 + F_2\cos(\omega t + \theta_2) \tag{6.24}$$

On the basis of the development described in Section 6.3, formulae can be developed for numerical simulations. Corresponding to equation (6.23), the "external excitation" rearranged for implementation of P-T method and its derivatives for ensuring the 4th order accuracy of the P-T method can be derived as the following.

$$f_{1i,[Nt]/N} = \left\{\alpha_1 x_1 x_2 + F_1\cos(\omega t + \theta_1)\right\}_{[Nt]/N} \tag{6.25}$$

$$f'_{1i,[Nt]/N} = \left\{\alpha_1(\dot{x}_1 x_2 + x_1\dot{x}_2) - F_1\omega\sin(\omega t + \theta_1)\right\}_{[Nt]/N} \tag{6.26}$$

$$f''_{1i,[Nt]/N} = \left\{\alpha_1(G_{11}x_2 + 2\dot{x}_1\dot{x}_2 + x_1 G_{12}) - F_1\omega^2\cos(\omega t + \theta_1)\right\}_{[Nt]/N} \tag{6.27}$$

$$f'''_{1i,[Nt]/N} = \left\{\begin{array}{l}\alpha_1(G_{21}x_2 + 3G_{11}\dot{x}_2 + 3\dot{x}_1 G_{12} + x_1 G_{22}) \\ + F_1\omega^3\sin(\omega t + \theta_1)\end{array}\right\}_{[Nt]/N} \tag{6.28}$$

and similarly, for equation (6.24)

$$f_{2i,[Nt]/N} = \left\{\alpha_2 x_1^2 + F_2\cos(\omega t + \theta_2)\right\}_{[Nt]/N} \tag{6.29}$$

$$f'_{2i,[Nt]/N} = \left\{ 2\alpha_2 \left(x_1 \dot{x}_1 \right) - F_2 \omega \sin(\omega t + \theta_2) \right\}_{[Nt]/N} \qquad (6.30)$$

$$f''_{2i,[Nt]/N} = \left\{ 2\alpha_2 \left(\dot{x}_1^2 + G_{11} x_1 \right) - F_2 \omega^2 \cos(\omega t + \theta_1) \right\}_{[Nt]/N} \qquad (6.31)$$

$$f'''_{2i,[Nt]/N} = \left\{ 2\alpha_2 \left(3G_{11}\dot{x}_1 + G_{21} x_1 \right) + F_2 \omega^3 \sin(\omega t + \theta_2) \right\}_{[Nt]/N} \qquad (6.32)$$

and in all the equations above

$$G_{11} = \alpha_1 x_1 x_2 + F_1 \cos(\omega t + \theta_1) - 2\mu_1 \dot{x}_1 - \omega_1^2 x_1 \qquad (6.33)$$

$$G_{12} = \alpha_2 x_1^2 + F_2 \cos(\omega t + \theta_2) - 2\mu_2 \dot{x}_2 - \omega_2^2 x_2 \qquad (6.34)$$

$$G_{21} = f_1' - 2\mu_1 G_{11} - \omega_1^2 \dot{x}_1 \qquad (6.35)$$

$$G_{22} = f_2' - 2\mu_2 G_{12} - \omega_2^2 \dot{x}_2 \qquad (6.36)$$

With these formulae, the numerical solution for the system can be obtained with the following equations in the time interval $[Nt]/N \leq t \leq ([Nt]+1)/N$.

$$x_{ji} = e^{-\mu_j \left(t - \frac{[Nt]}{N} \right)} \left\{ B_{j1} \cos\left[\xi_j \left(t - \frac{[Nt]}{N} \right) \right] + B_{j2} \sin\left[\xi_j \left(t - \frac{[Nt]}{N} \right) \right] \right\}$$

$$+ A_{j1} + A_{j2}\left(t - \frac{[Nt]}{N} \right) + A_{j3}\left(t - \frac{[Nt]}{N} \right)^2 + A_{j4}\left(t - \frac{[Nt]}{N} \right)^3$$

$$i = 1, 2, 3, ..., \qquad j = 1, 2 \qquad (6.37)$$

where

$$\xi_j^2 = \omega_j^2 - \mu_j^2$$

$$A_{j4} = \frac{1}{6\omega_j^2} f'''_{ji,[Nt]/N}$$

$$A_{j3} = \frac{1}{\omega_j^2}\left(\frac{1}{2} f''_{ji,[Nt]/N} - 6\mu_j A_{j4}\right)$$

$$A_{j2} = \frac{1}{\omega_j^2}\left(f'_{ji,[Nt]/N} - 4\mu_j A_{j3} - 6A_{j4}\right)$$

$$A_{j1} = \frac{1}{\omega_j^2}\left(f_{ji,[Nt]/N} - 2\mu_j A_{j2} - 2A_{j3}\right)$$

$$B_{j1} = x_{ji}\left(\frac{[Nt]}{N}\right) - A_{j1}$$

$$B_{j2} = \frac{1}{\xi_j}\left(\dot{x}_{ji}\left(\frac{[Nt]}{N}\right) + \mu_j B_{j1} - A_{j2}\right)$$

To determine the numerical solution for the system over the entire time range considered, i.e., $t \geq 0$, the solutions of all the time intervals are combined with satisfaction of the following conditions of continuity.

$$x_{ji}\left(\frac{[Nt]}{N}\right) = x_{ji-1}\left(\frac{[Nt]}{N}\right) \quad \text{and} \quad \dot{x}_{ji}\left(\frac{[Nt]}{N}\right) = \dot{x}_{ji-1}\left(\frac{[Nt]}{N}\right) \quad (6.38)$$

The behaviors of the 2-DOF nonlinear system with the quadratic nonlinearity can thus be investigated on the basis of the solutions. Numerical simulation for this system is carried out with the solution developed.

Some results of free and forced oscillatory response of the system are illustrated in Figure 6.1 and Figure 6.2. For the free system, Figure 6.1(a) shows that, in the absence of damping, the energy in the system continuously exchange between the two modes of oscillation. In the presence of damping, however, Figure 6.1(b) shows that the energy still continues to be exchanged between the two modes but is gradually dissipating. For the forced oscillation of the 2-DOF system, one may clearly see the saturation phenomenon. As can be observed from Figure 6.2(a) and (b), the displacement x2 increases as F2 increases from zero, while x1 remains zero.

When F_2 reaches 0.31, x_2 reaches its maximum value and further increase in F_2 will not produce further increase in x_2 and will cause x_1 to increase, as shown in Figure 6.2(c) and (d). The x_2 mode is saturated as expected. The other behaviors of this system can also be simulated with the solutions developed. The solutions developed match well with that presented in the references (Nayfeh 1979, Lee and Hsu 1994).

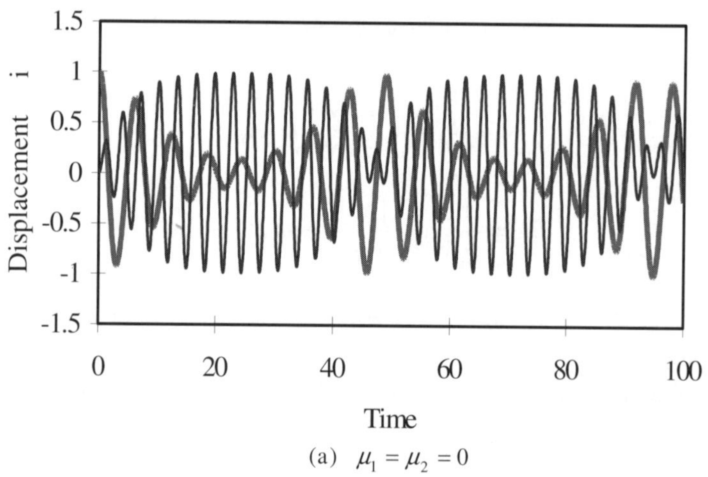

Time

(a) $\mu_1 = \mu_2 = 0$

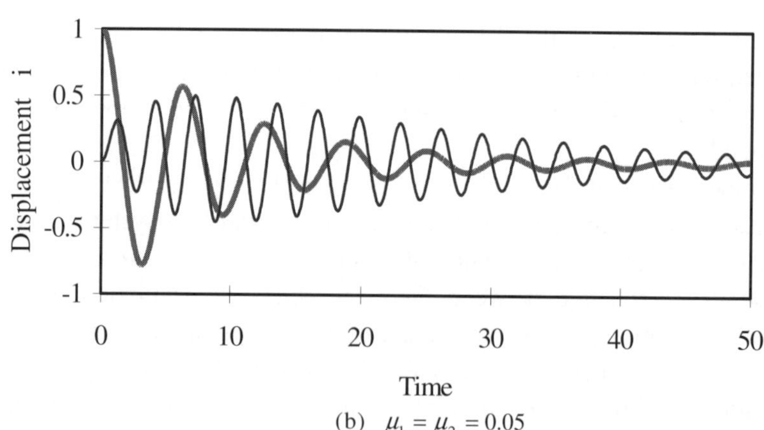

Time

(b) $\mu_1 = \mu_2 = 0.05$

Figure 6.1.   Free oscillations of 2-DOF system with quadratic; non-linearity ———— x_1; ———— x_2, $F_1 = F_2 = 0$, $\omega_1 = 1$, $\omega_2 = 2\omega_1$, $\alpha_1 = 0.5$, $\alpha_2 = 1$, $x_1(0) = 1$, $x_2(0) = 0$.

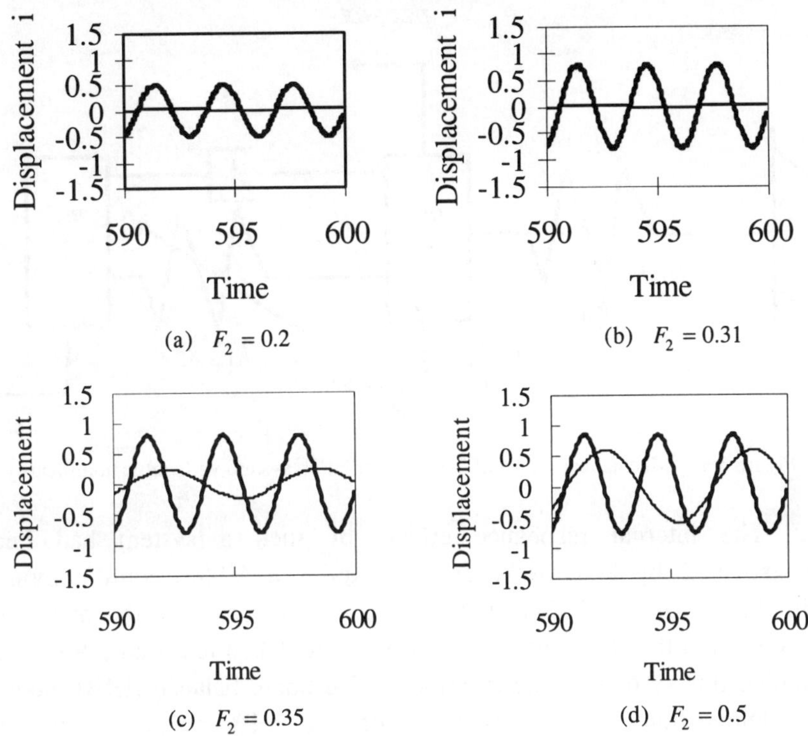

(a) $F_2 = 0.2$

(b) $F_2 = 0.31$

(c) $F_2 = 0.35$

(d) $F_2 = 0.5$

Figure 6.2.   Forced oscillations of 2-DOF system with quadratic; x_1 ———— , x_2 ▬▬▬ non-linearity, $F_1 = 0, \omega_1 = 1, \omega_2 = 2\omega_1, \omega = \omega_2, \alpha_1 = 0.5, \alpha_2 = 1, \mu_1 = \mu_2 = 0.1$.

Example 2. *Nonlinear systems with linear coupling and general damping terms*

P-T method can also be employed for solving the nonlinear dynamic problems with linear coupling and general damping forms with the formulae developed above. Consider a two-degree-of-freedom system consisting of two concentrated masses and two springs with a linear damper, under a harmonic excitation as shown in Figure 6.3. One of the springs is linear with the stiffness coefficient k_{10} and another one is a cubic nonlinear spring. The restoring force is defined as

$$f = k_{12}(x_1 - x_2) + k_3(x_1 - x_2)^3 \qquad (6.39)$$

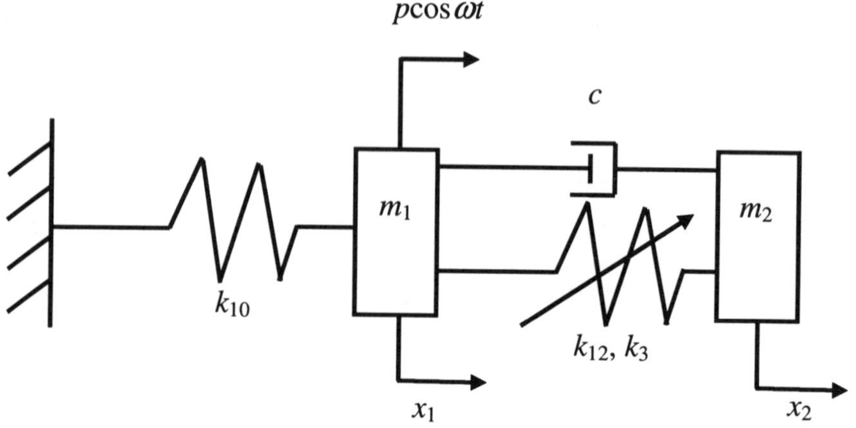

Figure 6.3.　Mechanical model of 2-DOF oscillation system with cubic nonlinearity.

The internal resonance effects of such a system had been investigated by Bajkowski and Szemplinska (1986) with computer simulation and analytical approaches, on the basis of the averaging method and the Ritz method. Later, Cheung, Chen and Lau (1990) also studied this system by the incremental harmonic balance (IHB) method to demonstrate the effectiveness of the IHB method. With the characteristics of this system, it makes a good example for elucidating the application of the methodology developed to solve for the nonlinear multi-degree-of-freedom problems with coupled linear terms and general damping terms.

The governing equations of the system can be expressed in the following matrix form.

$$\begin{bmatrix} m_1 & 0 \\ 0 & m_2 \end{bmatrix}\begin{bmatrix} \ddot{x}_1 \\ \ddot{x}_2 \end{bmatrix} + \begin{bmatrix} c & -c \\ -c & c \end{bmatrix}\begin{bmatrix} \dot{x}_1 \\ \dot{x}_2 \end{bmatrix} + \begin{bmatrix} k_{10}+k_{12} & -k_{12} \\ -k_{12} & k_{12} \end{bmatrix}\begin{bmatrix} x_1 \\ x_2 \end{bmatrix}$$
$$= \begin{bmatrix} p\cos \omega t - k_3\left(x_1 - x_2\right)^3 \\ k_3\left(x_1 - x_2\right)^3 \end{bmatrix} \tag{6.40}$$

In the above equation, x_1 and x_2 are the displacements of the concentrated masses m_1 and m_2, k_{10}, k_{12}, k_3, are the coefficients of linear stiffness of the springs, c is the coefficient of damping, p designates the amplitude of the external excitation, and ω denotes the frequency of the excitation.

Based on the development in Section 6.3.3, the matrices as expressed in Equation (6.14) can be given by

$$\tilde{M} = \begin{bmatrix} 0 & 0 & \gamma & 0 \\ 0 & 0 & 0 & 1 \\ \gamma & 0 & \zeta & -\zeta \\ 0 & 1 & -\zeta & \zeta \end{bmatrix} \tag{6.41}$$

$$\tilde{K} = \begin{bmatrix} -\gamma & 0 & 0 & 0 \\ 0 & -1 & 0 & 0 \\ 0 & 0 & \omega_{20}^2(k^2\gamma+1) & -\omega_{20}^2 \\ 0 & 0 & -\omega_{20}^2 & \omega_{20}^2 \end{bmatrix} \tag{6.42}$$

$$Z = \begin{bmatrix} \dot{x}_1 \\ \dot{x}_2 \\ x_1 \\ x_2 \end{bmatrix} \tag{6.43}$$

$$\tilde{G} = \begin{bmatrix} 0 \\ 0 \\ F\cos\omega t - \mu\omega_{20}^2(x_1-x_2)^3 \\ \mu\omega_{20}^2(x_1-x_2)^3 \end{bmatrix} \tag{6.44}$$

where the normalized parameters are defined as follows.

$$\gamma = \frac{m_1}{m_2}, \quad \mu = \frac{k_3}{k_{12}}, \quad k^2 = \sqrt{\frac{k_{10}/m_1}{k_{12}/m_2}}, \quad \omega_{20}^2 = \frac{k_{12}}{m_2}, \quad \zeta = \frac{c}{m_2}, \quad F = \frac{p}{m_2} \tag{6.45}$$

The coefficients corresponding to that in equations from (6.46) to (6.22) on an ith time interval $[Nt]/N \le t \le ([Nt]+1)/N$ can be expressed as the following.

$$f_{ji,[Nt]/N} = \left\{ P_{3j}F\cos\omega t - (P_{3j}-P_{4j})\mu\omega_{20}^2 G_{j1}^3 \right\}_{[Nt]/N} \tag{6.46}$$

$$G_{j1} = \sum_{k=1}^{4}\left[\left(P_{3k} - P_{4k}\right)y_k\right] \tag{6.47}$$

$$f'_{ji,[Nt]/N} = \left\{-P_{3j}F\omega\sin\omega t - \left(P_{3j} - P_{4j}\right)\mu\omega_{20}^{2}\left[3G_{j1}^{2}G_{j2}\right]\right\}_{[Nt]/N} \tag{6.48}$$

$$G_{j2} = \sum_{k=1}^{4}\left[\left(P_{3k} - P_{4k}\right)\dot{y}_k\right] = \sum_{k=1}^{4}\left[\left(P_{3k} - P_{4k}\right)\left(-a_k y_k + f_k\right)\right] \tag{6.49}$$

$$f''_{ji,[Nt]/N} = \left\{\begin{array}{l} -P_{3j}F\omega^{2}\cos\omega t \\[2mm] -\left(P_{3j} - P_{4j}\right)\mu\omega_{20}^{2}\left[6G_{j1}G_{j2}^{2} + 3G_{j1}^{2}G_{j3}\right] \end{array}\right\}_{[Nt]/N} \tag{6.50}$$

$$G_{j3} = \sum_{k=1}^{4}\left[\left(P_{3k} - P_{4k}\right)\ddot{y}_k\right]$$

$$= \sum_{k=1}^{4}\left\{\left(P_{3k} - P_{4k}\right)\left[-a_k\left(-a_k y_k + f_k\right) + f'_k\right]\right\} \tag{6.51}$$

$$f'''_{ji,[Nt]/N} = \left\{\begin{array}{l} P_{3j}F\omega^{3}\sin\omega t - \left(P_{3j} - P_{4j}\right)\mu\omega_{20}^{2} \\[2mm] \left[6G_{j2}^{3} + 18G_{j1}G_{j2}G_{j3} + 3G_{j1}^{2}G_{j4}\right] \end{array}\right\}_{[Nt]/N} \tag{6.52}$$

$$G_{j4} = \sum_{k=1}^{4}\left[\left(P_{3k} - P_{4k}\right)\dddot{y}_k\right]$$

$$= \sum_{k=1}^{4}\left(\left[P_{3k} - P_{4k}\right]\left\{-a_k\left[-a_k\left(-a_k y_k + f_k\right) + f'_k\right] + f''_k\right\}\right) \tag{6.53}$$

With the solutions developed above, numerical simulations can be performed conveniently. The influence of different system parameters on the nonlinear behaviors of the system is studied. The oscillation amplitudes of both of the concentrated masses under different frequencies of excitation are calculated with the solutions expressed above. The oscillatory amplitudes of the masses are plotted in Figures 6.4, 6.5 and 6.6, the curves are for the responses of the system at its steady state. The numerical values used for the simulations are as the following: $\zeta = 0.2$, $\gamma = 1.2$, $k^2 = 6.95$, and $\omega_{20}^2 = 2$.

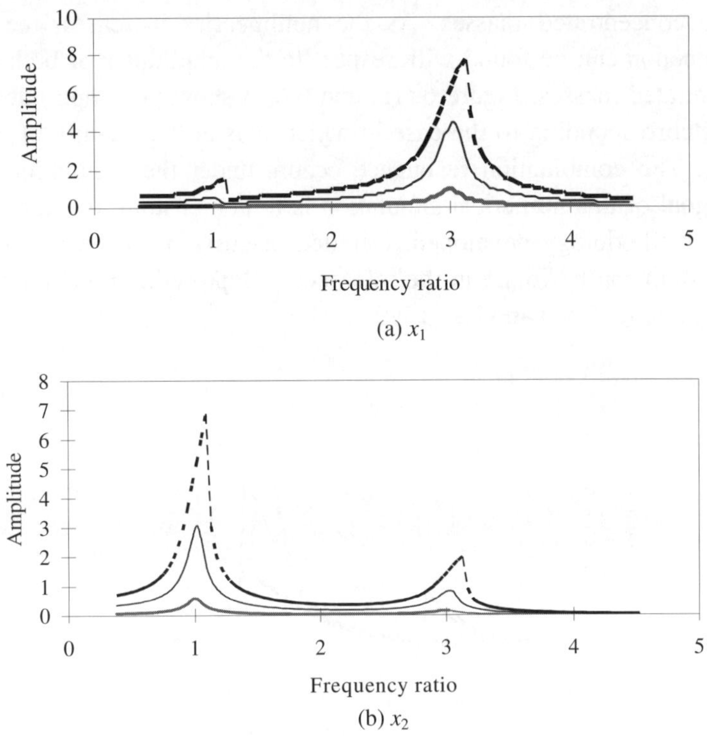

(a) x_1

(b) x_2

Figure 6.4.   Amplitude-frequency curves, $\mu = 0.01$; ▬▬ $F = 1$, ▬▬ $F = 5$, •••••• $F = 10$.

The parameters used in the numerical calculations are so selected that the second natural frequency, $\omega_2 = 3.978$, is three times of the first natural frequency, $\omega_1 = 1.326$. With these parameters, an internal resonance can be expected. The horizontal axes of these figures show the frequency ratio of the excitation frequency ω, to the first natural frequency, ω_1, of the system. Figure 6.4 illustrates the amplitudes of both of the concentrated masses corresponding to several values of the amplitude of the external excitation. As can be observed from the figure, the primary resonances occur when the excitation frequency is close to that of the first or the second natural frequency of the system. Figure 6.5 exhibits the influence of the nonlinearity of the system which consists of a spring of nonlinear stiffness. The nonlinearity represented by the parameter μ is plotted in the figure with respect to the amplitudes of both of the concentrated masses. As the nonlinearity increases, the jump phenomenon can be found with respect to the amplitudes of both of the concentrated masses. Figure 6.6(a) and 6.6(b) show the amplitudes of x_1 and x_2 corresponding to the case in which ω is at the vicinity of $2\omega_1$ or ω_2-ω_1. The combination resonance occurs under these situations. The main goal of this numerical simulation is to demonstrate the application of the methodology developed. Detailed discussion on the behavior of this system can be found in the references (Bajkowski and Szemplinska 1986, Cheung, Chen and Lau 1990).

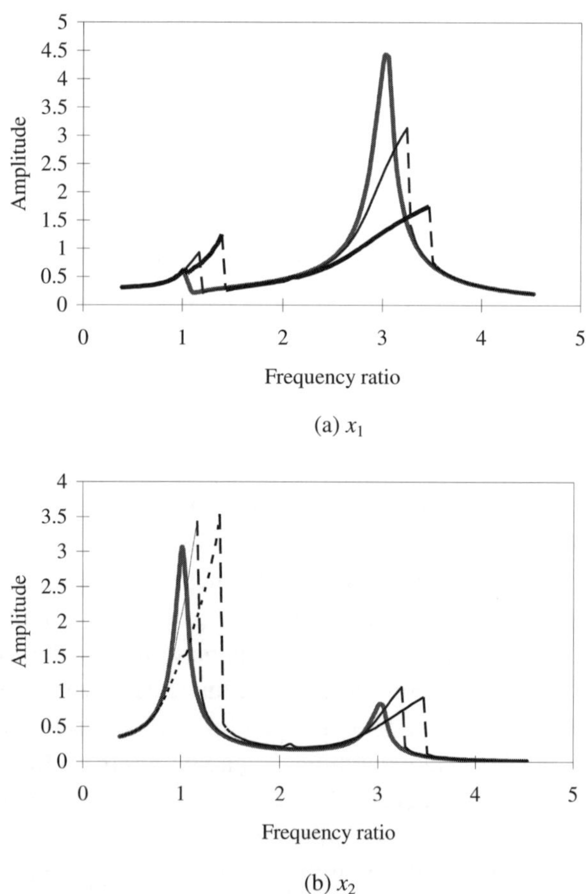

(a) x_1

(b) x_2

Figure 6.5.　Amplitude-frequency curves, $F = 5$; Nonlinearity: ▬▬ 0.01, ▬▬0.1, ▪▪▪▪▪▪ 0.5.

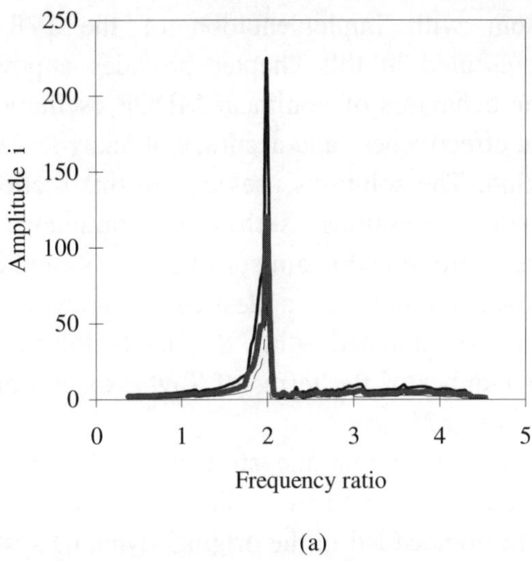

(a)

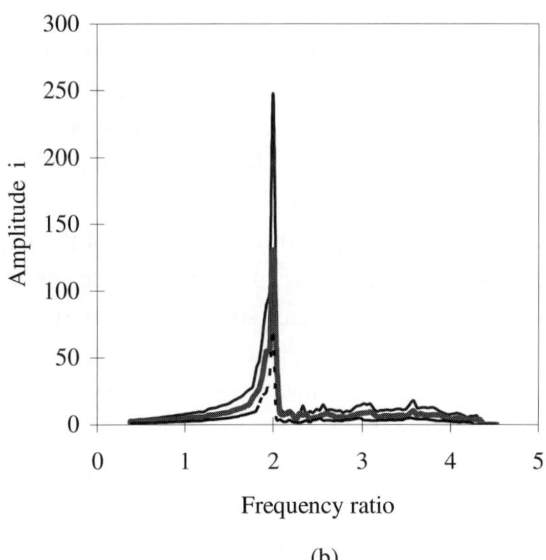

(b)

Figure 6.6.   Amplitude-frequency curves; (a) x_1; (b) x_2; ________ Nonlinearity: 0.1, $F = 50$; ________ Nonlinearity: 0.3, $F = 30$; .... Nonlinearity: 0.9, $F = 15$.

The examples discussed above demonstrate the procedures for numerically simulating the responses of nonlinear systems of multi-degree-of-freedom with implementation of the P-T method. The methodology presented in this chapter provides a powerful tool for investigating the behaviors of nonlinear MDOF oscillation systems and in validating the effectiveness and accuracy of analytical solutions of the systems such kind. The solutions provided in this section may also be used as approximate solutions as they are continuous on each time intervals and the entire time domain considered. As can be seen from the discussions above, solutions of a desired accuracy of the numerical calculations can be obtained with the methodology developed by utilizing proper number of the terms of Taylor expansion and a proper value of the parameter N.

It should be noticed that the strategy of the approach with the procedures presented above is to maintain as much as possible the physical properties embedded in the original dynamic system, in solving the multi-degree-of-freedom nonlinear dynamic problems with utilization of the P-T method. Reliable numerical and semi-analytical solutions with higher accuracy can be expedited. Same as that of single dimensional systems discussed in the previous chapters, there is a linear continuous dynamic system corresponding to each of the time intervals for the multi-degree-of-freedom nonlinear dynamic systems. In utilizing the methodology of implementing the P-T method, therefore, one should first of all arrange the terms in the governing equations of the dynamic system in such a way that as many as possible the terms are remaining unchanged to which an analytical solution is available in the time interval. In performing this, only those terms that may cause difficulties in obtaining the analytical solutions to the governing equation in the time interval should be manipulated.

The common nonlinear dynamic systems such as systems without linear coupling terms; the systems with linear coupling terms and proportional damping terms; and the systems with linear coupling terms and general damping terms are discussed in this chapter with the solution developments and numerical simulations. However, the MDOF nonlinear systems of the other types may also be solved numerically and semi-analytically through the similar procedures as described in the previous sections. All in all, the governing equations of the MDOF systems should be rearranged to the form of equations (6.7) or (6.13) for the P-T method to be applied. The more terms in the governing equations maintain unchanged, the higher accuracy will be the solutions.

References

Bajkowski, J. and Szemplinska-Stupnicka, W., "Internal Resonance Effects-Simulation versus Analytical Methods Results," Journal of Sound and Vibration, Vol. 104, pp. 259-275, 1986.

Cheung, Y. K., Chen, S. H. and Lau, S. L., "Application of the Incremental Harmonic Balance Method to Cubic Non-Linearity Systems," Journal of Sound and Vibration, Vol. 140, No. 2, pp. 273-286, 1990.

Dai, L. and Singh, M. C., "A New Approach with Piecewise Constant Arguments to Approximate and Numerical Solutions of Oscillatory Problems," Journal of Sound and Vibration, Vol. 263, No. 3, pp. 535-548, 2003.

Dai, L, Xu, L and Han, Q, "Semi-Analytical and Numerical Solutions of Multi-Degree-of-Freedom Nonlinear Oscillation Systems with Linear Coupling," Communications in Nonlinear Science and Numerical Simulations, Vol. 11, pp. 831-844, 2006.

Friedman, M. and Kandel, A., Fundamentals of Numerical Analysis, CRC Press, London, 1994.

Iserles, A., Ramaswami, G. and Sofroniou M., "Runge-Kutta Methods for Quadratic Ordinary Differential Equations," BIT—Numerical Mathematics, Vol. 38, pp. 315-336, 1998.

Kapitaniak, T., Controlling Chaos: Theoretical and Practical Methods in Nonlinear dynamics, New York, Academic, 1996.

Kelly, S. G, Fundamentals of Mechanical Vibrations, McGraw-Hill, New York, 1993.

Ku, Y.H., Analysis and Control of Nonlinear Systems, Ronald Press, New York, 1958.

Luo, A. C. J. and Han, R. P. S., "A Quantitative Stability and Bifurcation Analyses of a Generalized Duffing Oscillator with Strong Nonlinearity," Journal of Franklin Institute," Vol. 334B, pp. 447-459, 1997.

Lee, W. K. and Hsu, C. S., "A Global Analysis of an Harmonically Excited Spring-Pendulum System with Internal Resonance," Journal of Sound and Vibration, Vol. 171, No. 3, pp. 335-359, 1994.

Nayfeh, A. H., Mook, D. T. and Marshall, L. R., "Nonlinear Coupling of Pitch and Roll Modes in Ship Motions," Journal of Hydronautics, Vol. 7, No. 4, pp. 145-152, 1973.

Nayfeh, A. H. and Mook, D. T., Nonlinear Oscillations, New York, Wiley-Interscience, 1979.

Nayfeh, A. H., Introduction to Perturbation Techniques, New York, John Wiley & Sons, 1981.

Strogatz, S. H., Nonlinear Dynamics and Chaos: with Applications to Physics, Biology, Chemistry, and Engineering, Westview Press, Cambridge, MA, 2000.

Weaver, W. Jr, Timoshenko, S. and Young, D. H., Vibration Problems in Engineering, 5th Ed., New York, John Wiley & Sons, 1990.

Zingg, D. W. and Chisholm, T. T., "Runge-Kutta Methods for Linear Ordinary Differential Equations," Applied Numerical Mathematics, Vol. 31, pp. 227-238, 1999.

CHAPTER 7

Periodicity-Ratio and Its Application in Diagnosing Irregularities of Nonlinear Systems

7.1. Introduction

In the area of nonlinear dynamics, the behaviors of a nonlinear dynamic system can be roughly classified into two categories of regular and irregular motions. Irregular motion of a dynamic system may include nonperiodic and chaotic responses. Among the irregular responses, chaotic responses of nonlinear systems are probably the most attractive phenomena among the new observations in the last decades. Chaos is known (Baker and Gollub 1990, Moon 1987) as "randomness" that is a result of the sensitivity of nonlinear systems to the initial conditions. In nonlinear dynamics, chaotic response is defined as the emergence of random like behavior generated from completely deterministic systems (Guckenheimer and Holmes 1983, Thompson and Stewart 1986). It is recognized (Guckenheimer and Holmes 1983) that dynamical systems governed by nonlinear differential and nonlinear difference equations can admit bounded, nonperiodic or random like motions even though no random quantities appear in the equations of motion. In other words, a nonlinear deterministic system may behavior chaotically under regular such as periodic excitations. It is also recognized, chaos can only occur in nonlinear dynamical systems (Baker and Gollub 1990, Moon 1987) to which analytical solutions are, in general, difficult to obtain. Investigation of chaotic behavior in the nonlinear systems therefore has mostly to rely upon approximate or numerical techniques, as described in Chapter 2.

233

In studying the properties of motion of a nonlinear dynamical system, graphical analysis with mathematical constructs is inevitable. Some mathematical constructs that are very useful in the study of linear, nonlinear and chaotic vibrations will first be discussed and then used in the consequent sections. Behavior of motion of nonlinear systems, such as periodicity, nonperiodicity and chaos is to be investigated with the help of some mathematical constructs.

In the studies of nonlinear dynamic systems, which are usually governed by nonlinear second-order differential equations or a system of differential equations, the criteria for distinguishing a chaotic motion from regular motions are crucial. Such criteria are also important for diagnosing the nonlinear and regular behaviors of the system to which a mathematical model is hard to develop or hard to solve but numerical or experimental data for the behaviors of the system are available. Moreover, the techniques providing high efficiency and accuracy in diagnosing and quantifying chaos are always demanded for investigating nonlinear dynamic systems especially the nonlinear systems in the fields of science and engineering.

Regular motion of a system subjected to periodic exertions is usually periodic. In contrast with regular motions, final states of chaotic vibrations are extremely nonperiodic. In this chapter, a criterion named Periodicity-Ratio is to be developed for diagnosing the behavior of the nonlinear systems, on the basis of the response of the systems and the corresponding mathematical constructs. The detailed description on the derivation of the Periodicity-Ratio together with the development of a methodology for diagnosing the irregular motions from the regular motions of a dynamic system will be presented. It will be shown in the following sections; the Periodicity-Ratio describes the degree of periodicity of motion and can be conveniently used to distinguish a nonperiodic motion from a regular vibration or oscillation and to diagnose whether or not a motion is chaotic without plotting any figures. The analyses on the behavior of nonlinear dynamic systems with the implementation of the Periodicity-Ratio will be demonstrated.

With employment of the Periodicity-Ratio, the diagrams which may be used to graphically describe the regions of periodic, quasiperiodic and chaotic behaviors of a system can be constructed allowing simultaneous comparison of the initial conditions and the varying parameters of nonlinear dynamical systems. A simple pendulum is used for introducing the concept of Periodicity-Ratio as it is rich in nonlinear characteristics even for undamped pendulum (Lu 2007). This pendulum will be further used for demonstrating the applications of the Periodicity-Ratio in determining the periodic and chaotic regions for the Duffing's equation that governs the pendulum system. The generation of the periodic-chaotic region diagrams with Periodicity-Ratio will be presented for Duffing's equation which represents the widely used system in nonlinear dynamics (Yamapi and Aziz-Alaoui 2007; Lazzouni *et al.* 2007). The diagrams of periodic and chaotic regions generated on the basis of periodicity of the motions provide more detailed information in comparison with the results given by Ueda (1980, 1979). The procedures of generating the data for the periodic chaotic region diagrams is presented which includes numerical solutions of dynamical systems, calculation of the points for Poincare maps, distinguishing quasiperiodic motions from chaotic vibrations, collection of data for periodic chaotic region diagrams corresponding to varying parameters of dynamical systems and initial conditions.

Based on the current literature, Lyapunov-Exponent is probably the most popular criterion used for determining whether a system is periodic or chaotic. However, it may not provide reliable results for all the cases in nonlinear systems. Moreover, Lyapunov-Exponent merely predicts whether a system is chaotic or periodic. For the behavior of a system that is neither periodic nor chaotic, Lyapunov-Exponent is not applicable. With an approach different from that of Lyapunov-Exponent, Periodicity-Ratio shows some advantages in implementing in nonlinear dynamics analyses. In this chapter, a thorough and systematic investigation on the characteristics of the Periodicity-Ratio and its implementation in the analyses of nonlinear dynamic systems is to be performed. A detailed comparison of the characteristics between the Periodicity-Ratio and Lyapunov-Exponent will also be presented.

Strictly speaking, quasiperiodic motion of a dynamics system is not periodic. However, it is not an extremely nonperiodic motion like chaos either. In analyzing the behaviors of a nonlinear system, quasiperiodic motion must be distinguished from periodic and chaotic motions. A numerical technique for identifying quasiperiodic motions and differentiating them from the other nonperiodic motions is introduced.

In this chapter, nonlinear behavior especially the chaotic behavior of dynamical systems will be studied numerically with the implementation of the Periodicity-Ratio. All the numerical calculations employed in solving for the nonlinear dynamic systems involved are performed with the piecewise constant procedure and P-T method presented in the previous chapters, due to their accuracy and reliability in numerical simulations.

7.2. Phase Trajectories of Periodic, Nonperiodic and Chaotic Behavior of Nonlinear Systems

Many analytical and computational tools may be employed in studying linear and nonlinear dynamical systems. Among these tools, the mathematical constructs such as phase trajectories and Poincare maps are very helpful graphical devices in analyzing the behavior of motion for dynamical systems. Some of the most useful mathematical constructs will be discussed in this section to provide an insight into the behavior of nonlinear dynamical systems, especially the chaos of completely deterministic systems.

In analyzing the linear and nonlinear vibration problems discussed so far, the response of a dynamical system is illustrated on the basis of observation of the amplitude of motion with time. The wave forms or time history of the motion provided by an analytical solution or numerical simulation may show periodic or nonperiodic behavior. A time history of a motion illustrates certain visible behaviors of a dynamic system. For instance, the wave form representing the motion of the continuous system in Figure 3.7, is periodic, and on the contrary, the motion exhibits no visible pattern, such as that of the piecewise constant system in the same figure. The wave form shown in Figure 3.22 demonstrates extremely nonperiodic pattern and it is a good example of the wave form of a chaotic case.

In addition to the observation of time history, there are many other tools available in graphical analysis of the properties of motion for linear and nonlinear dynamical systems. Among them the form of a phase trajectory in phase space is one of the most important tools. With employment of phase trajectories, the states of linear and nonlinear motions may be displayed and the periodic and nonperiodic motions of dynamical systems may be distinguished graphically. In a two dimensional space, the state of a one dimensional vibration system can be specified by its position x and first derivative or velocity $\dot{x}$ in a phase plane; hence, its phase trajectory is defined as a planar figure in the $x - \dot{x}$ plane. Graphically, when a motion of a dynamical system is exhibited in the three dimensional $t - x - \dot{x}$ space as shown in Figure 7.1, the corresponding phase trajectory can be obtained by projecting the spiral orbits in the $t - x - \dot{x}$ space onto the $x - \dot{x}$ plane.

If a motion of a dynamic system is perfectly periodic, obviously, the orbits projected on the $x - \dot{x}$ plane will be overlapped completely. At a steady state, the phase trajectory corresponding to a periodic motion is therefore a closed curve with a fixed shape no matter how large the time range is covered in the steady state. Figure 7.2 shows an example of a steady state phase trajectory of a periodic vibration.

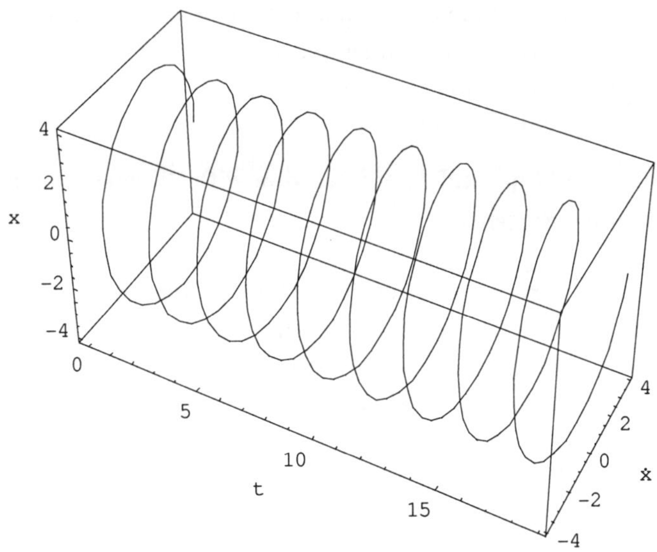

Figure 7.1.   Visualization of a motion of a dynamical system in $x - \dot{x} - t$ space.

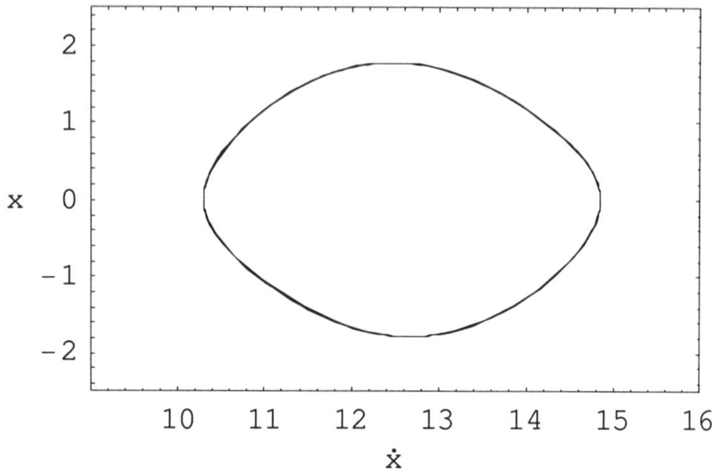

Figure 7.2.   Phase trajectory for $\ddot{x} + \theta\dot{x} + \omega^2 \sin x = A\cos\Omega t$, $\theta = 0.5$, $\omega = 1$, $A = 0.8$, $\Omega = 2/3$, $d_0 = 3$, $v_0 = 4$.

A phase trajectory of nonperiodic motion, on the other hand, will not form a closed curve. Extreme nonperiodic or chaotic motions have phase trajectories that tend to fill up a section of the phase space, though the trajectories may be bounded. A phase trajectory of chaotic motion is illustrated in Figure 5.4. One may observe from the figure that there is almost no curve overlapping with the others.

7.3.   Poincare Maps and Their Relation with Piecewise Constant Dynamic Systems

Although the wandering of orbits in phase trajectories is a clue to nonperiodic or chaotic motions, the information provided by the continuous wave forms and phase trajectories is usually not sufficient enough in analyzing the nonlinear dynamical systems, especially for the case of chaos. One may then employ a modified phase trajectory called Poincare map. A Poincare map is a device invented by Henri Poincare (Poincare 1946). Poincare map simplifies phase trajectories of a complicated dynamical system and demonstrate the main characteristics of the response of the system. Poincare map can be constructed by viewing a phase trajectory stroboscopically in such a way that the motion of the system is observed periodically. The construction of a Poincare map consists of sectioning the spiral orbits shown in Figure 7.1 at a regular time interval and projecting the points of intersection of the orbits at the sections on the $x - \dot{x}$ plane. As such, the intersected points, instead of curves, are shown on the phase plane as a stroboscopic picture, the Poincare map.

As an example, consider the following nonlinear driven pendulum governing equation:

$$\ddot{x}(t) + \theta \dot{x}(t) + \omega^2 \sin x(t) = A \cos \Omega t \qquad (7.1)$$

The motion corresponding to the pendulum may either be periodic or nonperiodic. For a periodic motion at a steady state with the phase trajectory as shown in Figure 7.3, the sectioning is done at intervals corresponding to the period of the external force, the Poincare map or stroboscopic picture shows a finite number of points as illustrated in Figure 7.4. Obviously, Poincare map corresponding to the system with phase trajectory shown in Figure 7.2 may have only one point visible since the motion will always come back to the same coordinate in the phase plane as t is increased by a period and the motion is perfectly periodic.

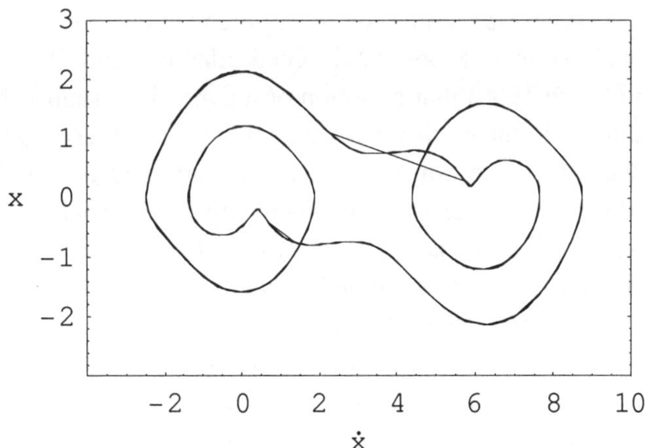

Figure 7.3. Phase trajectory $\ddot{x} + \theta \dot{x} + \omega^2 \sin x = A \cos \Omega t$, $\theta = 0.5$, $\omega = 1$, $A = 1.098$, $\Omega = 2/3$, $d_0 = -2.5$, $v_0 = 0$.

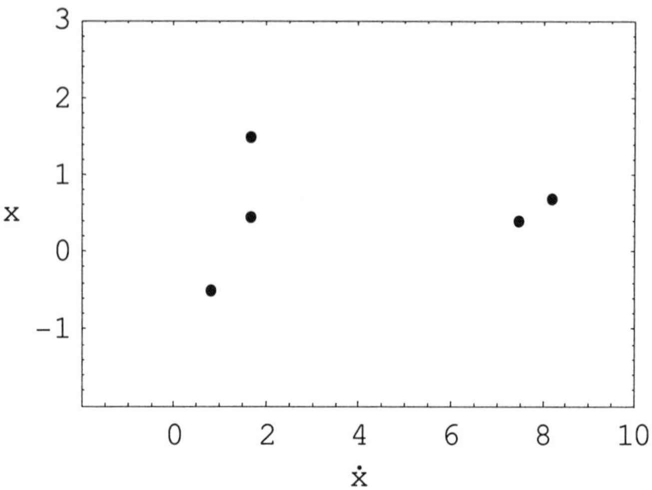

Figure 7.4.   Poincare map corresponding to the phase trajectory in Figure 7.3.

In comparing with the information provided by Figures 7.3 and 7.4, it can be seen that the information illustrated in Figure 7.4 is much simpler. If only the state of motion of nonlinear system is concerned, a Poincare map is clear and handy. One may also gain from the Poincare map Figure 7.4 that the period of this periodic case is 5T, where T is either the period of the external excitation acting on the dynamic system or the period used to sectioning for generating the Poincare map.

It is well known (Moon 1987, Guckenheimer and Holmes 1983, Baker and Gollub 1990) that a motion of a dynamic system is harmonic or periodic if the Poincare map corresponding to the motion consists of a finite number of visible points. In fact, Poincare map is a widely used tool in nonlinear dynamics to distinguish between a periodic motion and a nonperiodic motion or chaos. By definition, chaos is described by some authors (Moon 1987, Thompson and Stewart 1986, Hoppensteadt 1993) as a paradoxied combination of randomness and structures; the irregular, sensitive to initial conditions and unpredictable time evolution of nonlinear systems. Although chaos is not a precisely defined concept (Hoppensteadt 1993), it is used here in this book to indicate the behavior of a motion that is highly irregular and usually unpredictable for a longer period of time. The Poincare maps of chaotic motions often appear as a cloud of unorganized points in phase space such as those shown in Figure 7.5.

If, however, the Poincare map appears in the phase plane can be connected as a continuous smooth curve or continuous closed curve, such as the curve in the Poincare map of Figure 7.6, the corresponding motion is then considered as quasiperiodic (Moon 1987). As there are a large number of points of quasiperiodic motion appearing in phase diagram (theoretically infinite) and they are not overlapping in general, the quasiperiodic motions should be distinguished from chaos.

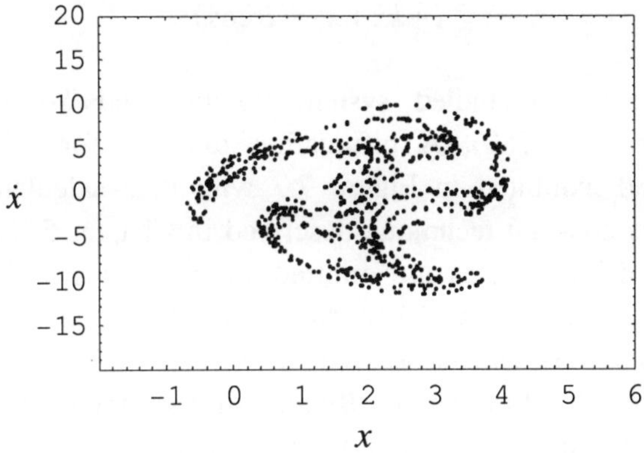

Figure 7.5. Poincare map of chaos for $\ddot{x} + k\dot{x} + x^3 = B\cos t$ as $k = 0.5$, $B = 23.9$, $d_0 = -2$, $v_0 = 0$.

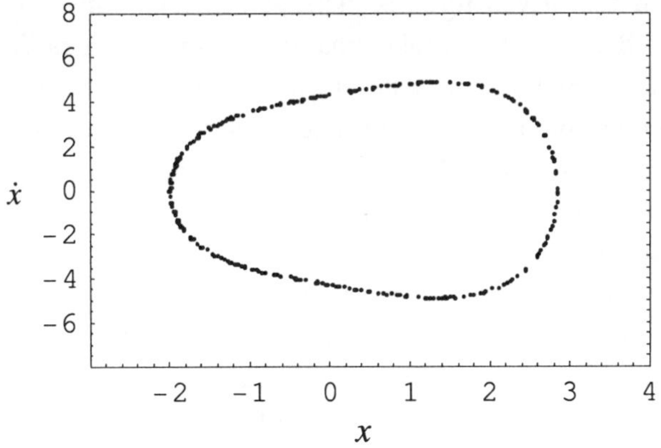

Figure 7.6. Poincare map of a quasiperiodic motion governed by $\ddot{x} + k\dot{x} + x^3 = B\cos t$ as $k = 0.00001$, $B = 1.8$, $d_0 = -2$, $v_0 = 0$.

7.4.   Bifurcation of Piecewise Constant Dynamic Systems

Behavior of a dynamical system can be viewed globally over a range of parameter values by a bifurcation diagram. As an example, a bifurcation diagram can be constructed by illustrating a graph of the first derivative of the solution or velocity $\dot{x}$ corresponding to a Poincare map versus one of the parameters in the governing equation of the dynamical system considered. A Duffing system governed by the following equation can be used for generating such a bifurcation diagram.

$$\ddot{x} + k\dot{x} + x^3 = B \cos t \qquad (7.2)$$

This is a well-studied system (Gwinn and Westervelt 1986, Blackburn *et al.* 1989). A bifurcation diagram for this system is generated and exhibited in Figure 7.7 with data calculated by using the piecewise constant technique described in Chapter 5. In Figure 7.7, the first 30 driven cycles are omitted so that the system is allowed to come to a steady state. When the amplitude of the driving force is small (say $B = 4.5$), as can be seen from Figure 7.7, there is only one point corresponding to the given B value. In fact, there can only be one point even if the time range is very large. This implies that the Poincare map in this case consists of only one visible point and the motion in this particular case is periodic or simple harmonic, as has been discussed in Chapter 2. If the driven force is increased to a certain range of B (say $B = 6.5$) the displacement x takes many number of values as seen in the diagram, the corresponding state shows nonperiodicity and the related motion can therefore be chaotic or quasiperiodic. For the cases of chaotic or quasiperiodic motions, theoretically, there should be infinite number of x values corresponding to a given B value.

In consideration of quasiperiodic cases, it is clear that the condition of infinite number of x values for a fixed B in a bifurcation diagram is not sufficient for a chaotic state to occur. There is no way to distinguish quasiperiodic motions from chaotic motions by merely using bifurcation diagram, since the numbers of x values are sufficiently large for both chaotic and quasiperiodic motions. Nevertheless, as indicated in Chapter 2 the bifurcation diagram provides a summary of the essential

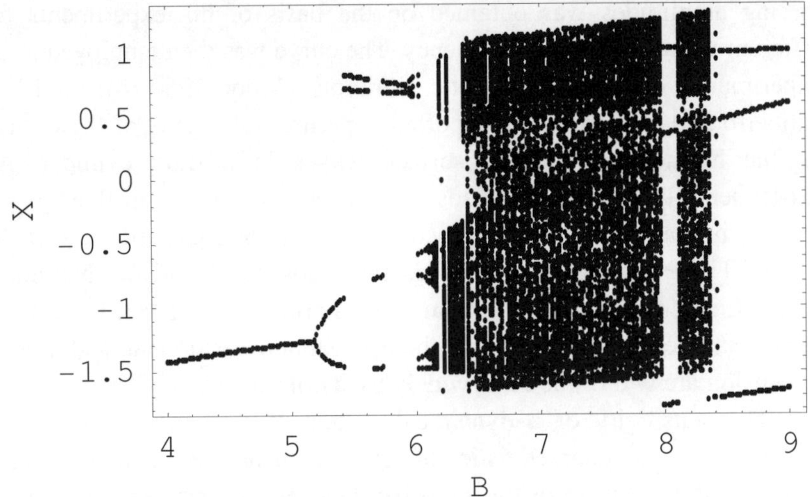

Figure 7.7.   Bifurcation diagram for a Duffing's equation governed by $\ddot{x} + k\dot{x} + x^3 = B \cos t$ where $k = 0.17$, initial conditions: $x_0 = -2$ and $\dot{x}_0 = 0$.

behavior of motion and is therefore a useful tool to overview the properties of motion for a nonlinear dynamical system.

7.5.   Derivation of Periodicity-Ratio

In investigating the motion of a nonlinear dynamical system, it is extremely important to set a criterion for chaos so that a chaotic motion can be distinguished from a regular motion. In the literature (Moon 1987), several methods are available for determining the onset of chaotic oscillations. Some predictive and diagnostic criteria for chaos can also be found in the field (Moon 1980 (a), Moon 1980 (b), Ciliberto and Gollub 1985). Among these, some are empirical methods that rely upon physical experiment results (Moon 1980 (a), Ciliberto and Gollub 1985) or data based on approximate mathematical models of the corresponding dynamical systems (Moon 1987, Moon 1980 (b)). The forced vibration of a buckled beam was studied by Moon (1987, 1980 (b)). He found that chaotic motion would occur when the amplitude of the load acting on the beam becomes sufficiently high. A curve representing the threshold of forcing amplitudes was obtained on the basis of his experiments for different values of forcing frequency. The curve was then employed as an experimental criterion for chaotic vibrations (Moon 1980 (b)). In 1985, Ciliberto and Gollub (1985) reported experimental results of their study on the harmonically driven surface waves in a fluid cylinder. An experimentally determined chaos diagram was presented in their paper. By the diagram, they distinguished chaos from periodic motions of the waves. There are also some criteria for chaos developed on theoretical bases (Lichtenberg and Lisberman 1983, Farmer *et al.* 1983, Wolf *et al.* 1985, Baierlein 1971). Among them, Lyapunov exponent and fractal dimension are two of the most widely used criteria.

The sensitivity of a dynamical system to a change in its initial conditions is a characteristic of chaotic behavior. The Lyapunov exponent is a measure of the sensitive dependence of a system upon its initial conditions (Wolf *et al.* 1985). The fractal dimension measures an extent to which the orbits in a phase space fill a certain subspace (Farmer *et al.* 1983). A fractal structure with noninteger dimensionality is considered as a hallmark of chaos. However, in a case that the fractal dimension is close to an integer, the other technique has to be employed to determine whether the case is chaotic. In practical applications, the Lyapunov exponent and fractal dimension are usually found by experiments or computer simulations.

For implementing Lyapunov-Exponent in analyzing the nonlinear behavior of dynamic systems, on the other hand, Lyapunov-Exponent may not provide reliable prediction for some odd cases in nonlinear dynamics, such as the cases in which the Lyapunov-Exponent close to zero. Furthermore, Lyapunov-Exponent can be used to determine whether a system is chaotic or periodic. However, it provides no information regarding nonperiodicity or quasiperiodicity. For the cases that the dynamic systems are neither periodic nor chaotic, Lyapunov-Exponent is not applicable. A new approach of diagnosing a dynamical system for its chaotic or periodic behavior was recently reported by Dai and Singh (1995, 1997). In this section, the approach will be further studied and the development of the approach will be presented in detail. As will be demonstrated in this section, with introduction of a new concept namely Periodicity-Ratio, periodic and nonperiodic motions can be conveniently distinguished, and consequently, chaotic motion can be distinguished from all of the other motions.

Complex appearance of various graphical representations of periodic and nonperiodic behavior naturally leads to a search for classifying the motions of nonlinear systems. Great efforts have been taken for associating the graphic representations with the types of motions. As discussed above, the Poincare map for a periodic oscillation in steady state consists only of a finite number of visible points. Since the motion is periodic, in a large enough time range allowing for the steady state case, each visible point in the Poincare map represents a large number of points overlapping each other. This can be seen graphically from Figures 7.8 and 7.9, which are generated by numerical simulations. Figure 7.8 illustrates a phase trajectory of a periodic motion to a nonlinear dynamical system.

The Poincare map corresponding to this motion is superposed on the phase trajectory and represented by the three diamonds (points). It should be noticed that the curves and dots in Figure 7.8 are the results of the superposition of a great number of curves in phase plan and a large number of points calculated for the Poincare map.

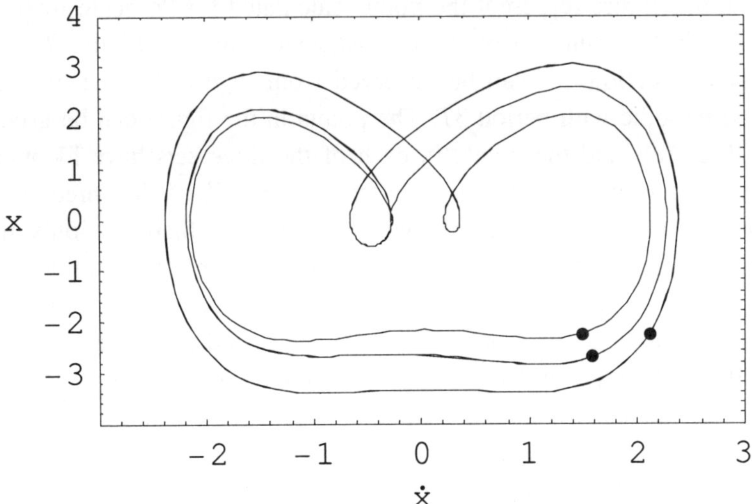

Figure 7.8.   Phase trajectory with corresponding Poincare map superposed on for a periodic solution of $\ddot{x} + k\dot{x} + x^3 = B\cos t$ as $k = 0.06$, $B = 3.0$, $d_0 = 3.0$, $v_0 = 4.0$.

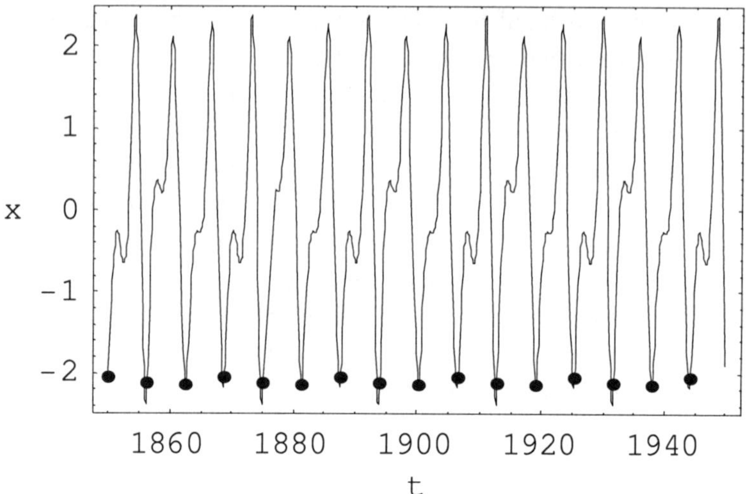

Figure 7.9. Wave form for a chaotic solution of $\ddot{x} + k\dot{x} + x^3 = B\cos t$ as $k = 0.06$, $B = 3.0$, $d_0 = 3.0$, $v_0 = 4.0$. Points for corresponding Poincare map are superposed on the wave form.

To illustrate this clearly, a small portion of the wave form diagram of the system is plotted in a $t - x$ plane as shown in Figure 7.9. The dots in this figure represent the points calculated for Poincare map. The distance in between any of the two adjacent points in Figure 7.9 is T, the system period. As can be observed from Figure 7.9, the motion is indeed periodic with period $3T$. The points in the figure can be grouped into three sets and the points in each of the three sets have identical x values and slopes $\dot{x}$. Therefore, after plotting all of the three sets of points on a phase plane as shown in Figure 7.8, there are only three points visible and the other points are overlapped by these three points. This should be the case no mater how large the time range is considered, provided that the system maintains the stable periodic motion.

If, on the other hand a motion is extremely random or chaotic, the points in the corresponding Poincare map must be spread evenly or unevenly over the phase plane. In this extreme case there may be a few or even no point overlapping any one of the other points in the corresponding Poincare map. As an example, the points calculated for a

Poincare map of a chaotic case is exhibited in Figure 7.10. As can be seen from the figure, in terms of x value and the slop (its first derivative), there is no point identical to any of the other point in the figure.

In view of the above discussion, a criterion for periodic and chaotic motions can be found on the basis of examining the overlapping points with respect to the total number of points in a Poincare map.

Consider a general dynamic system subjected to an external excitation of period T. Assume that the solution of this system is $x(t)$ and the behavior of the system is completely periodic. This periodic system represented by the solution of the system's governing equation should be a multiple of T in general; instead of a specific solution say simple harmonic solution. For such a general system, its periodic solution must satisfy the following expression.

$$x(t_0 + jT) = x(t_0) \tag{7.3}$$

where t_0 is a given time, and j is the number of periodic points visible in the Poincare map.

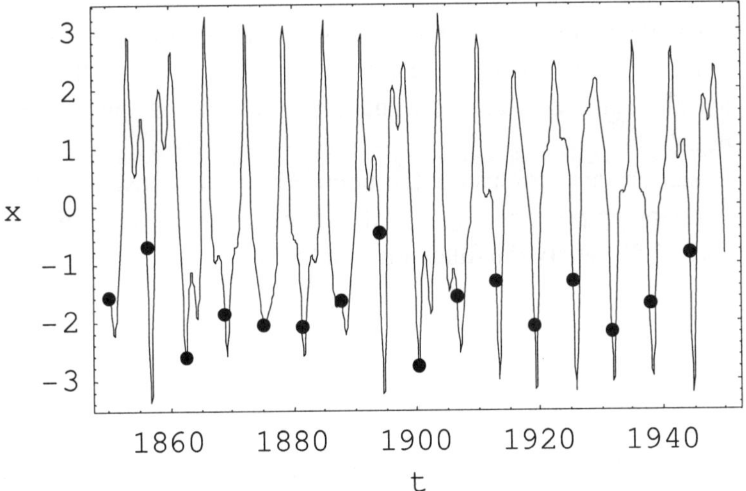

Figure 7.10. Wave form for a chaotic solution of $\ddot{x} + k\dot{x} + x^3 = B\cos t$ as $k = 0.05$, $B = 7.5$, $d_0 = 3.0$, $v_0 = 4.0$. Points for corresponding Poincare map are superposed on the wave form.

For a complete periodic system, no matter how large the time range is, there should only be a finite number of j points appearing in the Poincare map for a real periodic system and all the other points overlap these j points, provided that the time range considered is large enough. The determination of whether or not a point in the Poincare map is an overlapping point is based on the judgment described by the following equations (Dai and Singh 1997).

$$X_{ki} = x(t_0 + kT) - x(t_0 + iT) \tag{7.4}$$

$$\dot{X}_{ki} = \dot{x}(t_0 + kT) - \dot{x}(t_0 + iT) \tag{7.5}$$

where k is an integer in the range of $1 \le k \le j$. Designating n as the total number of points generated for the Poincare map within the time range considered, regardless whether they are overlapping points or not. i in the above equation is then an integer satisfying $1 \le i \le n$.

Equation (7.4) is to verify whether the displacements of the two points i and k are identical; whereas equation (7.5) determinates whether the first derivatives or the velocities of the system are equal for the two points. With the definitions given above, the following can be stated.

(1) On a Poincare map, a point $(x_i, \dot{x}_i)$ is said to be an overlapping point of the kth point represented by $(x_k, \dot{x}_k)$ only if the ith point has the same displacement and same velocity as those of the kth point.

(2) Based on this conclusion, the point $(x_i, \dot{x}_i)$ is considered to be an overlapping point of $(x_k, \dot{x}_k)$ if and only if the expressions of (7.4) and (7.5) satisfy the following conditions:

$$X_{ki} = 0 \tag{7.6}$$

and

$$\dot{X}_{ki} = 0 \tag{7.7}$$

On the other hand, points under consideration will not be overlapping points if

$$X_{ki} \ne 0, \ \ \text{and/or} \ \ \dot{X}_{ki} \ne 0. \tag{7.8}$$

(3) Otherwise if either of the conditions given above is not satisfied, the two points are not considered as overlapped points with each other.

Making use of the equations (7.6) and (7.7), the total number of points which overlap the kth point in the corresponding Poincare map can be calculated by the following equation:

$$\zeta(k) = \left\{ \sum_{i=k}^{n} Q(X_{ki}) \cdot Q(\dot{X}_{ki}) \right\} \cdot P\left(\sum_{i=k}^{n} \left[Q(X_{ki}) \cdot Q(\dot{X}_{ki}) \right] - 1 \right) \qquad (7.9)$$

where $Q(y)$ and $P(z)$ are two step functions represented by:

$$Q(y) = \begin{cases} 1, & if \ y = 0 \\ 0, & if \ y \neq 0 \end{cases} \qquad (7.10)$$

and

$$P(z) = \begin{cases} 0, & if \ z = 0 \\ 1, & if \ z \neq 0 \end{cases} \qquad (7.11)$$

In the above two equations, y and z are the arguments of the functions Q and P respectively. $\zeta(k)$ calculated by employing equation (7.9) gives the total number of points coincident with the kth point including the kth point itself in the Poincare map. Therefore, there are $\zeta(k)$ points in the Poincare map having the same values of displacement and velocity as that of the kth point. In the case that a pair of points are superposed, a visible point and a point overlapped by the visible point are calculated by equation (7.9) as two overlapping points. For a single visible point without any point overlapping it, equation (7.9) guarantees that this single point does not count as an overlapping point. The sum $\zeta(k)$ thus represents a set of points with the identical coordinates in the phase plane. The first summation in equation (7.9) calculates the number of overlapping points of the kth point and the function P assures $\zeta(k) = 0$ in the case that there is no point overlapping the kth point among the total number of n points of a Poincare map.

Once the number of points overlapping the kth point is calculated by equation (7.9), the total number of overlapping points corresponding to the j visible points in the Poincare map can be determined by summing up all these points overlapping with the j visible points. Designating N as the total number of overlapping points, and the equation developed for determining N can be expressed by

$$N = \zeta(1) + \sum_{k=2}^{n} \zeta(k) \cdot P\left(\prod_{l=1}^{k-1} \{X_{kl} + \dot{X}_{kl}\} \right) \qquad (7.12)$$

in which Π is the symbol for multiplication and $P(\cdot)$ the step function as defined previously. This equation ensures that the duplicate counting in the calculations for N or missing of any overlapping point is prevented.

In utilizing equation (7.12) in computing N, only the overlapping points should be counted. Therefore, the number of overlapping points should definitely be no more than the total number of points generated for the corresponding Poincare map. If the response of a dynamic system is completely periodic, all the points for the Poincare map must be overlapping points and the corresponding N in this case can then be simply expressed in the following form.

$$N = \sum_{k=1}^{j} \zeta(k) \qquad (7.13)$$

Equation (7.13) implies that there are j sets of points involved in the Poincare map for this periodic system, and the points in the same set are identical in terms of displacement and velocity. In this case, the points in the corresponding Poincare map are all periodic points and they are all overlapping points as well. This implies that there are only a fixed number of points in the Poincare map regardless the time range considered, once the first group of visible points is determined.

Based on our numerical analyses for the nonlinear dynamic systems in engineering field, it is found that the overlapping points calculated are in general periodic points for the pure periodic cases or the cases with behavior close to periodic. Therefore, the ratio of the total number of overlapping points N and the total number of calculated points n can be used as an index to describe the periodicity of a dynamic system.

It should be notice however, in some cases; overlapping points may not necessarily be periodic points which really represent the periodic motions of a nonlinear dynamic system. For these cases, determination of the periodic points and differentiation of the periodic from the general overlapping points are necessary. Moreover, in practice, numerical calculations or approximate calculations are unavoidable for most nonlinear dynamic analyses. Distinguishing periodic points from the general overlapping points is therefore important for the reliability of the calculations involved in this section. Nevertheless, in numerically analyzing the behavior of a nonlinear dynamic system, determining the periodic points and distinguishing between the periodic and general overlapping points can be a demanding task. The following are practically useful for numerically generating reliable and accurate periodic points, and they also provide advantages for the numerical analyses on the behavior of nonlinear dynamic systems.

(a) Assume there are M sets of points visible on the surface of a Poincare map. Among the M groups of points, j groups of points contain considerable number of overlapping points. Some of them may contain only one or two overlapping points, no mater how large the time range is used in the computation. With these few overlapping point samples, it is impossible to determine whether or not they are periodic points. Therefore, eliminating these groups of points from the M groups of points will increase the reliability of the calculation without affecting the accuracy of the determination for the total number of periodic points. In this case, there are $p = M - j$ groups of points to be ignored from the determination of the periodic points. However, if the system considered is a pure periodic system, all the points for the Poincare map are periodic points, there exist only j groups of overlapping points and therefore p in this case must be zero.

(b) Determine the average time span of the overlapping points in one of the j groups calculated. Consider the kth group among the j groups. Assuming there are q overlapping points in the kth group, the average time span of the overlapping points in this group can be expressed as

$$\eta = \frac{t_{k,q} - t_{k,1}}{q - 1} \tag{7.14}$$

where $t_{k,q}$ represents the time corresponding to the qth point in the kth group. If a group of overlapping points are all periodic points, then its period is the corresponding η. This then becomes a case to which the same displacement and velocity will occur after every η time units.

(c) Identification of the variance of a group of overlapping points is also helpful for increasing the reliability of the calculations. For the kth group, following equation can be used to determine the variance ρ of the overlapping points.

$$\rho^2 = \frac{\sum_{i}^{q-1}(t_{k,i+1} - t_{k,i} - \eta)^2}{q-1} \qquad i = 1,2,3,...q \qquad (7.15)$$

As can be seen from this equation, the variance value ρ must be zero, if a group of overlapping points contains only periodic points.

For the case that the overlapping points of kth group among the j groups of overlapping points are not all periodic points, distinguishing periodic points from the other nonperiodic points is important and grouping the periodic points with identical time span is more significant, since they are closely related to the periodic characteristics of a dynamic system. In order to compute the periodic points that have identical time spans, following formulas are developed for numerical determination of the periodic points. Consider all the overlapping points in the kth group. The time span between the ith point in the kth group and the $i+1$th point in the same group can be defined as

$$T_{k,i} = t_{k,i+1} - t_{k,i} \qquad (7.16)$$

and the time span between the $i+2$th point and the $i+1$th point is then

$$T_{k,i+1} = t_{k,i+2} - t_{k,i+1} \qquad (7.17)$$

Such that the overlapping points i, $i+1$ and $i+2$ in the kth group are periodic points only if the following equation is satisfied.

$$T_{k,i} - T_{k,i+1} = 0 \qquad (7.18)$$

With the above equations such defined, the number of periodic points with identical time spans as that of the ith and $i+1$th points in the kth group can be determined by employing the following formula.

$$\xi(i) = \left\{\sum_{h=i}^{q-1} Q(T_{k,i} - T_{k,h+1})\right\} \cdot P\left(\sum_{h=i}^{q-1} Q(T_{k,i} - T_{k,h+1}) - 1\right)$$

$$+ P\left(Q\left(\sum_{h=i}^{q-1} Q(T_{k,i} - T_{k,h+1})\right)\right) \tag{7.19}$$

where $Q(\cdot)$ and $P(\cdot)$ are the same step functions as described in equations (7.10) and (7.11) respectively. Note that q value in the above equation can be different from one group of points to another group of points. The overall number of periodic points in the kth group can thus be calculated by

$$\phi(k) = \xi(1) + \sum_{i=2}^{q-1} \xi(i) \cdot P\left[\prod_{h=1}^{i-1} \{T_{k,i} - T_{k,h}\}\right] \tag{7.20}$$

Therefore, there are $\phi(k)$ periodic points in the kth group of points. The spans of the periodic points, or the possible periods of the corresponding periodic motion can be determined by equation (7.16).

Designate NPP as the total number of periodic points among the entire n points. NPP is therefore the sum of the entire periodic points of all the j groups of overlapping points, such that

$$\text{NPP} = \sum_{k=1}^{j} \phi(k) \tag{7.21}$$

For a case of fixed n calculated for a given nonlinear system, larger number of periodic points such determined amount the n points means closer behavior of the system to periodic motion. Therefore, a ratio of NPP, the total number of periodic points, to n, the total number points generated for Poincare map, can be introduced to describe the characteristics of the system. This ratio, designated by γ, is defined as the Periodicity-Ratio by which the dynamic characteristics of the system can be described. Theoretically, the mathematical expression of the

Periodicity-Ratio can be given as

$$\gamma = \lim_{n \to \infty} \frac{\text{NPP}}{n} \tag{7.22}$$

If the behavior of a system in a steady state is periodic, the points in the corresponding Poincare map must all be overlapping points. Accordingly, the value of the Periodicity-Ratio, γ, should simply be unity. For a chaotic response of a system, on the other hand, the number of overlapping points should be zero or insignificant in comparing with n. This is to say, γ approaches zero for chaos.

With the definition of the Periodicity-Ratio, γ is clearly a quantified description of periodicity for a dynamic system. This is to state that γ indicates quantitatively how close the response of a dynamic system is to a perfect periodic motion. For example, a motion with γ equals to 0.9 is more close to a periodic motion in comparing with a motion to which γ equals to 0.8. Contrastively, a motion with γ approaching zero will show no periodic behavior at all, and therefore is a perfectly non-periodic motion. When γ takes a value such that $0 < \gamma < 1$, it implies that some points in the Poincare map are overlapping points while the others are not. Therefore, the corresponding motion is nonperiodic or a combination of periodic and nonperiodic motions. Nonperiodic cases in between chaos and periodic motions may include the intermittent chaos in which chaotic motions occur between periods of regular motion such as the cases reported by some of the authors in this field (Manneville and Pomeau 1980, Tsons 1992, Gwinn and Westervelt 1986).

Obviously, γ can be introduced here as a criterion to distinguish a periodic motion from a nonperiodic motion and to differentiate regular motion from chaos. γ thus describes the nature of periodicity of an oscillatory motion and is a measure of a degree of periodicity for a dynamical system. As the value of γ tends to unity the closer becomes its behavior to a periodic oscillatory motion. It is because of this that γ is called Periodicity-Ratio.

It should be noted, however, the expression shown in equation (7.22) is theoretical, as it requires an infinitely large number of n for a perfect measurement of γ and the time range considered must be $t \in [0, \infty)$ such that t will be sufficient for a perfect γ. This implies that the Periodicity-Ratio γ can be precisely calculated only in the cases for which the analytical solutions corresponding to the dynamical systems are available. For most nonlinear dynamic systems, however, the calculation for the Periodicity-Ratio has been done on a numerical basis with the aid of a computer, as analytical solutions for these systems are not available. An application of the formulae indicated above for numerically finding out the Periodicity-Ratio is presented in a computer program included in Appendix C. Since $Q(y)$ and $P(z)$ in equations (7.10) and (7.11) are step functions, the numerical calculation for γ can be conveniently carried out by a computer program.

Since numerical analysis is extremely important for finding solutions of nonlinear dynamic systems, numerical determination for the Periodicity-Ratio is almost unavoidable. In numerically determining for γ, therefore, a sufficiently large n should be used in performing the actual numerical calculation for γ in the practice of numerical calculation. In computing the Periodicity-Ratio, errors caused by numerical calculation and by the mathematical models of numerical purpose should also be considered. Furthermore, in numerically calculating for γ, all of the n points must be compared to see whether they are overlapping points or not. Once a point is counted as an overlapping point, it should not be counted again in the numerical calculations. The P function in equation (7.20) is used to prevent this duplicate counting in the calculation for the number of periodic points in a given group of overlapping points. With an approach such developed, ϕ in equation (7.20) and NPP in equation (7.21) and therefore γ in (7.22) only counts the periodic points.

It should also be noted that a motion with Periodicity-Ratio equals to zero may not necessarily be a chaotic motion. By the definition of Periodicity-Ratio, a perfect quasiperiodic motion also has a Periodicity-Ratio of zero. In this case, another technique, to be described in the following section, can be employed for distinguishing the quasiperiodic motion from chaos.

Periodicity-Ratio is applied to analyze the behavior of a nonlinear oscillatory system governed by equation (7.1). This nonlinear equation of motion has no analytical solution in engineering applications. With the nonlinear term in the equation, chaos may occur. Numerical results are obtained by implementation of the piecewise constant techniques developed in the previous chapters. Periodicity-Ratio is utilized in analyzing the characteristics of the motion of the nonlinear system. Figure 7.11 shows an example of the relationship between the amplitude of the external force and the corresponding Periodicity-Ratio values for this oscillatory system. In the figure, the motions with Periodicity-Ratio equals or very close to one are considered as periodic oscillation and those with Periodicity-Ratio equals or nearly equals to zero are regarded as extreme nonperiodic or chaotic oscillations. The motions in between these two extreme cases are neither periodic nor chaotic. This diagram is therefore useful in nonlinear dynamics analysis for diagnosing the state of motion for a given dynamic system. It is called the Periodicity-Ratio diagram in this book. In short, it will be referred as the P-R diagram hereafter.

As can be observed from Figure 7.11, the advantages of having such P-R diagram are the following.

1. The completely periodic and extremely nonperiodic responses of a nonlinear dynamic system can be clearly identified from the diagram.
2. Paths from periodic motion to a nonperiodic can be found from the diagram, and vise versa.
3. The diagram shows how close a nonlinear periodic motion to either completely periodic or perfectly nonperiodic motion. This is shown by the distance of a point in the figure to the top boundary ($\gamma = 1.0$) or the bottom boundary ($\gamma = 0.0$).
4. The amplitude of the external force is used as a variable in constructing Figure 7.11. The other system parameters such as the stiffness coefficient ω of the system, or initial conditions, may also be used as variables in constructing the diagrams, as desired.

The Periodicity-Ratio diagram in Figure 7.11 describes the state of motion with respect to only one parameter of the dynamical system. If different degree of grey is employed to represent the different values of γ, more system parameters can be considered simultaneously and a more global aspect of the motion can be presented for the system. As an example, a Duffing's equation is investigated and the states of motions represented by Periodicity-Ratios are presented in the Periodicity-Ratio diagram shown in Figure 7.12.

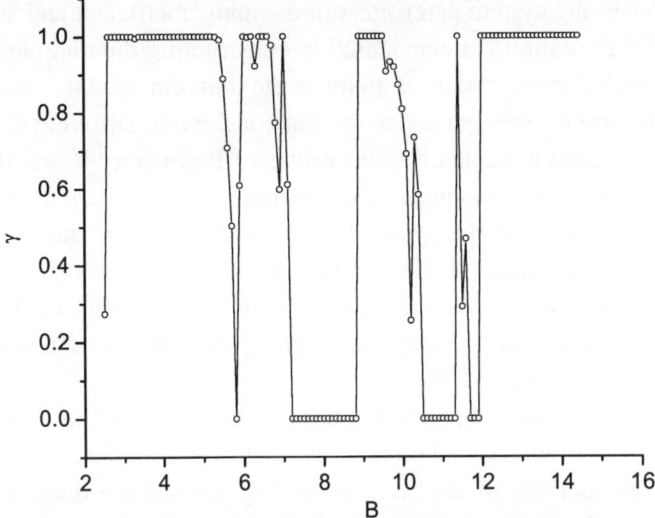

Figure 7.11. Periodicity-ratio versus amplitude of forcing function for a dynamical system governed by $\ddot{x} + \theta\dot{x} + \omega^2 \sin x = A\cos\Omega t$ as $\theta = 0.5$, $\omega = 1$, $d_0 = -2.5$, $v_0 = 0$.

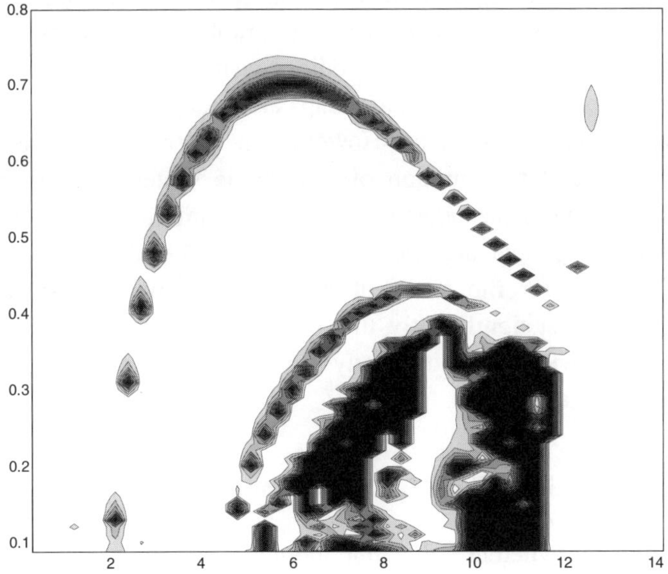

Figure 7.12. Periodicity-ratio diagram for $\ddot{x} + k\dot{x} + x^3 = B\cos t$.

Two varying system parameters of damping coefficient and amplitude of external excitation are considered in constructing the diagram. With a diagram such constructed, each point in the diagram actually represents a state of motion for the dynamical system considered. Different degrees of grey in the figure indicate different values of Periodicity-Ratio, therefore different degrees of periodicity. The distribution of the Periodicity-Ratios of different value for the system is thus illustrated by the diagram in regarding to the values of the system parameters and initial conditions. The state of motion, such as periodic or nonperiodic motions, can therefore be predicted by using the diagram with give system parameters and initial conditions.

The periodic and nonperiodic motions, as illustrated in Figure 7.12, are clearly distributed based on their degrees of periodicity. It can be seen from the comparison of the bifurcation diagram and the diagram shown in Figure 7.12, the former considers the relation in between $\dot{x}$ and a single parameter in the corresponding dynamical system whereas the latter relates the states consisting both the x and $\dot{x}$ values corresponding to the two varying system parameters. One may notice that it is the states of motion that Poincare maps really reveal. This is to say that the Poincare maps may be used to graphically identify at least the following states of motions: periodic, quasiperiodic and chaotic motions. This may explain why Poincare maps are widely used in the analyses for nonlinear dynamic systems. However, in order to diagnose for one of these three states of motion of a dynamic system, a Poincare map is necessary, unless the other theoretical or numerical techniques such as Lyapunov-Exponent are used. With use of Periodicity-Ratio, however, the diagrams for the states of motion of a dynamic system can be generated. Most significantly, the state of motion can also be numerically determined without plotting any figures, with implementation of the Periodicity-Ratio.

7.6.   Distinction of Quasiperiodic Motion from Chaos

In diagnosing whether a dynamic motion is chaotic or not, one may now simply concentrate on the states of motion with Periodicity-Ratios close or equal to zero where chaos is likely to occur. It is known (Moon 1987) that the Poincare map for a quasiperiodic motion is a smooth curve as shown in Figure 7.6. This indicates a state of motion to which the motion is not perfectly periodic but not unpredictable or random like either. From Figure 7.6, it can be seen that the points in the Poincare map form a closed curve. It is also found that the number of overlapping points is negligibly small as compared to the total number of points in the Poincare map. This results in that the corresponding Periodicity-Ratio for this case tends to zero. Since quasiperiodic motions may also have zero value Periodicity-Ratio, it is necessary to separate the quasiperiodic motions from all the motions with Periodicity-Ratios equal or close to zero. After the quasiperiodic motions are separated, the remaining states of motion to which the corresponding Periodicity-Ratio values are zero can be identified. As such, the chaotic regions can be determined for a Periodicity-Ratio diagram.

The least squares method is a well-known technique in fitting a curve or function to a set of data. Also, polynomials can be easily manipulated in numerically determining a function of curve for given numerical data. The least squares method for fitting the polynomial function of finite terms to the given data that do not plot linearly is very common (Draper and Smith 1981). In fitting a curve to a given set of data, usually, the curve generated by the least squares method with polynomials of finite terms does not pass all the data available, through all the points corresponding to the given data. However, there are cases in which the points of given data are distributed in such a way that they may be continuously connected to form a single smooth curve (Forsythe 1957, Kennedy and Gentlel 1980). The degree of fitness may also be numerically determined.

With comprehension of the above properties of the curve fitting, the least squares method is suitable for determining whether a motion is quasiperiodic or not, by evaluating the fitness of the given data to a single smooth curve. With the data of the points calculated for a Poincare map, a polynomial function of finite terms can always be constructed by the least square method. Theoretically, if all of the points of a Poincare map are located right on the curve generated by the polynomial function, then, according to the definition, the corresponding motion must be quasiperiodic. In other words, with the data of the points forming a smooth curve on a phase diagram or Poincare map, the polynomial function for the fitting curve can be so constructed such that all the points generated for the corresponding Poincare map satisfy the function. With this understanding, quasiperiodic motion is a state of response of a system to which all the data calculated for Poincare map are perfectly fitting with a smooth curve described by a polynomial function. One may therefore examine fitness of the data for the cases to which the Periodicity-Ratios are zero. Should the data fit perfectly with a smooth curve of polynomial function, the data must form a smooth curve which represents a quasiperiodic motion. The polynomial functions themselves are not significant in diagnosing quasiperiodic cases from chaos. It is the fitness that matters. Therefore, with the fitness evaluated, there is no need to plot any curves in distinguishing a quasiperiodic motion from chaotic motions, among all the cases to which the Periodicity-Ratios are zero.

In numerically determining the quasiperiodic motions on a computer, if the difference between the given data of Poincare map and the curve produced by the least squares method is within a predetermined limit representing a high fitness level, the corresponding motion is considered as quasiperiodic. This technique for determining quasiperiodic motions is used in this book. A computer program is included in Appendix C for the application of distinguishing the quasiperiodic motions from chaotic ones. In order to increase the accuracy of the numerical calculation, a polynomial of 10th order is applied in the program. In numerically diagnosing for the quasiperiodic cases, the data of Poincare map for the cases of zero Periodicity-Ratio should first be determined. These data should then be examined with the technique discussed above for distinguishing the quasiperiodic cases from all the chaotic cases. After the quasiperiodic cases are identified, a periodic chaotic region diagram can be conveniently plotted.

7.7. Comparison of Periodicity-Ratio and Lyapunov-Exponent

As discussed above, criteria for distinguishing chaos from the other types of response of a system are extremely important in nonlinear dynamics.

As described in Chapter 2, several methods such as Power Spectral, Fractals, Poincare Maps, Entropy, and Lyapunov-Exponent for diagnosing chaotic behavior of dynamic systems can be found in literature. The prevalent methods commonly used in the current nonlinear dynamics analyses are the Lyapunov-Exponent method and fractal dimension (Nayfeh and Mook 1988, Siegwart 1991, Weaver *et al.* 1990), though Lyapunov-Exponent is probably the most practical technique for diagnosing chaotic systems.

As described in Chapter 2, in considering that one of the important characteristics of a nonlinear dynamic system is its sensitivity to initial conditions, Lyapunov-Exponent measures the sensitive dependence of the system upon its initial conditions. With Lyapunov-Exponent, the average exponential rates of divergence or convergence of close orbits of a vibrating object in the phase space of a dynamic system (Lakshmanan and Rajasekar 2003) is quantitatively determined. This exponent is used as a criterion for determining whether a motion of a dynamic system is chaotic or periodic. The criterion is therefore widely used in nonlinear analyses and many research works on determination and application of Lyapunov-Exponents can be seen in the literature. Wolf *et al.* (1985) gave a method for determining Lyapunov-Exponents from time series. Rong *et al.* (2002) investigated the principal resonance of a stochastic Mathieu oscillator to random parametric excitation and gave the conclusion that the instability of the stochastic Mathieu system depends on the sign of the maximum Lyapunov-Exponent. Lyapunov-Exponents were also found to be used to analyze the numerical characteristic (Shahverdian and Apkarian 2007). Some other researchers (Terzic and Kandrup 2003) presented a scheme for the semi-analytic estimates of Lyapunov-Exponents in lower-dimensional systems.

Lyapunov-Exponents are mainly determined by computer simulations with numerical approaches, though experimentally collected data can also be used. Research on the numerical determination of Lyapunov-Exponents is continuously undertaking with the investigations on non-linear dynamics. Several new numerical techniques have been found in the current literature. Some innovative methods for the calculation of the Lyapunov-Exponent of continuous dynamical systems were reported recently by Lu *et al.* (2005) and He *et al.* (1999). Nevertheless, reliability of Lyapunov-Exponents depends on the accuracy and stability of the numerical results and numerical methods used.

Although Lyapunov-Exponent is probably the most popular criterion in diagnosing chaotic and periodic responses of a dynamic system, however, it has some short comes in implemented in nonlinear analyses.

1. As mentioned in Section 7.5, Lyapunov-Exponent is not suitable for identifying the behavior that is neither periodic nor chaotic.
2. Lyapunov-Exponent may not provide reliable predictions for certain cases such as that when its value is close to zero.
3. Lyapunov-Exponent can not be used to distinguish between quasiperiodic and periodic cases.
4. Lyapunov-Exponent is usually determined numerically in nonlinear analysis practice. The accuracy and reliability of the Lyapunov-Exponent calculated highly depend on the accuracy and reliability of the numerical calculations and the numerical methods used (see Chapter 2).
5. Each Lyapunov-Exponent calculated may be used to diagnose chaos in one direction only and the values of the Lyapunov-Exponents in different directions are usually different for a given nonlinear system. It is the largest Lyapunov-Exponents to be used for the diagnoses. In utilizing the Lyapunov exponent for analyzing a nonlinear dynamic system, for example, the dimensions of the system are important. Taking a driven pendulum system for example, three Lyapunov-Exponents should be used to describe the three dimensions of the phase space (θ, ω, ϕ); where θ is the phase angle, ω is phase velocity and ϕ is the initial phase angle. Since the orbits of the pendulum are the solutions to a set of the differential equations governing the dynamic system, the calculation of Lyapunov-Exponents is rather complicated. On a chaotic attractor, the directions of divergence and contraction are locally defined, and the calculation must be constantly adjusted for obeying this condition. For a dynamic system to which the expression of Lyapunov function can not be properly defined, the Lyapunov-Exponent approach is then not valid for determining whether or not this system is periodic or chaotic. The usual test for chaos with utilization of Lyapunov-Exponent is the calculation for the largest Lyapunov-Exponent.

In comparing with Lyapunov-Exponent, Periodicity-Ratio describes the periodicity of a nonlinear dynamic system, in addition to diagnosing chaos from regular motions. The characteristics of the Periodicity-Ratio in diagnosing the regular and irregular motions need to be identified for applying it in the analysis of dynamic systems. Moreover, for systematically analyzing the behavior of a nonlinear dynamic system with the implementation of the Periodicity-Ratio, a comparison of the Periodicity-Ratio with the other criteria such as Lyapunov-Exponent is necessary.

To demonstrate the practicality of Periodicity-Ratio and compare the characteristics of Periodicity-Ratio with that of the Lyapunov-Exponent approach, it is convenient to start with the numerical simulations of a driven pendulum system governed by the following equation, as its rich linear and nonlinear behaviors are well known to readers in the areas of nonlinear differential equations and nonlinear dynamics (Baker and Gollub 1990). This system is illustrated in Figure 1.3(a) with the governing equation expressed as the following:

$$\frac{d^2x}{dt^2} + \frac{1}{Q}\frac{dx}{dt} + \sin x = F\cos(\omega_D t) \qquad (7.23)$$

In the above equation, x represents the phase angle of the pendulum, $1/Q$ is the damping coefficient, F and ω_D designate respectively the amplitude and frequency of the external excitation acting on the pendulum system (notice the difference between the figure and the equation on the variable).

It is well known that the system such defined is linear if the angle x is small. For the linear case, analytical solution is available and the behavior of the linear system is well known. The nonlinear behavior governed by equation (7.23) is closer to that of reality and physically more meaningful. However, numerical method has to be used for solving this nonlinear problem. This nonlinear system is solved numerically with implantation of the P-T method described previously, in considering that the P-T method provides accurate results with good convergence (Dai and Singh 2003) in comparing with the commonly used numerical methods such as Runge-Kutta method.

Making use of the system parameters: $Q = 2$ and $\omega_D = 2/3$, and taking the value F in the range from 1.0 to 1.99, numerical simulations for this specific system are performed. With the numerical data generated for the solution by the P-T method, all data for Poincare maps should firstly be determined. The Periodicity-Ratio values can then be determined. Lyapunov-Exponents are also determined separately on the basis of the numerical solutions. Two numerical results of Lyapunov-Exponent calculated are plotted in Figures 7.13 and 7.14 for exhibiting the cases of chaotic and periodic responses of the system respectively. In the two figures, the "number of orbits" designates the number of drive cycles of the external excitation of the system.

For the purpose of comparison, the numerical solutions of the system are found for a set of F values in the range of 1.0 to 1.99. All the other system parameters and the initial conditions are maintained as constants in calculating for the Periodicity-Ratio and Lyapunov-Exponent values. With the determinations of the Periodicity-Ratio and Lyapunov-Exponent values, the states of motion of the system can be determined correspondingly.

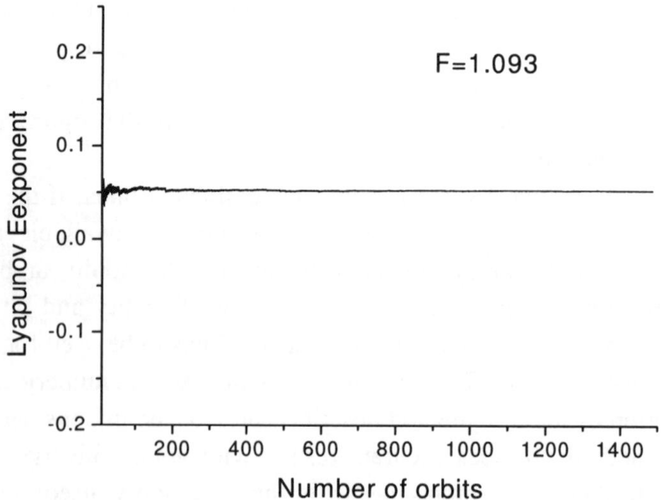

Figure 7.13.   Lyapunov-Exponent corresponding to a chaotic response of a driven pendulum, $Q = 2$ and $\omega_D = 2/3$.

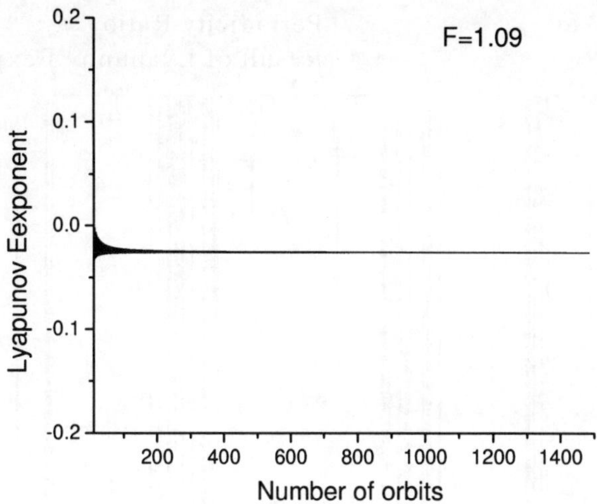

Figure 7.14. Lyapunov-Exponent corresponding to a periodic response of a driven pendulum, $Q = 2$ and $\omega_D = 2/3$.

The states of motion determined with implementations of both the Periodicity-Ratio and Lyapunov-Exponent are shown in Figure 7.15 for the purpose of comparison. The curve generated by the Periodicity-Ratio approach can be used to interpret the characteristics of the system. When the value reaches 1.0, the system is periodic; whereas the value reaches 0.0, the system is completely nonperiodic therefore chaos may occur. As described previously, the cases of neither perfectly periodic nor extremely nonperiodic can be presented by the Periodicity-Ratio with values in between 0 to 1. This provides the capability of describing the characteristics of the motion for the nonlinear dynamic systems in a much broad manner.

The Lyapunov-Exponent, on the other hand, merely diagnoses whether a system is periodic or chaotic, with the given system parameters and initial conditions. In other words, only two states of motion are considered by Lyapunov-Exponent. For the sake of clarity and quantitative comparison between Periodicity-Ratio and Lyapunov-Exponent, in the present research, the curve of Lyapunov-Exponent approach shown in Figure 7.15 consists of only 1 and 0 to represent the Lyapunov-Exponent values for periodic and chaotic motions respectively.

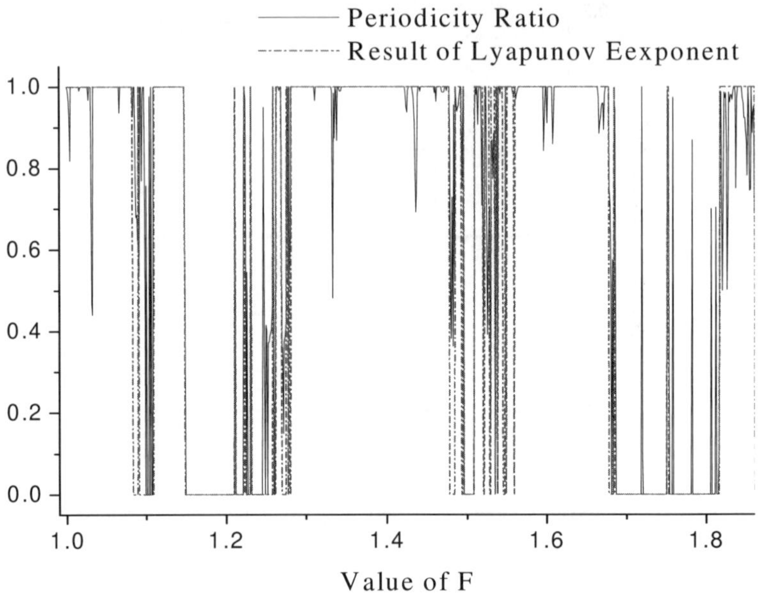

Figure 7.15.   Computation results of driven pendulum, $q = 2$ and $\omega_D = 2/3$.

As can be seen from the figure, the results of Periodicity-Ratio and Lyapunov-Exponent well agree with each other in most of the regions considered, especially when the regions in which the system is perfectly periodic or chaotic. However, for certain regions in Figure 7.15, the results of the two approaches do not agree with each other. The characteristics of the results generated by both the Periodicity-Ratio and Lyapunov-Exponent approaches can be classified into the following categories:

(a) *Perfectly agreed region*

These regions can be easily found from Figure 7.15, such as the regions with $F = 1.11 \sim 1.26$ and $F = 1.35 \sim 1.42$. In the region of $F = 1.23 \sim 1.245$, for example, both methods predict chaos. Indeed, the motions of the system in this region are chaotic based on the numerical simulations performed for the pendulum system. Motion in the region from $F = 1.35$ to $F = 1.42$, on the other hand, is a periodic motion region as predicted by both the Periodicity-Ratio and Lyapunov-Exponent approaches. Per the numerical simulations performed for the motion in this region, it is also found that the motion of the pendulum is indeed periodic.

Figure 7.16 shows the trajectory of the system in phase space with value of $F = 1.410$. With this F value, the system should be periodic as per the prediction made by both the Periodicity-Ratio and Lyapunov-Exponent approaches. As can be observed from the figure, the motion of the pendulum is really periodic and in actual fact the Periodicity-Ratio for this case is 1.0. For the periodic cases in the vicinity region of this F value, the motion of the system is actually very stable. For the motion in this region, therefore, the properties of motion can be easily determined numerically with only a few cycles of the motion or a few iterations in numerical calculations, if the Periodicity-Ratio approach is implemented.

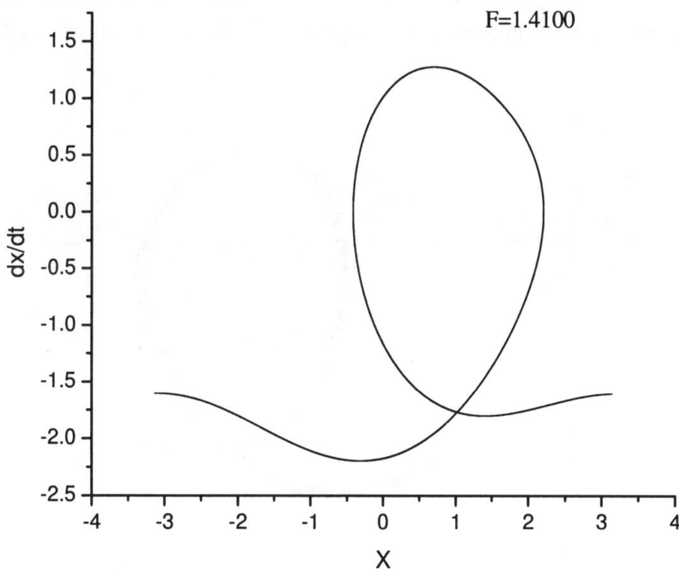

Figure 7.16.   Trajectory of driven pendulum in phase space, $Q = 2$ and $\omega_D = 2/3$.

(b) *Nearly agreed region*

In some regions, as can be found from Figure 7.15, the results of Periodicity-Ratio and Lyapunov-Exponent are not perfectly identical but very close, such as the region in which $F = 1.42 \sim 1.44$ and $F = 1.27 \sim 1.28$. In the region corresponding to $F = 1.42 \sim 1.44$, for example, the calculated Periodicity-Ratio value is close to 1 (> 0.75), the motion in this region can be considered as periodic as only a small number or insignificant number of points not overlapping with the others. Figure 7.17 shows the trajectory of the driven pendulum in the phase space with $F = 1.4244$.

This is a case slightly different from that shown in Figure 7.16 in which $F = 1.410$ whereas the other parameters and initial conditions are kept as the same as that of Figure 7.16. As can be seen from Figure 7.17, the motion for this case is close to periodic, but the behavior of the system in this case with $F = 1.4244$ is not perfectly periodic as its period of motion varies with time and the pattern of its phase diagram is slightly varying from that of the perfect periodic case shown in Figure 7.16. Nevertheless, as can be seen from Figure 7.17, the behavior of the motion

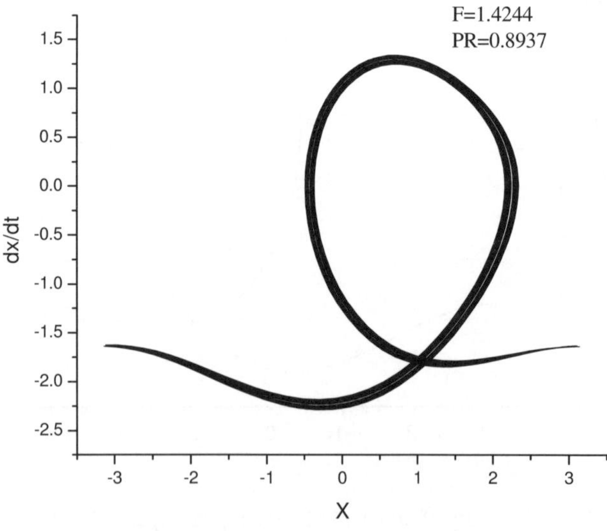

Figure 7.17.   Trajectory of driven pendulum in phase space, $q = 2$, $\omega_D = 2/3$ and $F = 1.4243$.

is similar to periodic in terms of period and pattern in comparing with that of the periodic case of Figure 7.16. As expected, the Periodicity-Ratio for this case is 0.8937, quantitatively exhibits its difference from a perfect periodic system to which the Periodicity-Ratio should be a unity.

For the same system in the region of $F = 1.27 \sim 1.28$, on the other hand, Lyapunov-Exponent approach predicts chaos and the corresponding Periodicity-Ratio is very close to zero. This can be seen from Figure 7.15. The small variation of γ from zero by the Periodicity- Ratio is the contribution of a few overlapping points picked during the numerical calculation for determining the ratio γ. However, the time spans between any two neighboring points of these points are random. The effects of these points on the characteristics of the motion are actually negligible as they do not represent periodically repeating values. To further increase the accuracy of the determination for γ, the strategies described previously can be employed. The motions in this region are chaotic.

(c) *Disagreed regions*

In some particular regions, the results from the two methods conflict with each other. For example, when $F = 1.093$, the system is predicted to be chaotic by Lyapunov-Exponent, whereas the calculated Periodicity-Ratio value declares that the system is very close to a perfect periodic. To check this out, one may simply find out the numerical solution for it and plot a wave form or phase diagram accordingly. The trajectory of the driven pendulum in the phase space generated by a numerical simulation for this specific case is shown in Figure 7.18. As can be seen from the figure, the system is periodic. The reason that Lyapunov-Exponent does not give the right prediction is that the value of Lyapunov-Exponent is usually not a fixed constant and it may vary around certain number under specified conditions, even when the system is in its steady state. In the case that the largest Lyapunov-Exponent is negative, but its absolute value is at the vicinity of zero, the computed Lyapunov-Exponent value may vary around zero, or the Lyapunov-Exponent shifts between positive and negative values with a varying system parameter considered. This may result with an unstable prediction for the behavior of a nonlinear system and the instability of the result may lead to improper predictions.

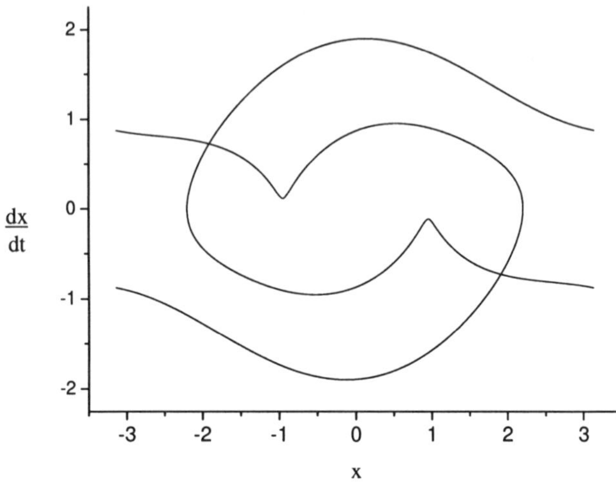

Figure 7.18.   Trajectory of driven pendulum in phase space, $q = 2$ and $\omega_D = 2/3$, $F = 1.093$.

With the numerical simulations for the dynamic system considered above, numerous irregular motions, which are neither periodic nor chaotic, can be found. These irregular motions are interesting and may attract the attention from the readers. Some responses of the system under the excitation are close to periodic but not completely periodic in the time duration considered, on the basis of the numerical calculations performed for the dynamic system with a set of fixed system parameters. A typical such case is selected from Figure 7.15, the point "a" indicated in the figure at which $F = 1.032$. The Periodicity-Ratio for this case is 0.4403. The phase diagram for this case is shown in Figure 7.19. As can be seen from the figure, the motion in this case is neither periodic nor chaotic. However, some regulation and patterns can still be identified. This case is therefore a case in a region of transition between periodic and chaotic responses. The response of the system in this region is neither chaotic nor perfectly periodic. The Periodicity-Ratio may also be useful for the analysis on the nonlinear systems which show mixed periodic and nonperiodic or chaotic behavior under fixed system parameters and initial conditions.

With the discussions above, one may see the application and advantages of Periodicity-Ratio to be used as an index for quantitatively

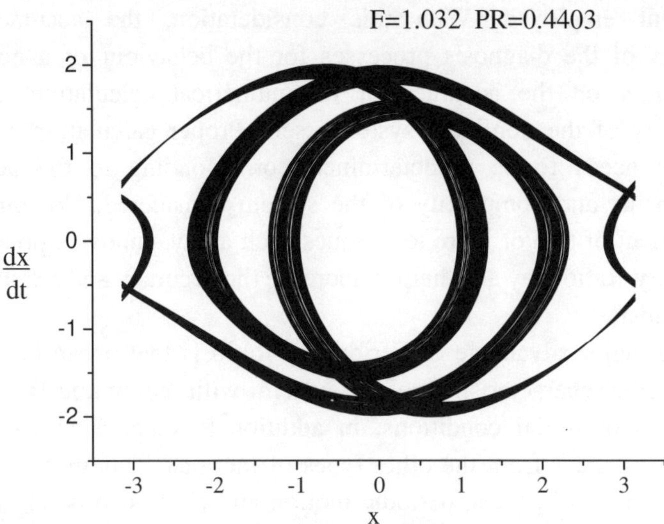

Figure 7.19.   Trajectory of driven pendulum in phase space, $Q = 2$ and $\omega_D = 2/3$, $F = 1.032$.

describing the periodicity of a nonlinear system's behavior. All the responses of a nonlinear system can be measured by the index to exam how close a system is to periodic cases, from perfect periodic behavior ($\gamma = 1$) to perfect non-periodic or chaos ($\gamma = 0$). This is actually where the terminology, Periodicity-Ratio, comes from. One of the advantages of implementing Periodicity-Ratio in nonlinear dynamic analysis is that the periodicity of a motion corresponding to a nonlinear dynamic system can be quantitatively described with a single numerical value γ. With the Periodicity-Ratio, it can be stated that there are infinite number of non-periodic and quasiperiodic cases in between perfect periodic and chaotic cases quantitatively described by Periodicity-Ratio from 0 to 1.

It should be noted that the values of Lyapunov-Exponent and Periodicity-Ratio are both determined by computer calculations in practice. Therefore, the accuracy of the numerical values is directly depending on the accuracy of the numerical calculations, including the accuracy requirements set for calculation, accumulative errors of the calculation, numerical method used and accuracy of the computational equipment employed. With this consideration, the accuracy and reliability of the diagnosis processes for the behaviour of a nonlinear system rely on the accuracy of the numerical calculation and the complexity of the nonlinear system itself. Proper calculation accuracy therefore needs to be predetermined corresponding to the accuracy requirements and complexity of the system considered. Combinatorial employment of two or more techniques such as Lyapunov-Exponent and Periodicity-Ratio may also help to increase the accuracy and reliability of the diagnosis.

One major advantage of Periodicity-Ratio is that it can be used to describe the characteristics of a system with continuously varying parameters or initial conditions, in addition to diagnosing chaos and periodic instances from the other types of motions. This means that the transition region between periodic motion and chaos can be clearly and quantitatively depicted. To illustrate the transition of the pendulum's motion from one type of behavior to another, a bifurcation diagram is useful. A bifurcation diagram for the pendulum system with F values varying from 1.255 to 1.270 is plotted in Figure 7.20.

Focusing on the region in which F takes the value between 1.261 and 1.2635 in Figure 7.20, one may observe the gradual changing

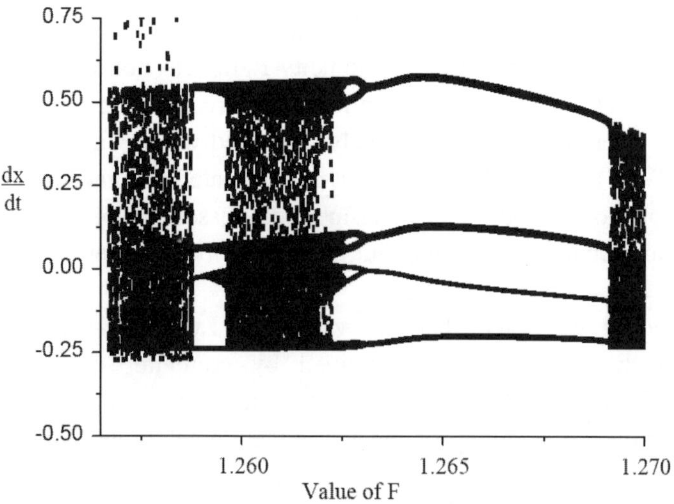

Figure 7.20.   Bifurcation diagram of the driven pendulum, $Q = 2$, $\omega_D = 2/3$.

behavior of the system. When $F = 1.263$, the behavior of the system seems periodic based on Figure 7.20. The trajectory of the motion in phase diagram for the system in this case indeed shows the periodic behavior as exhibited in Figure 7.21(a), with multi periods. The Poincare map of this case is plotted in Figure 7.21(b) and the corresponding Periodicity-Ratio in this case is 0.9613, indicating this is almost a perfect periodic case with only a few odd points slightly out of track.

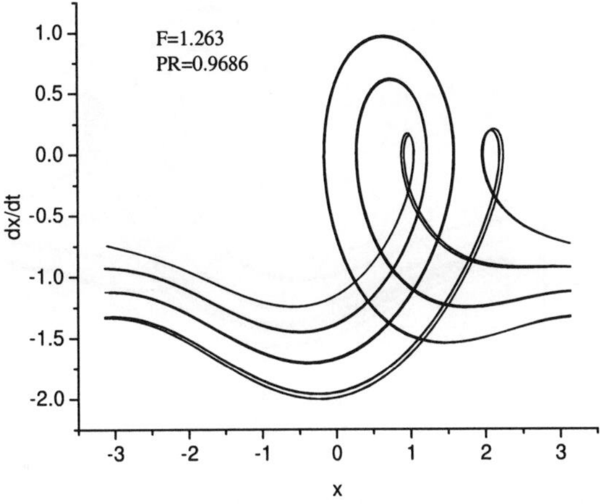

Figure 7.21(a).   Phase diagram of the driven pendulum with $F = 1.263$, $Q = 2$, $\omega_D = 2/3$.

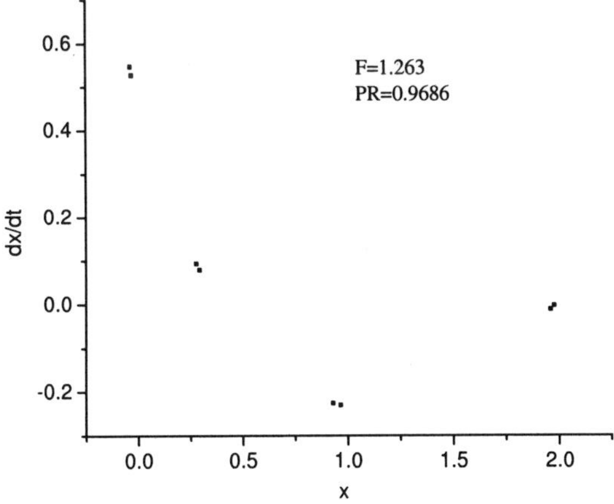

Figure 7.21(b).   Poincaré map of the driven pendulum with $F = 1.263$, $Q = 2$, $\omega_D = 2/3$.

Reducing *F* to 1.2625 gives a trajectory of the pendulum's motion in the phase space with a nearly periodic appearance, as shown in Figure 7.22(a). Indeed, the Periodicity-Ratio for this case is 0.8232 corresponding to a nearly periodic motion. The Poincare map corresponding to this case is shown in Figure 7.22(b).

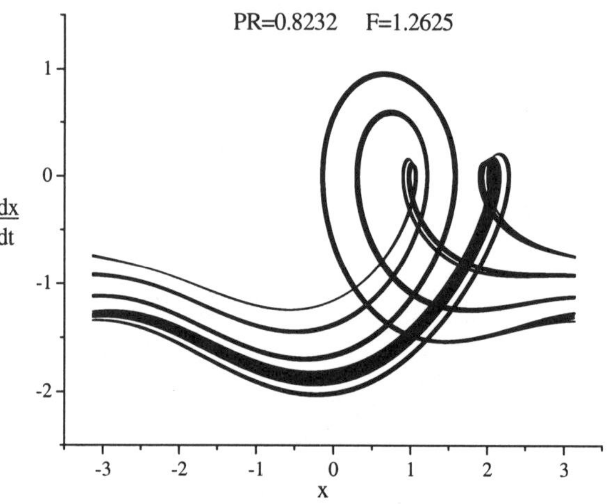

Figure 7.22(a). Phase diagram of the driven pendulum with $F = 1.2625$, $Q = 2$, $\omega_D = 2/3$.

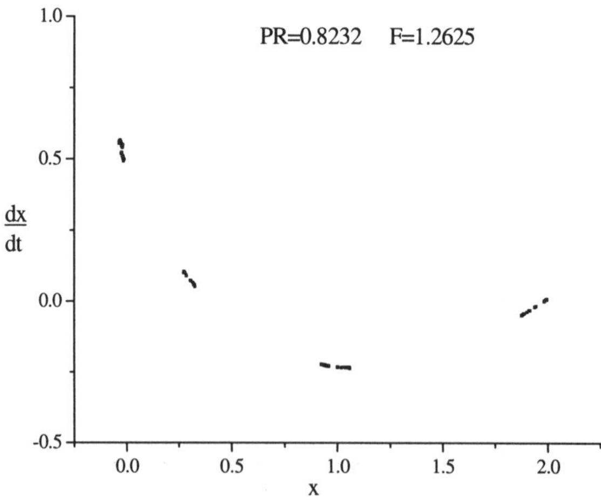

Figure 7.22(b). Poincaré map of the driven pendulum with $F = 1.2625$, $Q = 2$, $\omega_D = 2/3$.

With Figures 7.22(a) and 7.22(b), one may see the similarities of these figures with the phase diagram and Poincare map showing in Figures 21(a) and 21(b) respectively. Although the period of the motion corresponding to Figures 22(a) and 22(b) is slightly varying with time, the pattern of motion is similar to that of Figures 21(a) and 21(b). The Periodicity-Ratios for the two cases correctly represent the behaviors of the motion for the dynamic system, as the behavior of the case in Figure 7.21 is closer to a perfect periodic motion and that of Figure 7.22 is slightly away towards a more random behavior.

Further reduce F to 1.2615. The behavior of the pendulum system shows some randomness. The phase diagram and Poincare map for this case are shown in Figures 7.23(a) and 7.23(b) respectively. The Periodicity-Ratio calculated for this case is 0.5886, indicating further off of the system from a perfect periodic case. The points in the Poincare map are unevenly distributed primarily along a curve. Similar pattern in the two diagrams can also be recognized in comparing with that of Figures 21 and 22, though the case is far from perfectly periodic. However, the response of the system is not chaotic either. The Periodicity-Ratio tells this information numerically in comparing with that of Figures 7.21 and 7.22.

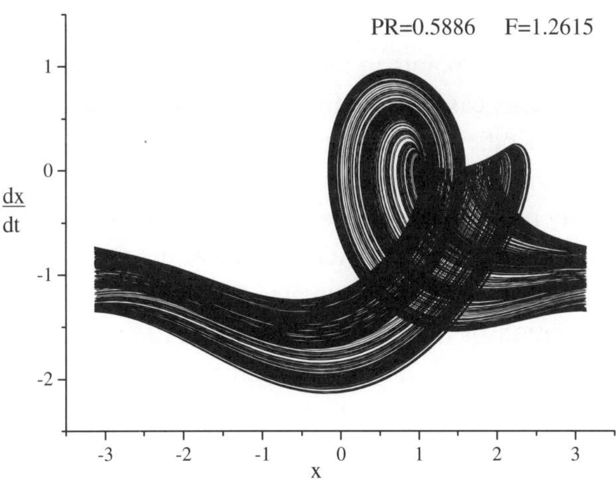

Figure 7.23(a).   Phase diagram of the driven pendulum with $F = 1.2615$, $Q = 2$, $\omega_D = 2/3$.

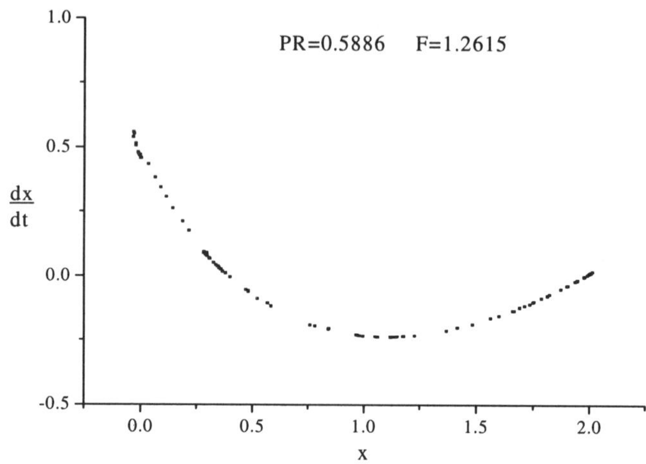

Figure 7.23(b).   Poincare map of the driven pendulum with $F = 1.2615$, $Q = 2$, $\omega_D = 2/3$.

As it is expected, the Periodicity-Ratio for chaotic case should be zero and the corresponding Poincare map of the system considered should therefore show highly irregular distribution of the Poincare points. The Periodicity-Ratio is almost a perfect zero for a case in which $F = 1.88$. Corresponding to the case of $F = 1.88$, Figures 7.24(a) and 7.24(b) are constructed for phase diagram and Poincare map respectively.

As can be seen from the figures that the phase diagram does show the irregularity and the points in the Poincare map are indeed spread in a random manner. This case is therefore a clear chaotic case.

Since the Periodicity-Ratio can be used as a single value index in diagnosing for periodic and nonperiodic motions of a nonlinear dynamic system, one may use it to determine the states of the motions and plot the states without showing the Periodicity-Ratio values. The periodic-chaotic region diagrams can therefore be created for nonlinear dynamic systems with implementation of the Periodicity-Ratio. A Periodic-chaotic region diagram is a diagram that shows the regions of periodic, quasiperiodic and chaotic responses of nonlinear systems.

The diagrams are therefore useful for analytically studying the behavior of a given dynamic system. One of the advantages of employing Periodicity-Ratio for the diagrams is that there is no need to plot any of

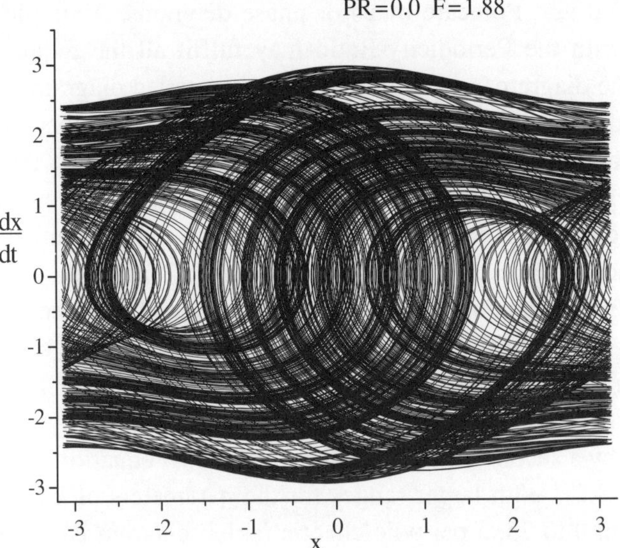

Figure 7.24(a).   Phase diagram of the driven pendulum in chaos, $Q = 2$, $\omega_D = 2/3$.

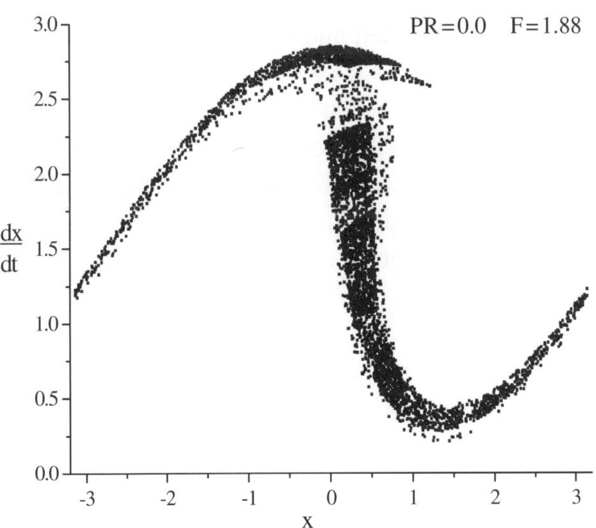

Figure 7.24(b).   Poincaré map of the driven pendulum in chaos, $Q = 2$, $\omega_D = 2/3$.

the wave curves, Poincare maps or phase diagrams. A simple computer program with the Periodicity-Ratio may fulfill all the requirements for plotting the diagrams, even for plotting the detailed diagrams illustrating the periodic, chaotic, quasiperiodic regions and the other nonperiodic instance regions for a nonlinear dynamic system. With the periodic-chaotic region diagrams, for example, one may identify a chaotic case in the areas of chaos from the periodic-chaotic region diagram and plot the corresponding phase diagram and Poincare map if desired.

A periodic-chaotic region diagram generated by Ueda (1980) for analyzing the Duffing's equation has been widely cited by the researchers in the areas of nonlinear phenomena studies. The Duffing's equation investigated by Ueda is for a typical nonlinear dynamic system governed by a second-order nonlinear differential equation in the form of equation (7.2). With larger ranges for the parameters of k from 0 to 0.8 and B from 0 to 25, a periodic-chaotic region diagram is generated with the employment of Periodicity-Ratio. The periodic-chaotic region diagram generated with Periodicity-Ratio is shown in Figure 7.25.

Much detailed information is provided by this diagram in comparing with that generated by Ueda (1980). Yet, with Periodicity-Ratio, much finer diagram can be generated and more parameters or larger ranges of parameters can be considered if so desired. Again, no Poincare maps or phase diagrams need to be plotted for generating the diagram. With employment of Periodicity-Ratio, some new phenomena, which were not reported by Ueda, can be found within the regions the same as that considered by Ueda. The application and interesting characteristics of the periodic-chaotic region diagram illustrated in Figure 7.25 are described below.

From the periodic-chaotic region diagrams generated with the Periodicity-Ratio, the transition areas between period zones and chaotic zones can be easily identified, as that shown in Figure 25. Identification of the transition areas is significantly important for analyzing the behavior of nonlinear dynamic systems. Existence of "transition" phenomena reflects the actual behavior of the physical problem itself regardless of the numerical methods used. It should be noticed that the periodic-chaotic region diagrams are usually independent of the initial conditions. Periodicity-Ratio thus provides the availability of revealing

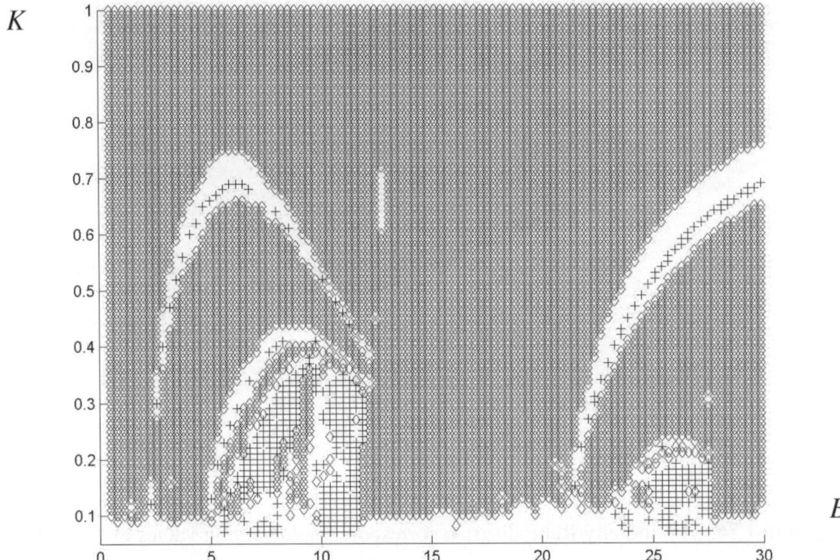

Figure 7.25.   Periodic-Chaotic region diagram of a Duffing's equation.

the gradual variation of a nonlinear system from one state to another different state, e.g., from periodic to chaotic, or from quasiperiodic to chaotic, or from one form of periodic instance to another form of periodic one. It therefore provides more information in comparing with the bifurcation diagrams on this aspect. Implementation of Periodicity-Ratio, which can be any number between zero and one corresponding to the nonperiodic motions between perfect periodic case and chaos, therefore provides a significant advantage over Lyapunov-Exponent, which provides merely two values to represent either periodic or chaotic cases.

As it is well known, all the system parameters and initial conditions affect the behavior of a nonlinear dynamic system. In many cases, it will be beneficial to the researchers in analyzing the nonlinear behavior of a system if a three-dimensional periodic-chaotic region diagram can be generated for two system parameters to be simultaneously considered. Periodicity-Ratio provides the availability for such diagrams to be plotted. Figure 7.26 illustrates such a diagram for the pendulum system in equation (7.23) with the varying parameters of Q and F. In the figure, PR is designated for Periodicity-Ratio.

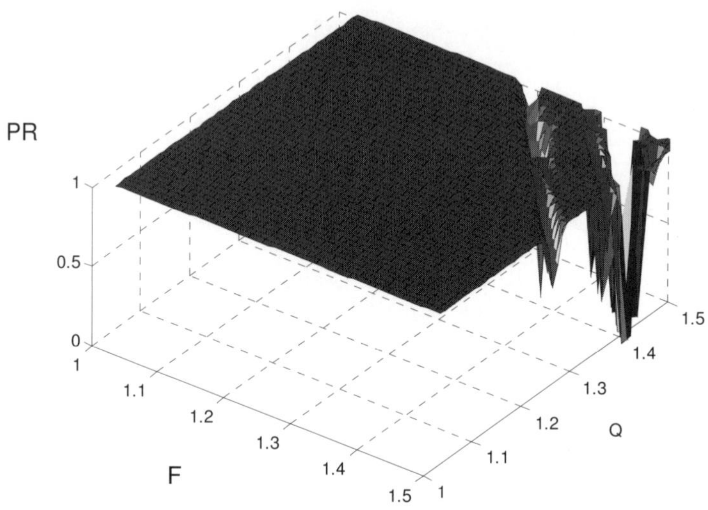

Figure 7.26.   Periodicity-Ratio diagram corresponding to varying values of Q and F.

7.8.   Characteristics of Periodicity-Ratio

Once the Periodicity-Ratio is developed and the approach with Periodicity-Ratio is presented, characteristics of Periodicity-Ratio need to be investigated thoroughly. The applications of Periodicity-Ratio and Lyapunov-Exponent in analyzing the behavior of nonlinear dynamic systems can be discussed and compared. Such a systematic approach for the characteristics of the Periodicity-Ratio and the comparison with that of Lyapunov-Exponent is significant for implementing Periodicity-Ratio as an index and criterion.

Based on the analyses and discussions in the previous sections, it can be concluded that Periodicity-Ratio is an effective tool to quantify the characteristic of a nonlinear dynamic system in addition to diagnosing chaos from regular motions and distinguishing periodic motions from the nonperiodic ones. The following characteristics are important and need to be summarized here for implementing the Periodicity-Ratio in nonlinear dynamics analyses.

(a) Periodicity-Ratio quantitatively describes the periodic behavior of a nonlinear dynamic system. The higher the Periodicity-Ratio value is, under the condition of $0 \leq \gamma \leq 1$, the more periodic is the system. The Periodicity-Ratio is one when the system is perfectly periodic. When Periodicity-Ratio approaches zero, the corresponding motion of the dynamic system can be chaotic.

(b) The application of Periodicity-Ratio for diagnosing chaos is to reveal the periodicity, irregularities and perfect non-periodicity of a nonlinear system.

(c) The Periodicity-Ratio makes it available for charactering a non-linear system with a single numerical value. Though, Lyapunov-Exponent is also a single value criterion, in practice however, the largest Lyapunov-Exponent needs to be identified first before it can be properly used for the diagnosis.

(d) For a dynamic system governed by nonlinear differential equations, there may exist infinite number of nonperiodic solutions, which are neither periodic nor chaotic. For these nonperiodic solutions, the corresponding Periodicity-Ratio values are in the range of $0 < \gamma < 1$.

(e) The response of a nonlinear system may switch intermittently between regular behavior and chaotic behavior. The intermittency has been identified numerically and experimentally for a number of cases (Platt *et al.* 1993, Hammer *et al.* 1994). Periodicity-Ratio provides a useful tool for analyzing the intermittent behaviors.

(f) In certain cases, the largest Lyapunov-Exponent is negative or positive and its absolute value varies around zero. For these cases, the Lyapunov-Exponent is no longer valid for diagnosing chaos, as it may lead to unstable or even wrong predictions. However, application of the Periodicity-Ratio is not restricted by these cases.

(g) Quasiperiodic phenomena can be efficiently diagnosed with the Periodicity-Ratio, since all the data for a Poincare map are determined in the process of computing for the Periodicity-Ratio. Moreover, with the Periodicity-Ratio, pure periodic cases are distinguished from quasiperiodic cases. However, these two types of motion can not be differentiated by Lyapunov-Exponent.

(h) With the Periodicity-Ratio, periodic-chaotic region diagrams can be generated. The diagrams provide clear graphical information for the regions of periodic, quasiperiodic and chaotic motions with various system parameters. Moreover, the diagrams present transition regions which represent nonperiodic and non-chaotic motions of a dynamic system considered. Such transitions can not be identified by using Lyapunov-Exponents, because Lyapunov-Exponents only justify whether or not a system is chaotic or periodic. More significantly, with the Periodicity-Ratio, the nonlinear behavior of the system in the transition regions can be quantitatively described. The transition regions also provide the information for the paths from one type of motion to another type of motion. It should be noted, in generating the periodic-chaotic region diagrams, there is no need for plotting any diagrams such as wave curves, phase diagrams or Poincare maps. Conventionally, however, plotting these diagrams is necessary for generating the periodic-chaotic diagrams.

(i) For describing the characteristics of a nonlinear dynamic system, if the frequency of the external excitation is known, the relation between the frequency of the excitation and the frequencies of the system can be obtained for periodic cases with the help of Periodicity-Ratio.

(j) Making use of the Periodicity-Ratio, different form of periodic phenomena can be identified and the period of a given periodic phenomenon of a nonlinear system can be easily determined during the process of determining for the Periodicity-Ratios.

(k) For the nonlinear dynamic systems to which the external excitations are nonperiodic or unknown, the Periodicity-Ratio can be used for analyzing the behavior of the systems and determining for the periods of the systems, provided that enough data regarding the motion of the systems are known.

7.9. Implementation of Periodicity-Ratio in Analyzing Nonlinear Dynamic Problems

For a nonlinear dynamical system, it is important to find as to under what initial conditions and for what values of the system parameters periodic, quasiperiodic or chaotic motion is possible. To describe the application of Periodicity-Ratio in diagnosing chaos of a nonlinear dynamic system with considerations of various parameter values and initial conditions, let us start with an example. Through the applications of the Periodicity-Ratio in the example, the readers may have a clearer picture of the methodology of implementing Periodicity-Ratio in analyses of nonlinear systems.

A nonlinear dynamic system governed by the Duffing's equation in the form shown in equation (7.2) demonstrates rich nonlinear properties and is therefore good for the description of applying of Periodicity-Ratio. This system has been systematically studied by Ueda via a graphical approach (Moon 1987, Ueda 1980). The findings of Ueda's study can be used for comparison with the approach of Periodicity-Ratio. Ueda found that a wide variety of periodic and chaotic behavior can be obtained with varying the two parameters of k and B of the governing equation. His diagram illustrating the regions of chaotic motions is widely cited (Moon 1987, Thompson *et al.* 1986). A periodic-chaotic region diagram similar to that provided by Ueda can be constructed with utilization of the Periodicity-Ratio, since periodic and chaotic motions can be diagnosed by the single-value criterion γ. In constructing the periodic-chaotic region diagram, the P-T method can be used to numerically solve Duffing's equation and to calculate the points for Poincare map. The complete procedure for numerically determining the Periodicity-Ratio is shown in the following flow chart, on the basis of the equations developed in Section 7.5.

With the data for Poincare map, the Periodicity-Ratio corresponding to different values of k and B in the Duffing's equation (7.2) can be determined by employing equation (7.22). Consequently, the nonperiodic cases with γ equals or very close to zero may then be differentiated from all of the other cases. The nonperiodic data such selected involving two cases of chaos and quasiperiodic cases, as described previously.

In order to distinguish chaotic cases from all the other states of motion including quasiperiodic cases, the procedures presented in Section 7.6 can be used.

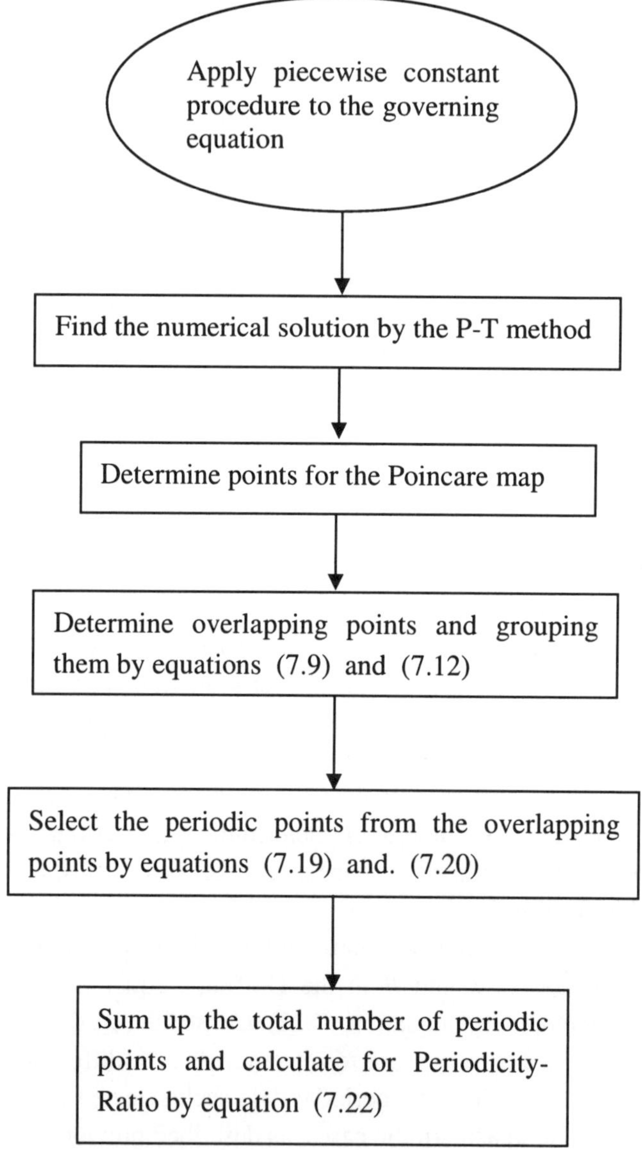

Figure 7.27. Flow chart for determining Periodicity-Ratio.

By utilizing Periodicity-Ratio and the other techniques discussed above, there are four cases of motion can be determined, namely periodic, quasiperiodic, nonperiodic and chaotic cases. With these states of motion determined, a detailed periodic-chaotic diagram can be constructed. For constructing such a diagram with the four states of motion demonstrated, the flow chart showing in Figure 7.28 can be employed.

Taking k and B in equation (7.2) as the two varying parameters of the system, Figure 7.25 is developed to show the periodic and chaotic regions determined by the corresponding Periodicity-Ratios calculated. This periodic-chaotic region diagram presents a global aspect of motion for the system. Each point in the diagram represents a state of motion. Therefore, the diagram allows a simultaneous comparison of the periodic and chaotic behavior of the system with varying system parameters and variety of initial conditions.

To find out the chaotic behavior of the system, for example, a chaotic case considered can be identified from the diagram and the phase diagram, Poincare map or wave form for the case can then be plotted if needed. Similar diagrams can be developed with considerations of the other variables. The dependence of chaos on initial conditions, for instance, can be shown in a chaotic region diagram under two different initial conditions. This diagram is plotted in Figure 7.29.

It can be seen from Figure 7.29, there are some regions which are initial condition dependent. Chaotic motion may not occur for some initial conditions.

To emphasize the regions in which the chaotic motions may depend upon initial conditions, chaos regions under various initial conditions are further investigated and the results are presented in Figure 7.30. The initial condition dependent regions shown in the figure are almost identical to those presented by Ueda.

The unique chaotic regions in Figure 7.30 also compare well with the results by Ueda (1980) except in two areas. The first area is a window around the point at $k = 0.25$ and $B = 11.4$ as shown in the figure. According to Ueda's diagram, the motion in this area is uniquely chaotic. However, based on the Periodicity-Ratios calculated for Figure 7.30, the motions in this area are found to be periodic or nonchaotic no matter what initial conditions are taken. As illustrated by a phase trajectory for the

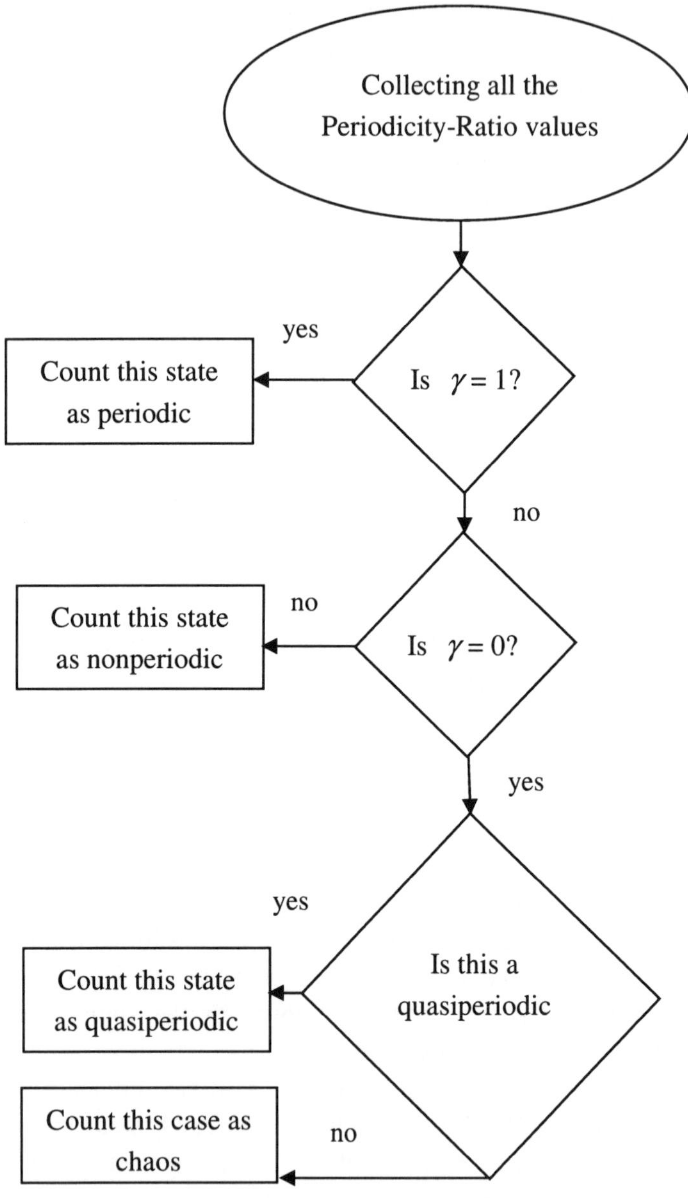

Figure 7.28.   Flow chart for determining data for periodic-chaotic diagram.

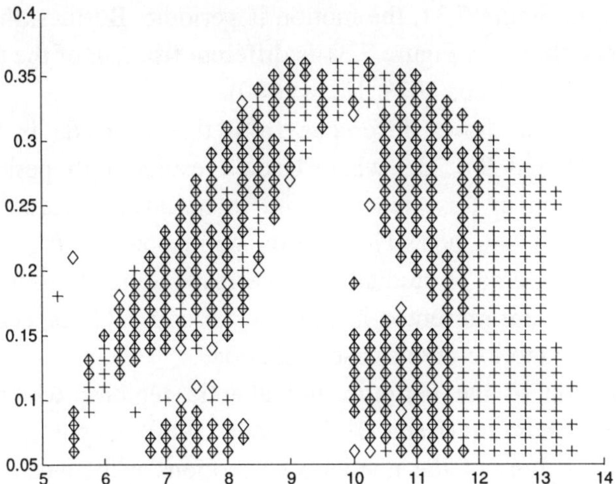

Figure 7.29. Chaotic region diagram for $\ddot{x} + k\dot{x} + x^3 = B\cos t$ under two initial conditions: diamonds for $d_0 = -2$, $v_0 = 0$; and crosses for $d_0 = 3$, $v_0 = 0$.

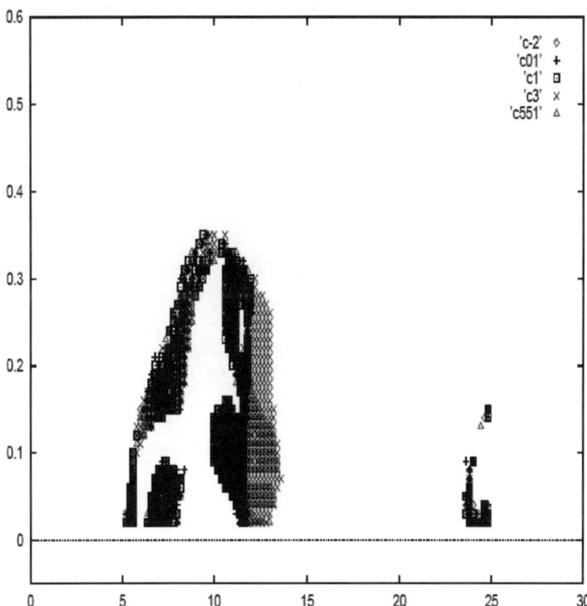

Figure 7.30. Chaotic region diagram of the system governed by $\ddot{x} + k\dot{x} + x^3 = B\cos t$ under various initial conditions. Initial condition: $'c-2'$: $d_0 = -2$, $v_0 = 0$; $'c01'$: $d_0 = 0$, $v_0 = 1$; $'c1'$: $d_0 = 1$, $v_0 = 0$; $'c3'$: $d_0 = 3$, $v_0 = 0$; $'c551'$: $d_0 = 5.5$, $v_0 = 1$.

above point in Figure 7.31, the motion is periodic. Besides, the shape of the trajectory shown in Figure 7.31 is different from all of the trajectories collected by Ueda (Ueda 1979, Ueda 1980).

Another area shows difference from that of Ueda is the region around $k = 0.5$ and $B = 23.9$ where Ueda's results imply periodic vibration. The motion in this area is found to be chaotic according to the Periodicity-Ratio obtained. The motion corresponding to $k = 0.5$ and $B = 23.9$ in this area is studied. The corresponding Poincare map and phase trajectory are presented in Figures 7.5 and 7.32 respectively. It is evident that the motion in this case is chaotic.

In addition to the periodic and chaotic regions, which are also reported by Ueda in (Ueda 1979, Ueda 1980), some quasiperiodic regions and regions of the nonperiodic motions in between chaos and periodic motions are found in constructing the periodic chaotic region diagrams. Although the quasiperiodic and nonperiodic regains were not reported by the Ueda and some other authors for this system, there are important to be noticed for the nonlinear analysis of dynamic systems. The quasiperiodic regions and the regions of the other motions which are neither periodic nor chaotic are represented by the unmarked regions in Figure 7.25.

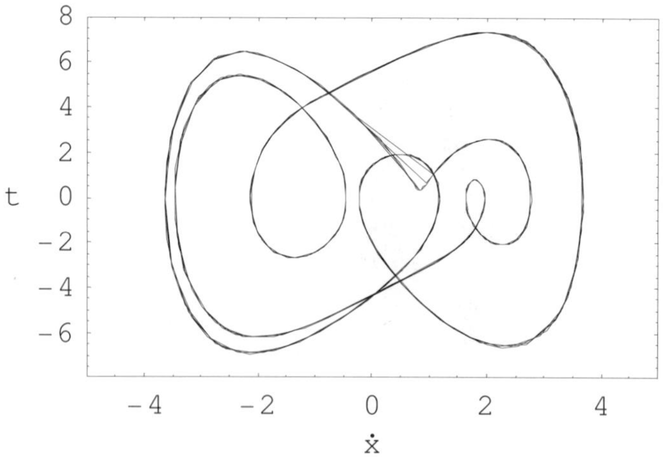

Figure 7.31.　Phase trajectory of a periodic case governed by $\ddot{x} + k\dot{x} + x^3 = B\cos t$. $k = 0.26$, $B = 11.4$, $d_0 = -2$, $v_0 = 0$.

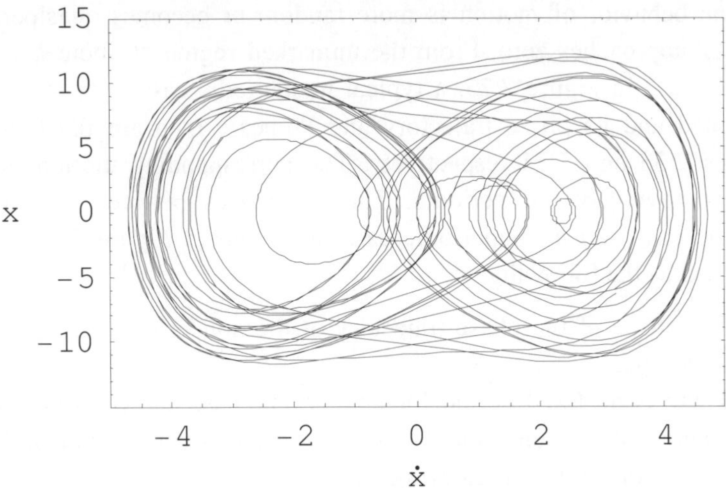

Figure 7.32. Phase trajectory of a chaotic case governed by $\ddot{x} + k\dot{x} + x^3 = B\cos t$. $k = 0.5$, $B = 23.9$, $d_0 = -2$, $v_0 = 0$.

A typical quasiperiodic case found is shown in Figure 7.6. As can be seen from the figure, a curve can be constructed to join all the points in the Poincare map for this quasiperiodic case and the curve is smooth and continuous. It can also observed from the figure that the number of overlapping points for this quasiperiodic case is negligibly small as compared to the total number of points appearing in the Poincare map. This leads to the zero value of the Periodicity-Ratio corresponding to this quasiperiodic case. Obviously, the behavior of a quasiperiodic case is different from that of a chaotic case. Since both a quasiperiodic case and a chaotic case have zero Periodicity-Ratio, it is necessary to distinguish the quasiperiodic cases from chaos. In distinguishing a quasiperiodic motion from chaos, the least-square method with combination of periodicity ratio can be used, through the procedure discussed previously. It should be noticed that there is no need to plot any figures for distinguishing the quasiperiodic and chaotic cases, in utilizing the procedure presented in Section 7.6.

Within the range of $0 < \gamma < 1$, corresponding to the regions of the nonperiodic motions in between chaos and periodic motions, the behavior of motion tends to be periodic as γ becomes close to unity, and the behavior of motion is more random or becomes quasiperiodic when γ approaches zero. From the unmarked region at about $k = 0.67$ and $B = 5.2$ in Figure 7.25, a typical nonperiodic case with $\gamma = 0.8$ is examined and its phase trajectory and Poincare map are illustrated in Figures 7.33 and 7.34 respectively. The corresponding motion in this case is obviously not periodic. The differences between a perfectly periodic case and a nonperiodic case as shown in Figure 7.33 are be identified as the following:

(a)   The shape of the phase trajectory of the nonperiodic case changes with time.

(b)   There is no fixed period for nonperiodic case and the time for an identical displacement and velocity to occur is not constant and may vary as the motion is carrying on.

(c)   The points appearing in the Poincare map are not all overlapped.

However, if variations of the phase trajectory are very slight, the corresponding Periodicity-Ratio should be close to a unity. Accordingly, the behavior of motion in this case is therefore almost periodic as indicated in the figures.

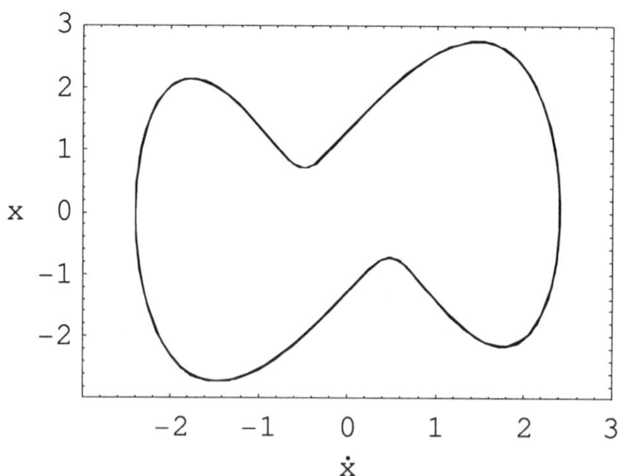

Figure 7.33.   Phase trajectory of a nonperiodic case ($\gamma = 0.8$) governed by $\ddot{x} + k\dot{x} + x^3 = B \cos t$. $k = 0.67$, $B = 5.2$.

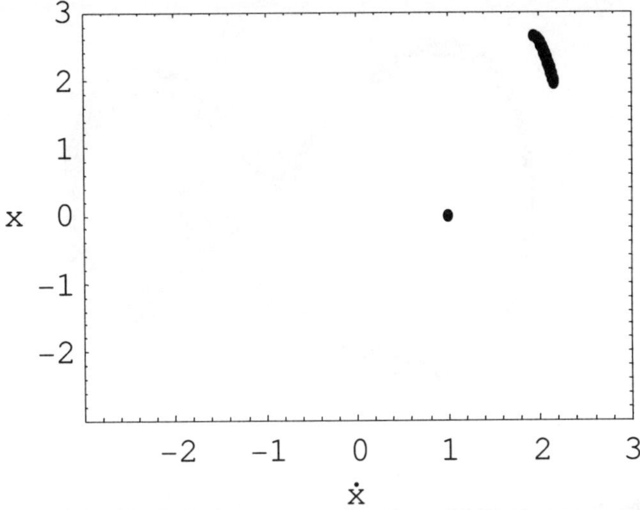

Figure 7.34. Poincare map of a nonperiodic case ($\gamma = 0.8$) governed by $\ddot{x} + k\dot{x} + x^3 = B \cos t$. $k = 0.67$, $B = 5.2$.

In comparison with the almost periodic case above, a case of highly nonperiodic with small Periodicity-Ratio ($\gamma = 0.17$) is taken from an unmarked region in Figure 7.25, around $k = 0.35$ and $B = 9.2$. Figure 7.35 is the phase trajectory corresponding to this case. It can be seen from the figure that the motion in this case is far from periodic. Nevertheless, the motion is not chaotic either since it is not completely random. By the corresponding Poincare map shown in Figure 7.36, the overlapping points at the two ends of the curve can be observed. Therefore, the behavior of the corresponding motion is nonperiodic and nonchaotic but shows more randomness.

With the discussion about the application of Periodicity-Ratio in analyses of nonlinear dynamic systems, it can be seen that the Periodicity-Ratio reveals the behavior of motion of nonlinear oscillatory systems and can be used as an effective criterion for periodic and chaotic motions. With the Periodicity-Ratio, chaotic motion can be conveniently diagnosed for a dynamical system and periodic chaotic region diagrams may be obtained such that a simultaneous comparison of the chaotic and periodic behavior with varying system parameters and various initial conditions becomes available. Making use of the Periodicity-Ratio, as

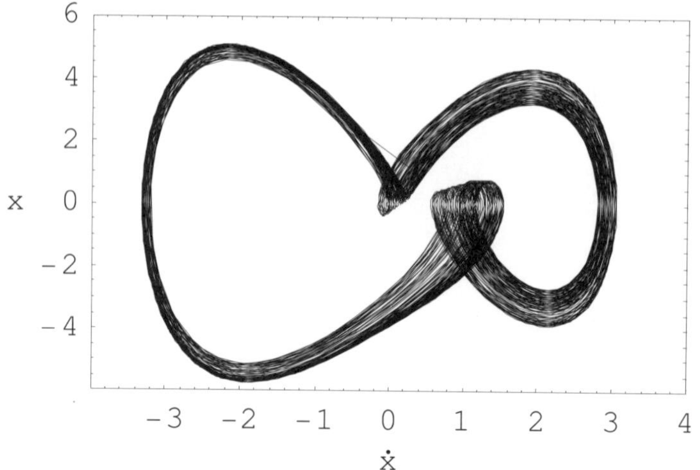

Figure 7.35. Phase trajectory of a nonperiodic motion governed by $\ddot{x} + k\dot{x} + x^3 = B\cos t$. $k = 0.35$, $B = 9.2$, $\gamma = 0.17$.

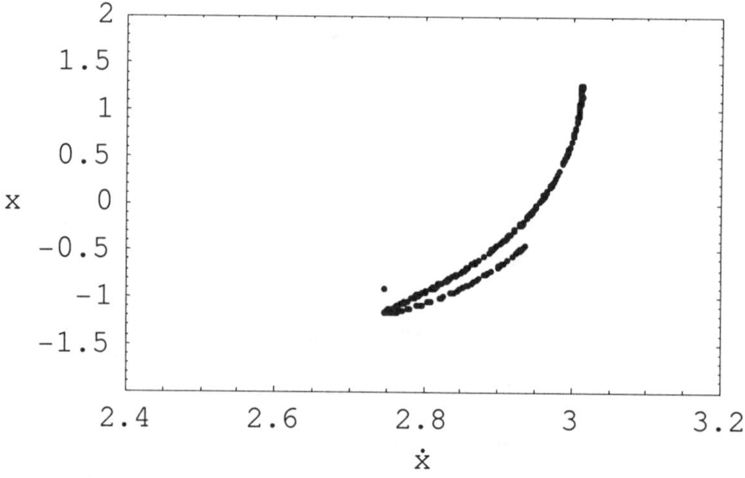

Figure 7.36. Poincare map of a nonperiodic motion for $\ddot{x} + k\dot{x} + x^3 = B\cos t$. $k = 0.35$, $B = 9.2$, $\gamma = 0.17$.

indicated in the above discussion and Appendix C, the diagnostic procedure for chaotic, quasiperiodic and periodic motions can be completed by a single computer program without plotting any of the wave forms, phase trajectories and Poincare maps. In analyzing the behavior of motion for nonlinear dynamical systems with implementtation of Periodicity-Ratio, it is significant and beneficial to have computers do all the work of "observing" the behavior of motion and distinguishing the periodic, nonperiodic, quasiperiodic and chaotic motions.

References

Baierlein, R., Atoms and Information Theory, W. H. Freeman and Co., San Francisco, 1971.

Baker, G. L. and Gollub, J. P., Chaotic Dynamics an Introduction, Cambridge University Press, Cambridge, 1990.

Blackburn, J. A., Vik, S., Wu, B. and Smith, H. J. T., "Driven Pendulum for Studying Chaos," Review of Scientific Instruments, Vol. 60, pp. 422-426, 1989.

Ciliberto, S. and Gollub, J. P., "Chaotic Mode Competition in Parametrically Forced Surface Waves," Journal of Fluid Mechanics, Vol. 158, pp. 381-398, 1985.

Dai, L. and Singh, M. C., "Periodicity-ratio in Diagnosing Chaotic Vibrations," Proceedings of 15th Canadian Congress of Applied Mechanics, Vol. 1, pp. 390-391, 1995.

Dai, L. and Singh, M. C., "Diagnosis of Periodic and Chaotic Responses in Vibratory Systems," Journal of Acoustic Society of America, Vol. 102, No. 6, pp. 3361-3371, 1997.

Dai, L. and Singh, M. C., "A New Approach with Piecewise-Constant Arguments to Approximate and Numerical Solutions of Oscillatory Problems," Journal of Sound and Vibration, Vol. 263, No. 3, pp. 535-548, 2003.

Draper, N. R. and Smith, H., Applied Regression Analysis, second edition, John Wiley and Sons, Inc., New York, 1981.

Farmer, J. D., Ott, E. and Yorke, J. A., "The Dimension of Chaotic Attractors," Physica, Vol. 7D, pp. 153-170, 1983.

Forsythe, G. E., "General and Use of Orthogonal Polynomials for Fitting Data with a Digital Computer," SIAM Journal on Applied Mathematics, Vol. 5, pp. 74-88, 1957.

Gwinn, E. G. and Westervelt, R. M., "Intermittent Chaos and Low-Frequency Noise in the Driven Damped Pendulum," Physical Review Letters, Vol. 55, No. 15, pp. 1613-1616, 1985.

Guckenheimer, J. and Holmes, P., Nonlinear Oscillations, Dynamical Systems, and Bifurcations of Vector Fields, Springer Verlag, New York, 1983.

Gwinn, E. G. and Westervelt, R. M., "Fractal Basin Boundaries and Intermittency in the Driven Damped Pendulum," Physical Review A, Vol. 33, pp. 4143-4155, 1986.

Hammer, P.W., Platt, N., Hammel, S.M., Heagy, J.F., and Lee, B.D., "Experimental Observation of On-Off Intermittency," Physical Review Letters, Vol. 73, pp. 1095-1098, 1994.

He, D., Xu, J., Chen, Y. and Tan, N., "A Simple Method for the Computation of the Conditional Lyapunov Exponents," Communications in Nonlinear Science and Numerical Simulation, Vol. 4, No. 2, pp. 113-117, 1999.

Hoppensteadt, F. C., Analysis and Simulation of Chaotic Systems, Springer-Verlag, New York, 1993

Kennedy, W. J., Jr. and Gentle, J. E., Statistical Computing, Marcel Dekker, New York, 1980.

Lakshmanan, M. and Rajasekar, S., Nonlinear Dynamics: Integrability, Chaos, and Patterns, Springer, New York, 2003.

Lazzouni, S.A., Bowong, S., Kakmeni, F.M.M., Cherki, B. and Ghouali, N., "Chaos Control Using Small-Amplitude Damping Signals of the Extended Duffing Equation," Communications in Nonlinear Science and Numerical Simulation, Vol. 12, No. 5, pp. 804-813, 2007.

Lichtenberg, A. J. and Lisberman, M. A., Regular and Stochastic Motion, Springer-Verlag, New York, 1983.

Lu, C., "Chaos of a Parametrically Excited Undamped Pendulum," Communications in Nonlinear Science and Numerical Simulation, Vol. 12, No. 1, pp. 45-57, 2007.

Lu, J., Yang, G., Oh, H. and Luo, A.J., "Computing Lyapunov-Exponents of Continuous Dynamical Systems: Method of Lyapunov Vectors," Chaos, Solitons and Fractals, Vol. 23, pp. 1879-1892, 2005.

Manneville, P. and Pomeau, Y., "Different Ways to Turbulence in Dissipative Dynamical Systems," Physica, Vol. 1D, pp. 219-226, 1980.

Moon, F. C., "Experimental Models for Strange Attractor Vibration in Elastic Systems," in New Approaches to Nonlinear Problems in Dynamics, Holmes, P. J. (ed.), pp. 487-495, 1980 (A).

Moon, F. C., "Experiments on Chaotic Motions of a Forced Nonlinear Oscillator: Strange Attractors," ASME Journal of Applied Mechanics, Vol. 47, pp. 638-644, 1980 (B).

Moon, F. C., Chaotic Vibrations, John Wiley & Sons, Inc., New York, 1987.

Nayfeh, A.H. and Mook, D.T., Nonlinear Dynamics, American Society of Mechanical Engineers, New York, NY: 1988.

Platt, N, Spiegel, E.A. and Tresser, C., "On-Off Intermittency: A Mechanism for Bursting," Physical Review Letters, Vol. 70, pp. 279-282, 1993.

Poincare, H., The Foundation of Science (Translated by G.B. Halsed), Science Press, Lancaster, 1946.

Rong, H., Meng, G, Wang, X., Xu, W. and Fang, T., "Invariant Measures and Lyapunov-Exponents for Stochastic Mathieu System," Nonlinear Dynamics, Vol. 30, pp. 313-321, 2002.

Shahverdian, A., Yu. and Apkarian, A.V., A Difference Characteristic for One-Dimensional Deterministic Systems," Communications in Nonlinear Science and Numerical Simulation, Vol. 12, No. 3, pp. 233-242, 2007.

Siegwart, D., Nonlinear Dynamics: Applications and Visualisation Techniques, IBM UK Scientific Centre, 1991.

Terzic, B. and Kandrup, H.E., "Semi-Analytic Estimates of Lyapunov Exponents in Lower-Dimensional Systems," Physics Letters, A311, 165, pp. 1-6, 2003.

Thompson, J. M. T. and Stewart, H. B., Nonlinear Dynamics and Chaos, John Wiley and Sons, Inc., New York, 1986.

Tsons, A. A., Chaos from Theory to Applications, Plenum Press, New York, 1992.

Ueda, Y., "Random Transitional Phenomena in the System Governed by Duffing's Equation," Journal of Statistical Physics, Vol. 20, pp. 181-196, 1979.

Ueda, Y., "Steady motions exhibited by Duffing's equation: a picture book of regular and chaotic motions, in New Approaches to Nonlinear Problems in Dynamics, Holmes, P. J. (ed.), SIAM, Philadelphia, pp. 311-322, 1980.

Weaver, W. Jr., Timoshenko, S., and Young, D.H., Vibration Problems in Engineering, John Wiley & Sons, Inc., New York, 1990.

Wolf, A., Swift, J. B., Swinney, H. L. and Vastano, J. A., "Determining Lyapunov Exponents from a Fine Series," Physica, Vol. 16D, pp. 285-317, 1985.

Yamapi, R. and Aziz-Alaoui, M.A., "Vibration Analysis and Bifurcations in the Self-Sustained Electromechanical System with Multiple Functions," Communications in Nonlinear Science and Numerical Simulation, Vol. 12, No. 8, pp. 1534-1549, 2007.

APPENDIX A

Mathematical Developments and Proofs

1. L'Hopital's Rule for Indeterminate

To determine for the limits of the indeterminate forms of 0/0 and ∞/∞, as that used in Chapters 3 and 4, L'Hopital's Rule is useful. The rule states, for two functions $f(x)$ and $F(x)$, if the limit of $f(x)/F(x)$ produces the indeterminate form 0/0 as x approaches c, then

$$\lim_{x \to c} \frac{f(x)}{F(x)} = \lim_{x \to c} \frac{f'(x)}{F'(x)} \tag{A.1}$$

L'Hopital's Rule is also valid for the limit of $f(x)/F(x)$ as x approaches c produces any one of the indeterminate forms of ∞/∞, $-\infty/\infty$, $\infty/-\infty$, or $-\infty/-\infty$.

The rule can be applied only if the following conditions are satisfied.

- The two functions $f(x)$ and $F(x)$ are differentiable on an open interval (a, b) containing c except possibly at c itself

- $\lim_{x \to a} \dfrac{f'(x)}{F'(x)}$ exists (or tends to infinity)

- Both $f'(x)$ and $F'(x)$ exist, and $F'(x) \neq 0$

L'Hopital's Rule can be continuously applied should the above conditions are satisfied, i.e.,

$$\lim_{x \to c} \frac{f(x)}{F(x)} = \lim_{x \to c} \frac{f'(x)}{F'(x)} = \lim_{x \to c} \frac{f''(x)}{F''(x)} = \cdots \tag{A.2}$$

297

The existence of the rule can be proved by the Cauchy Mean Value Theorem.

2. Proof of the Existence of for Following Relationship

$$\begin{bmatrix} \cos\omega & \dfrac{\sin\omega}{\omega} \\ -\omega\sin\omega & \cos\omega \end{bmatrix}^{[t]} = \begin{bmatrix} \cos(\omega[t]) & \dfrac{\sin(\omega[t])}{\omega} \\ -\omega\sin(\omega[t]) & \cos(\omega[t]) \end{bmatrix} \tag{A.3}$$

By the definition, $[t]$ is an integer. We may therefore replace $[t]$ by n where $n \in Z$. The above relationship is obviously correct as $n = 1$. For $n = 2$, one may have

$$\begin{bmatrix} \cos\omega & \dfrac{\sin\omega}{\omega} \\ -\omega\sin\omega & \cos\omega \end{bmatrix}^{2} = \begin{bmatrix} \cos\omega & \dfrac{\sin\omega}{\omega} \\ -\omega\sin\omega & \cos\omega \end{bmatrix}\begin{bmatrix} \cos\omega & \dfrac{\sin\omega}{\omega} \\ -\omega\sin\omega & \cos\omega \end{bmatrix}$$

$$= \begin{bmatrix} \cos^2\omega - \sin^2\omega & \dfrac{2\sin\omega\cos\omega}{\omega} \\ \dfrac{-2\sin\omega\cos\omega}{\omega} & -\sin^2\omega + \cos^2\omega \end{bmatrix} = \begin{bmatrix} \cos 2\omega & \dfrac{\sin 2\omega}{\omega} \\ -\omega\sin 2\omega & \cos 2\omega \end{bmatrix} \tag{A.4}$$

As such, the existence of the relationship can be proved by mathematical induction. Assume that the relationship exist as $n = k$ and $k > 2$, i.e.,

$$\begin{bmatrix} \cos\omega & \dfrac{\sin\omega}{\omega} \\ -\omega\sin\omega & \cos\omega \end{bmatrix}^{k} = \begin{bmatrix} \cos(k\omega) & \dfrac{\sin(k\omega)}{\omega} \\ -\omega\sin(k\omega) & \cos(k\omega) \end{bmatrix} \tag{A.5}$$

Therefore, the existence of the relationship would be proved should the relationship is also true as $n = k + 1$, via the implementation of the assumption above.

$$\begin{bmatrix} \cos\omega & \dfrac{\sin\omega}{\omega} \\ -\omega\sin\omega & \cos\omega \end{bmatrix}^{k+1} = \begin{bmatrix} \cos\omega & \dfrac{\sin\omega}{\omega} \\ -\omega\sin\omega & \cos\omega \end{bmatrix}^{k} \begin{bmatrix} \cos\omega & \dfrac{\sin\omega}{\omega} \\ -\omega\sin\omega & \cos\omega \end{bmatrix}$$

$$= \begin{bmatrix} \cos(k\omega) & \dfrac{\sin(k\omega)}{\omega} \\ -\omega\sin(k\omega) & \cos(k\omega) \end{bmatrix} \begin{bmatrix} \cos\omega & \dfrac{\sin\omega}{\omega} \\ -\omega\sin\omega & \cos\omega \end{bmatrix}$$

$$= \begin{bmatrix} \cos(k\omega)\cos\omega - \sin(k\omega)\sin\omega & \dfrac{\cos(k\omega)\sin\omega + \sin(k\omega)\cos\omega}{\omega} \\ -\omega\sin(k\omega)\cos\omega - \omega\sin\omega\cos(k\omega) & -\sin(k\omega)\sin\omega + \cos(k\omega)\cos\omega \end{bmatrix}$$

$$= \begin{bmatrix} \cos(k+1)\omega & \dfrac{\sin(k+1)\omega}{\omega} \\ -\omega\sin(k+1)\omega & \cos(k+1)\omega \end{bmatrix} \tag{A.6}$$

The existence of the relationship shown in equation (A.3) is therefore proved.

3. **Derivation of the Limits Shown in Equations (3.22), (3.23) and (3.24) as ω Approaches Zero**

For equation (3.22),

$$\lim_{\omega\to 0}\left\{\left(1-\frac{\beta}{\omega^2}\right)\cos[\omega(t-[t])] + \frac{\beta}{\omega^2}\right\}$$

$$= \lim_{\omega\to 0}\left\{\cos[\omega(t-[t])] + \frac{\beta}{\omega^2}(1-\cos[\omega(t-[t])])\right\}$$

$$= 1 + \beta\lim_{\omega\to 0}\frac{1-\cos[\omega(t-[t])]}{\omega^2} \tag{A.7}$$

Implementing L'Hopital's rule to obtain

$$\lim_{\omega\to0}\frac{1-\cos[\omega(t-[t])]}{\omega^2}=\lim_{\omega\to0}\frac{(t-[t])\sin[\omega(t-[t])]}{2\omega}$$
$$=\lim_{\omega\to0}\frac{(t-[t])^2\cos[\omega(t-[t])]}{2}=\frac{1}{2}(t-[t])^2 \qquad (A.8)$$

Thus, the following limit is correct

$$\lim_{\omega\to0}\left\{\left(1-\frac{\beta}{\omega^2}\right)\cos[\omega(t-[t])]+\frac{\beta}{\omega^2}\right\}=1+\frac{\beta}{2}(t-[t])^2 \qquad (A.9)$$

This may also be proved by using the following trigonometry identity

$$\cos\upsilon=1-\frac{\upsilon^2}{2} \qquad (A.10)$$

Therefore,

$$\lim_{\omega\to0}\left\{\left(1-\frac{\beta}{\omega^2}\right)\cos[\omega(t-[t])]+\frac{\beta}{\omega^2}\right\}$$
$$=\lim_{\omega\to0}\left\{\left(1-\frac{\beta}{\omega^2}\right)-\frac{\omega^2-\beta}{\omega^2}\frac{[\omega(t-[t])]^2}{2}+\frac{\beta}{\omega^2}\right\}$$
$$=\lim_{\omega\to0}\left\{1-\omega^2\frac{[(t-[t])]^2}{2}+\frac{\beta[(t-[t])]^2}{2}\right\}$$
$$=1+\frac{\beta[(t-[t])]^2}{2} \qquad (A.11)$$

For equation (3.23),

$$\lim_{\omega\to0}\left\{\left(1-\frac{\beta}{\omega^2}\right)\cos\omega+\frac{\beta}{\omega^2}\right\}$$
$$=\lim_{\omega\to0}\left\{\left(1-\frac{\beta}{\omega^2}\right)\left(1-\frac{\omega^2}{2}\right)+\frac{\beta}{\omega^2}\right\}$$
$$=1+\frac{\beta}{2} \qquad (A.12)$$

The above derivation utilizes the trigonometry identity as well. For equation (3.24),

$$\lim_{\omega \to 0}\left\{\frac{\beta - \omega^2}{\omega}\sin\omega\right\}$$

$$= \lim_{\omega \to 0}\left\{\beta\frac{\sin\omega}{\omega} - \omega\sin\omega\right\}$$

$$= \beta \qquad\qquad (A.13)$$

4. Matrix Manipulations

For an $n \times n$ square matrices $[A]$ with $|A| \neq 0$ and n eigenvalues $\lambda_1, \lambda_2, \ldots, \lambda_n$, one may have $[A]$ to be diagonalized as

$$[A] = [B][D][B]^{-1} = [B]\begin{bmatrix} \lambda_1 & 0 & \cdots & 0 \\ 0 & \lambda_2 & \cdots & 0 \\ \vdots & \vdots & \vdots & \vdots \\ 0 & 0 & \cdots & \lambda_n \end{bmatrix}[B]^{-1} \qquad (A.14)$$

This can be obtained from equation (B.19) of Appendix B with considerations that $[B]^{-1}[B] = [B][B]^{-1} = I$.

With this conclusion, the following can be proved.

$$[A]^k = [B][D]^k[B]^{-1} \qquad\qquad (A.15)$$

where the exponent k is a positive integer.

Expand $[A]^k$ such that

$$[A]^k = \underbrace{([B][D][B]^{-1})([B][D][B]^{-1})\cdots([B][D][B]^{-1})}_{\text{multiplication of } 3k \text{ square matrices}} \qquad (A.16)$$

By the associative law of matrix multiplication and utilizing $[B]^{-1}[B] = [B][B]^{-1} = I$, equation (A.15) can be obtained. Also, the multiplication of k diagonal matrix $[D]$ containing n eigenvalues can have the following expression.

$$[D]^k = \begin{bmatrix} \lambda_1 & 0 & \cdots & 0 \\ 0 & \lambda_2 & \cdots & 0 \\ \vdots & \vdots & \vdots & \vdots \\ 0 & 0 & \cdots & \lambda_n \end{bmatrix}^k$$

$$= \underbrace{\begin{bmatrix} \lambda_1 & 0 & \cdots & 0 \\ 0 & \lambda_2 & \cdots & 0 \\ \vdots & \vdots & \vdots & \vdots \\ 0 & 0 & \cdots & \lambda_n \end{bmatrix}\begin{bmatrix} \lambda_1 & 0 & \cdots & 0 \\ 0 & \lambda_2 & \cdots & 0 \\ \vdots & \vdots & \vdots & \vdots \\ 0 & 0 & \cdots & \lambda_n \end{bmatrix} \cdots \begin{bmatrix} \lambda_1 & 0 & \cdots & 0 \\ 0 & \lambda_2 & \cdots & 0 \\ \vdots & \vdots & \vdots & \vdots \\ 0 & 0 & \cdots & \lambda_n \end{bmatrix}}_{\text{multiplication of } k \text{ square matrices}}$$

$$= \begin{bmatrix} \lambda_1^k & 0 & \cdots & 0 \\ 0 & \lambda_2^k & \cdots & 0 \\ \vdots & \vdots & \vdots & \vdots \\ 0 & 0 & \cdots & \lambda_n^k \end{bmatrix} \qquad (\text{A.17})$$

by the definition of the multiplication of the matrices.

APPENDIX B

Theory of Matrices

1. Definitions

For a given matrix with m rows n columns in the following form

$$[A] = \begin{bmatrix} a_{11} & a_{12} & \cdots & a_{1n} \\ a_{21} & a_{22} & \cdots & a_{2n} \\ \vdots & \vdots & \cdots & \vdots \\ a_{m1} & a_{m2} & \cdots & a_{mn} \end{bmatrix} \tag{B.1}$$

is known as a $m \times n$ matrix. The scalar values are the elements of the matrix, and each element is designated by a_{ij} with row subscript i and column subscript j respectively.

For a square matrix ($n \times n$ matrix) $[A]$, all the elements a_{ij} of which $i = j$ are known as diagonal elements. A square matrix is called identity matrix if all the diagonal elements are 1's and the other elements are zeros. An identity matrix is usually designated by I. Some important properties of identity matrix are shown below.

$$[A]I = I[A] = [A] \tag{B.2}$$

$$I^n = I, \qquad n = 1, 2, 3, \ldots. \tag{B.3}$$

2. Transpose of Matrices

The transpose of the matrix corresponding to the matrix $[A]$ above is denoted as the following.

$$[A]^T = \begin{bmatrix} a_{11} & a_{12} & \cdots & a_{m1} \\ a_{21} & a_{22} & \cdots & a_{m2} \\ \vdots & \vdots & \cdots & \vdots \\ a_{1n} & a_{2n} & \cdots & a_{mn} \end{bmatrix} \qquad (B.4)$$

3.   Manipulations of Matrices

- Addition and subtraction

  Addition and subtraction of matrices can be performed only if they are conformable for the operation, i.e., the matrices involved have the identical rows and columns. The sum of two $m \times n$ matrices $[A]$ and $[B]$ can be given by

$$[C] = [A] \pm [B] \qquad (B.5)$$

where the elements of the sum

$$c_{ij} = a_{ij} \pm b_{ij} \qquad (B.6)$$

- Multiplication

  The multiplication of a $m \times p$ matrix $[A]$ and a $p \times n$ matrix $[B]$ can be defined as

$$[C] = [A][B] \qquad (B.7)$$

where the elements of the product

$$c_{ij} = \sum_{k=1}^{p} a_{ik} b_{kj} \qquad (B.8)$$

4.   Determinants

For an $n \times n$ square matrix, its determinant can be defined as

$$\det[A] = |A| = \begin{vmatrix} a_{11} & a_{12} & \cdots & a_{1n} \\ a_{21} & a_{22} & \cdots & a_{2n} \\ \vdots & \vdots & \cdots & \vdots \\ a_{n1} & a_{n2} & \cdots & a_{nn} \end{vmatrix} \qquad (B.9)$$

The determinant can be determined by the following summation

$$|A| = \sum_{i=1}^{n} a_{ij} C_{ij} \tag{B.10}$$

in which

$$C_{ij} = (-1)^{i+j} \left| M_{ij} \right| \tag{B.11}$$

where $\left| M_{ij} \right|$ known as a cofactor is a determinant obtained by removing row I and column j for $|A|$.

5. Matrix Inversion

Inversion of a square matrix $[A]$ denoted by $[A]^{-1}$ is defined as

$$[A]^{-1} = \frac{adj[A]}{|A|} \tag{B.12}$$

where $adj[A]$ is known as adjoint of $[A]$ and defined as $adj[A] = [C]^{T}$. The square matrix $[C]$ is a cofactor matrix with the elements defined in (B.11). Several important properties of inverse of matrices can be listed as the following.

$$[A][A]^{-1} = [A]^{-1}[A] = I \tag{B.13}$$

$$([A][B])^{-1} = [B]^{-1}[A] \tag{B.14}$$

6. Eigenvalues and Eigenvectors

Assume $[A]$ is an $n \times n$ square matrix, $[X]$ is a matrix with only one column such that

$$\begin{bmatrix} a_{11} & a_{12} & \cdots & a_{1n} \\ a_{21} & a_{22} & \cdots & a_{2n} \\ \vdots & \vdots & \vdots & \vdots \\ a_{1n} & a_{2n} & \cdots & a_{nn} \end{bmatrix} \begin{bmatrix} x_1 \\ x_2 \\ \vdots \\ x_n \end{bmatrix} = \lambda \begin{bmatrix} x_1 \\ x_2 \\ \vdots \\ x_n \end{bmatrix} \tag{B.15}$$

This equation has nontrivial solutions if and only if the following condition is satisfied.

$$\begin{vmatrix} a_{11}-\lambda & a_{12} & \cdots & a_{1n} \\ a_{21} & a_{22}-\lambda & \cdots & a_{2n} \\ \vdots & \vdots & \vdots & \vdots \\ a_{n1} & a_{n2} & \cdots & a_{nn}-\lambda \end{vmatrix}=0 \tag{B.16}$$

The above equation with respect to λ is known as characteristic equation of matrix $[A]$. The roots of the equation are called eigenvalues of the matrix. Corresponding to each of the eigenvalues, there will be solutions for $[X]\neq 0$. The matrix composes the solutions of $[X]$ corresponding to the eigenvalue is known as an eigenvector of $[A]$.

7.   Similarity of Matrices

For two $n\times n$ square matrices $[A]$ and $[B]$ are called similar if there exists a nontrivial matrix $[R]$ such that

$$[B]=[R]^{-1}[A][R] \tag{B.17}$$

8.   Diagonalization of Matrices

For an $n\times n$ square matrices $[A]$ and $|A|\neq 0$. Assume the matrix has n eigenvalues λ_1, $\lambda_2,\ldots$, λ_n, and the corresponding eigenvectors can be used to construct a matrix in which the eigenvectors arrayed as columns, such that

$$[B]=\begin{bmatrix} b_{11} & b_{12} & \cdots & b_{1n} \\ b_{21} & b_{22} & \cdots & b_{2n} \\ \vdots & \vdots & \vdots & \vdots \\ b_{n1} & b_{n2} & \cdots & b_{nn} \end{bmatrix} \tag{B.18}$$

One may have

$$[B]^{-1}[A][B] = [D] = \begin{bmatrix} \lambda_1 & 0 & \cdots & 0 \\ 0 & \lambda_2 & \cdots & 0 \\ \vdots & \vdots & \vdots & \vdots \\ 0 & 0 & \cdots & \lambda_n \end{bmatrix} \tag{B.19}$$

where $[D]$ is the diagonal matrix with the eigenvalues of $[A]$.

APPENDIX C

Computer Programs for Analyses of Dynamics

This appendix provides three computer program listings which may be used in their present or modified versions for solving the differential equations of dynamic systems and generating data for plotting the graphs discussed in this book. The listings in this appendix are in Fortran. The first program solves a nonlinear differential equation of an oscillatory system with the piecewise-constant procedure and provides a two-dimensional phase plane representation of the system. The second program generates data for a Poincare map by using the P-T method and diagnoses whether a motion is chaotic, quasiperiodic or periodic so that a periodic-quasiperiodic-chaotic region diagram can be constructed.

COMPUTER PROGRAM 1

```
C   PROGRAM: PIECEWISE-CONSTANT METHOD
C
C   PURPOSE:
C         This program is for calculating data to draw phase
C         trajectories of a driven Froude pendulum system. The
C         program first solves the equation of motion of Froude
C         pendulum described by
C               d2U/dt2+MU*AL*dU/dt+MU*BE*
C               (dU/dt)**3+D*SINU=FT+Fcos(t)
C         where MU, AL (Alpha), BE (Beta), D and F are
C         constants, [Nt] is the greatest integer function, U is the
C         angular displacement of the pendulum and dU/dt is the
C         angular velocity. The values of U and dU/dt are
```

```
C            recorded into a file 'pha.da'. A phase plane can be plotted
C            with the data in the file.
C
C   TYPE: MAIN PROGRAM
```
■■
```
C
C        VARIABLE DEFINITION:
C        TMIN: minimum time for a stable state
C        TMAX: time length
C        DN: local initial angular displacement
C        VN: local initial angular velocity
C        T: time
C        TN: time step
C        EPS: accuracy (epsilon)
C        PHI: initial phase
C        PI:   constant ACOS(-1.0)
C        N: parameter that controls accuracy and time step

C        VARIABLE DECLARATION
         REAL DN,VN,AL,BE,D,F,TMIN,TMAX,T,MU
         REAL X,V,TN,XNEW,VNEW,TNEW,XD,VD,THALF
         REAL XN,VN
         REAL D1,D2,DELTA,PHI,EPS,PI,W1,W2,N,A,B

C        OPEN A DATA FILE FOR SAVING THE DATA OF
C        U AND dU/dt
         OPEN (7, FILE='PHA.DA')

C        DATA INPUT
         READ (12,*) AL,BE,MU,D,F,DN,VN,TMIN,TMAX,PHI
         READ(12,*) EPS,N

C        INPUT DATA VERIFICATION
         WRITE(*,*) 'A=',AL,'   B=',BE,'   D=',D,'   F=',F,'
PHI',PHI
         WRITE(*,*)  'DN=',DN,'        VN=',VN,'        TMIN',TMIN,'
```

```fortran
TMAX',TMAX
      WRITE(*,*) 'EPS',EPS,'    N',N


C       DATA INITIATION
C         T=0.0
          X=DN
          V=VN
          A=AL*MU
          B=BE*MU
          PI=ACOS(-1.0)
          TN=1.0/N


C       CALCULATE U AND V WHILE TIME t IS LESS THAN
C       TMAX
222 IF(T.LT.TMAX) THEN
      CALL PC(N,X,V,T,XNEW,VNEW,A,B,F,D)


C DOUBLE N VALUE
      CALL PC(2.0*N,X,V,T,XD,VD,A,B,F,D)
      THALF=T+TN/2.0
      CALL PC(2.0*N,XD,VD,THALF,XN,VN,A,B,F,D)
C       COMPARE U AND V CALCULATE BY TN AND TN/2
   D1=ABS(XN-XNEW)
   D2=ABS(VN-VNEW)
   DELTA=MAX(D1,D2)


C       RECORD U AND V WHILE REQUIRED ACCURACY
C       (epsilon) IS SATISFIED
   IF(DELTA.LT.EPS) THEN
     IF(T.GT.TMIN) THEN


C   RECORD DATA U AND dU/dt
      WRITE(7,110) X,V
    ENDIF
   X=XNEW
```

```fortran
      V=VNEW
      T=T+TN


C REDUCE N VALUE
      N=N/(0.95*(EPS/DELTA)**0.25)
      TN=1.0/N
      ELSE


C EXPAND N VALUE
      N=N/(0.95*(EPS/DELTA)**0.2)
      TN=1.0/N
      ENDIF
      GO TO 222
      ENDIF
110      FORMAT(1X,F15.8,5X,F15.8)
         STOP
         END



C SUBROUTINE PC
C............................................................
C SUBRUTINE DESCRIPTION:
C This subroutine is used for calculating angular displacement and its
C first derivative. Calculation is based on the numerical method,
C The piecewise constant technique described in the context.
C
C TYPE: SUBRUTINE
C............................................................
         SUBROUTINE PC(N,X,V,T,XNEW,VNEW,A,B,F,D)


C   DESCRIPTION OF VARIABLES
C   LAM:   LAMBDA
C   X:      LOCAL INITIAL DISPLACEMENT
C   V:      LOCAL INITIAL VELOCITY
C   XNEW: CALCULATED ANGULAR DISPLACEMENT
C   VNEW: CALCULATED ANGULAR VELOCITY
C   F:      FORCE AMPLITUDE
```

```
C   A:      Alpha
C   B:      Beta

C       VARIABLE DECLARATION AND DATA INITIATION
        REAL LAM,X,V,XNEW,VNEW,F,T,N,THETA,A,B,D
        PI=ACOS(-1.0)
        LAM=D*SIN(X)+B*(V**3)-F*COS(T)
        THETA=A/N

C       RECURRENCE RELATIONS
XNEW=X+V*(1.0/A-(1.0/A)*EXP(-THETA))+LAM*(1.0/(A**2)
     +          -(1.0/A**2)*EXP(-THETA)-1.0/(A*N))
VNEW=EXP(-THETA)*V+LAM*((1.0/A)*EXP(-THETA)-1.0/A)

        IF(ABS(XNEW).GT.PI)
XNEW=XNEW-2.0*PI*ABS(X)/X
        RETURN
     END
```

COMPUTER PROGRAM 2

▮▪▪▪

```
C    PROGRAM: P-T METHOD
C
C  PURPOSE:
C       This program for numerically solving the governing
C       equation in nonlinear dynamics, such as
C               d2U/dt2+AL*dU/dt+BE*U=F*U**3
C       with the 4th-order P-T method
C
C  TYPE: MAIN PROGRAM
C
```
▪▪
```
C
C       VARIABLE DEFINITION:
C       TMIN: minimum time for a stable state
```

```
C       TMAX: time length
C       CN: local initial displacement
C       DN: local initial velocity
C       T: time
C       PI:   constant ACOS(-1.0)
C       N: parameter that controls accuracy and time step
C       TSTEP: time step


C       VARIABLE DEFINITION:
        REAL CN,DN,AL,BE,F,TMIN,TMAX,T,A,C,XI,N


C        OPEN A DATA FILES FOR SAVING THE DATA
        OPEN(2, FILE='pa4nl.da')
        OPEN(3, FILE='xt4nl.da')


C        DATA INPUT
        READ (12,*) AL,BE,F,CN,DN,TMIN,TMAX
        READ (12,*) TSTEP
C        DATA VERIFICATION
       WRITE(*,*) 'A=',AL,'   B=',BE,'   F=',F
       WRITE(*,*) '       Cn=',CN,'       Dn=',DN,'       TMIN',TMIN,'
TMAX',TMAX
       WRITE(*,*) '   TSTEP',TSTEP


C         DATA INITIATION
     T=TMIN
     X=CN
     V=DN
     PI=ACOS(-1.0)
        N=1.0/TSTEP
      C=AL/2.0
      XI=SQRT(BE-C*C)
     WRITE(*,*) '   XI',XI
        NUM=3
```

```
C          CALCULATIONS
222        IF(T.LT.TMAX) THEN
       CALL PT4(X,V,N,XNEW,VNEW,T,BE,C,XI,F)
           IF(NUM.GE.3) THEN

C          RECORD DATA
             WRITE(2,110) X,V
             WRITE(3,110) T,X
             NUM=1
             ENDIF
             NUM=NUM+1
             X=XNEW
             V=VNEW
             T=T+TSTEP
             GO TO 222
             ENDIF
            WRITE(3,*) '-999    -999'


110        FORMAT(1X,F15.8,5X,F15.8)
           STOP
100        END
```

```
!..........................................................
C
C SUBROUTINE PT4
C
C SUBROUTINE DESCRIPTION
C This subroutine is for calculating the displacement and its
C derivatives with the 4th-order P-t method.
C
Type: subroutine
C
!..........................................................


      SUBROUTINE PT4(X,V,N,XNEW,VNEW,T,BE,C,XI,F)


C    VARIABLE DESCRIPTION
```

```
C    A1,A2,A3,A4: COEFFICIENTS (SEE EQUATIONS (5.84)
C    TO (5.87))
C    B1,B2: COEFFICIENTS (SEE EQUATIONS IN (5.88))
C    XNEW: CALCULATED DISPLACEMENT
C    VNEW: CALCULATED VELOCITY

C        VARIABLE DECLARATION AND DATA INITIATION
         REAL N,EAA,A1,A2,A3,A4,B1,B2,XAA
         REAL E1,E2,F1,F2,F3
         F0=F*X**3
         X2D=F0-2.0*C*V-BE*X
         F1=3.0*F*X*X*V
         X3D=F1-2.0*C*X2D-BE*V
         F2=6.0*F*X*V*V+3.0*F*X*X*X2D
         F3=6.0*F*V**3+18.0*F*X*V*X2D+3.0*F*X*X*X3D
         A4=(F3/6.0)/BE
         A3=(0.5*F2-6.0*C*A4)/BE
         A2=(F1-4.0*C*A3-6.0*A4)/BE
         A1=(F0-2.0*C*A2-2.0*A3)/BE
         B1=X-A1
         B2=(V+C*B1-A2)/XI
         EAA=EXP(-C/N)
         XAA=EAA*(B1*COS(XI/N)+B2*SIN(XI/N))

C        RECURRENCE RELATIONS
      XNEW=A1+A2/N+A3/(N*N)+A4/(N**3)+XAA
      VNEW=-C*XAA+EAA*(-XI*B1*SIN(XI/N)
     +      +XI*B2*COS(XI/N))+A2+2.0*A3/N+3.0*A4/(N*N)
      RETURN
      END
```

COMPUTER PROGRAM 3
■■

```
C   PROGRAM: GENERATION OF PERIODIC-CHAOTIC-
C   QUASIPERIODIC DIAGRAMS
```

```
C
C   PURPOSE:
C           This program is for calculating data to draw periodic-
C           quasiperiodic-chaotic region diagrams for a driven Froude
C           pendulum system. By using the 2ed-order P-T method, the
C           program first solves the equation of motion of the Froude
C           pendulum described by
C                   d2U/dt2+AL*dU/dt+BE*(dU/dt)**3+
C                   DE*SINU=FT+Fcos(t)
C           where AL (Alpha), BE (Beta), DE, FT and F are constants.
C           U the angular displacement of the pendulum.
C           The calculated data are stored in the files of 'peri', 'chao',
C           'quas1' and 'quas2' for the periodic, chaotic and
C           quasiperiodic motions respectively. The periodicity ratio
C           diagrams can then be plotted with the data.
C
C   TYPE: MAIN PROGRAM
```
∎∎∎

```
C           VARIABLE DEFINITION:
C           TMIN: minimum time for a stable state
C           TMAX: time length
C           DN: local initial angular displacement
C           VN: local initial angular velocity
C           T: time
C           TN: time step
C           EPS: accuracy (epsilon)
C           PHI: initial phase
C           PI:   constant ACOS(-1.0)
C           N: parameter that controls accuracy and time step
C           RC: periodicity ratio for chaos
C           RP: periodicity ratio for periodic motion
C           NDEG: degree of the fitting polynomial
C           NOBS: number of data pairs
C           B: coefficients of the polynomial
```

```
C       SSPLOY: sequential sums of squares
C       STAT: statistics

C       VARIABLE DECLARATION
        INTEGER I,COUN,NDEG,NOBS,NHAL,N
        PARAMETER (NDEG=10,NOBS=650)
        REAL B(NDEG+1),SSPOLY(NDEG+1),STAT(10)
        REAL XSEMU(330),VSEMU(330)
        REAL XSEML(330),VSEML(330)
        REAL DN,VN,AL,BE,DE,F,TMIN,TMAX,T,FT
        REAL X,V,TN,XNEW,VNEW,TNEW,XH,VH
        REAL TSHALF,RC,RP
        REAL D1,D2,DELTA,PHI,EPS,PI,W1,W2,XQ(NOBS)
        REAL VQ(NOBS)
        EXTERNAL RCURV

C       OPEN DATA FILES FOR SAVING THE DATA
        OPEN(2,FILE='peri')
        OPEN(3,FILE='chao')
        OPEN(18,FILE='quas1')
        OPEN(8,FILE='quas2')
        OPEN(9,FILE='pc3d')
        DE=1.0

C       DATA INPUT
        READ (12,*)DN,VN,TMIN,TMAX,PHI,RC,RP,BE,FT,N
        READ(12,*) EPS,EPSP,AMIN,AMAX,ASTEP,FMIN,
        READ(12,*) FMAX,FSTEP,RQUA

C       INPUT DATA VERIFICATION
        WRITE(*,*) 'BE',BE,'DE',DE,'FT',FT
        WRITE(*,*) '     DN=',DN,'     VN=',VN,'     TMIN',TMIN,'
     TMAX',TMAX
        WRITE(*,*) 'EPS',EPS,'  N',N,' RC',RC,  'RP',RP,'PHI',PHI
        WRITE(*,*) 'AMIN',AMIN,'     AMAX',AMAX,'     ASTEP',
     ASTEP
```

```
      WRITE(*,*) 'RQUA',RQUA
      WRITE(*,*)   'AL',AL,'       FMIN',FMIN,'      FMAX',FMAX,'
FSTEP',FSTEP

C       CALCULATE FOR X AND V WITH TIME T
      DO 102 F=FMIN,FMAX,FSTEP
        DO 101 AL=AMIN,AMAX,ASTEP

C       DATA INITIATION
    T=TMIN
    X=DN
    V=VN
        COUN=1
    TN=1.0/N
    PI=ACOS(-1.0)

C       CALCULATE U AND V WHILE TIME T IS LESS THAN
C       TMAX
222 IF(T.LT.TMAX) THEN
      CALL PT2(X,V,TN,XNEW,VNEW,T,F,AL,BE,DE,FT)
      TSHALF=TN/2.0

C       DOUBLE N VALUE
      CALL PT2(X,V,TSHALF,XH,VH,T,F,AL,BE,DE,FT)
      CALL      PT2(XH,VH,TSHALF,XN,VN,T+TSHALF,F,AL,BE,
DE,FT)

C       COMPARE X AND V CALCULATE BY TN AND TSHALF
    D1=ABS(XN-XNEW)
    D2=ABS(VN-VNEW)
    DELTA=MAX(D1,D2)
    IF(DELTA.LT.EPS) THEN

C       CALCULATION FOR THE POINTS IN A POINCARE
C       MAP
      IF(T.GT.TMIN) THEN
```

```
          TNEW=T+TN
          W1=AMOD(PHI-T,2.0*PI)+2.0*PI
          W2=AMOD(TNEW-PHI,2.0*PI)
          IF((W1.LT.TN).AND.(W2.LT.TN)) THEN
           TS=W1
                CALL PT2(X,V,TS,XP,VP,T,F,AL,BE,DE,FT)
                  IF(ABS(XP).GT.PI) XP=XP-2.0*PI*ABS(XP)/XP
                  XQ(COUN)=XP
                  VQ(COUN)=VP
                  COUN=COUN+1
              ENDIF
         ENDIF
       X=XN
       V=VN
       T=T+TN

C EXPAND STEP SIZE
            TN=TN*0.95*(EPS/DELTA)**0.25
            IF(ABS(X).GT.PI) X=X-2.0*PI*ABS(X)/X
      ELSE

C REDUCE STEP SIZE
        TN=TN*0.95*(EPS/DELTA)**0.2
        ENDIF
     GO TO 222
     ENDIF
110     FORMAT(1X,F15.8,5X,F15.8)

C       FIND OUT IF THERE IS A VALUE AMONG XQ(I),
C       I=1,2,3..., EQUALS TO A GIVEN XQ(KJ)
        K=COUN-1
        I=1
        NEP=0
        NDP=0
777     IF(I.LT.K) THEN
        J=1
```

```
666       IF(J.LT.K) THEN
             IF(I.EQ.J) GO TO 555
          DX=ABS(XQ(I)-XQ(J))
          DV=ABS(VQ(I)-VQ(J))
            IF((DX.LT.EPSP).AND.(DV.LT.EPSP)) THEN
              NEP=NEP+1
              GO TO 999
            ENDIF
555          J=J+1
             GO TO 666
          ENDIF
999          I=I+1
             GO TO 777
          ENDIF
           RAEQ=REAL(NEP)/REAL(K-1)
           WRITE(9,*) F,AL,RAEQ


C    RECORD AL AND F VALUES OF PERIODIC CASES FOR
C    PERIODIC-C QUASIPERIODIC-CHAOTIC REGION
C    DIAGRAM
           IF(RAEQ.GT.RP) WRITE(2,*) F,AL


C    DISTINGUISH QUASIPERIODIC MOTIONS FROM CHAOS
           IF(RAEQ.LT.RC) THEN

           CALL RCURV(COUN-1,XQ,VQ,NDEG,B,SSPOLY,STAT)
           STATT=STAT(5)
           IF (STAT(5).GT.RQUA) THEN


C    RECORD AL AND F VALUES OF QUASIPERIODIC CASES
C    FOR PERIODIC-QUASIPERIODIC-CHAOTIC REGION
C    DIAGRAM
             WRITE(18,*) F,AL
           ELSE
             VMA=VQ(1)
```

```
         DO 12 I=1,COUN-1
           VMA=MAX(VMA,VQ(I))
12       CONTINUE
         VMI=VQ(1)
         DO 21 J=1,COUN-1
           VMI=MIN(VMI,VQ(J))
21       CONTINUE
          SDIF=(VMA-VMI)/2.0
          VHAL=VMA-SDIF
          NHALL=1
          NHALU=1
          DO 24 K=1,COUN-1
            IF(VQ(K).GT.VHAL) THEN
              VSEMU(NHALU)=VQ(K)
              XSEMU(NHALU)=XQ(K)
              NHALU=1+NHALU
            ELSE
              VSEML(NHALL)=VQ(K)
              XSEML(NHALL)=XQ(K)
              NHALL=1+NHALL
            ENDIF
24       CONTINUE
         CALL
RCURV(NHALU-1,XSEMU,VSEMU,NDEG,B,SSPOLY,STAT)
         STATU=STAT(5)
        WRITE(*,*)'STATU',STAT(5)
         CALL
RCURV(NHALL-1,XSEML,VSEML,NDEG,B,SSPOLY,STAT)
         STATL=STAT(5)
        WRITE(*,*)'STATL',STAT(5)
         RQUAUL=RQUA-4.0
         IF    ((STATU.GE.RQUA+1).AND.(STATL.GE.RQUA+1))
THEN

C    RECORD AL AND F VALUES FOR QUASIPERIODIC
```

```
C    CASES WITH CLOSED-CURVE POINCARE MAP
             WRITE(8,*) F,AL
          ELSEIF
((STATT.GE.RQUAUL).AND.(STATU.GE.RQUAUL)
      +              .AND.(STATL.GE.RQUAUL)) THEN

C    RECORD AL AND F VALUES FOR QUASIPERIODIC CASES
WITH
C    OPEN-CURVE POINCARE MAP
             WRITE(18,*) F,AL
          ELSE
C    RECORD AL AND F VALUES FOR CHAOS
             WRITE(3,*) F,AL
          ENDIF

          ENDIF
          ENDIF
C    END OF DISTINCTION

101     CONTINUE
102     CONTINUE
        STOP
        END
```

■■

```
C
C SUBROUTINE PT2
C
C SUBRUTINE DESCRIPTION:
C This subroutine is used for calculating the angular displacement U
C and its first derivative. Calculation is based on the numerical
C method, P-T method of second order.
C
C TYPE: SUBRUTINE
C
```

■■

```fortran
      SUBROUTINE PT2(X,V,TN,XNEW,VNEW,T,F,AL,BE,DE,FT)
        REAL X,V,TN,XNEW,VNEW,T,F,AL,BE,DE
        REAL AC,ACC,B1,B2,B3,D1,D2
      B3=-(3.0*BE*(2.0*V*AC(X,V,T,F,AL,BE,DE,FT)**2
     +     +V*V*ACC(X,V,T,F,AL,BE,DE))+DE*(
     +
AC(X,V,T,F,AL,BE,DE,FT)*COS(X)-V*V*SIN(X)))/(6.0*AL)
      B2=-(3.0*BE*V*V*AC(X,V,T,F,AL,BE,DE,FT)
     +     +DE*V*COS(X)+6.0*B3)/(2.0*AL)
      B1=(FT-BE*V**3-DE*SIN(X)-2.0*B2)/AL
      D2=(B1-V+F*(SIN(T)+AL*COS(T))/(1.0+AL*AL))/AL
      D1=X-D2-F*(AL*SIN(T)-COS(T))/(1.0+AL*AL)
      XNEW=D1+D2*EXP(-AL*TN)-F*COS(T+TN)/(1.0+AL*AL)
     +     +F*AL*SIN(T+TN)/(1.0+AL*AL)+B1*TN+B2*TN**2
     +     +B3*TN**3
      VNEW=-AL*D2*EXP(-AL*TN)+F*SIN(T+TN)/(1.0+AL*AL)
     +     +F*AL*COS(T+TN)/(1.0+AL*AL)+B1+2.0*B2*TN
     +     +3.0*B3*TN**2
      RETURN
      END


C FUNCTION SUBPRGRAM AC
        REAL FUNCTION AC(X,V,T,F,AL,BE,DE,FT)
        REAL X,V,T,F,AL,BE,DE,FT
      AC=-AL*V+FT+F*COS(T)-BE*V**3-DE*SIN(X)
      RETURN
      END
C FUNCTION SUBPRGRAM ACC
        REAL FUNCTION ACC(X,V,T,F,AL,BE,DE)
        REAL X,V,T,F,AL,BE,DE
      ACC=-F*SIN(T)-DE*V*COS(X)-(AL+3.0*BE*V*V)*AC
      RETURN
      END
```

INDEX